D0793761
046234

Acoustic Communication in Birds

Volume 1

COMMUNICATION AND BEHAVIOR

AN INTERDISCIPLINARY SERIES

Under the Editorship of ***Duane M. Rumbaugh,***
Georgia State University and Yerkes Regional Primate Research Center of Emory University

DUANE M. RUMBAUGH (ED.), LANGUAGE LEARNING BY A CHIMPANZEE: THE LANA PROJECT, 1977

ROBERTA L. HALL AND HENRY S. SHARP (EDS.), WOLF AND MAN: EVOLUTION IN PARALLEL, 1978

HORST D. STEKLIS AND MICHAEL J. RALEIGH (EDS.), NEUROBIOLOGY OF SOCIAL COMMUNICATION IN PRIMATES: AN EVOLUTIONARY PERSPECTIVE, 1979

P. CHARLES-DOMINIQUE, H. M. COOPER, A. HLADIK, C. M. HLADIK, E. PAGES, G. F. PARIENTE, A. PETTER-ROUSSEAUX, J. J. PETTER, AND A. SCHILLING (EDS.), NOCTURNAL MALAGASY PRIMATES: ECOLOGY, PHYSIOLOGY, AND BEHAVIOR, 1980

JAMES L. FOBES AND JAMES E. KING (EDS.), PRIMATE BEHAVIOR, 1982

DONALD E. KROODSMA AND EDWARD H. MILLER (EDS.), ACOUSTIC COMMUNICATION IN BIRDS, VOLUME 1: PRODUCTION, PERCEPTION, AND DESIGN FEATURES OF SOUNDS, 1982. VOLUME 2: SONG LEARNING AND ITS CONSEQUENCES, 1982

Acoustic Communication in Birds

Volume 1

Production, Perception, and Design Features of Sounds

Edited by

DONALD E. KROODSMA
Department of Zoology
University of Massachusetts
Amherst, Massachusetts

EDWARD H. MILLER
Vertebrate Zoology Division
British Columbia Provincial Museum
Victoria, British Columbia, Canada

Taxonomic Editor
HENRI OUELLET
Vertebrate Zoology Division
Museum of Natural Sciences
National Museums Canada
Ottawa, Ontario, Canada

1982

ACADEMIC PRESS
A Subsidiary of Harcourt Brace Jovanovich, Publishers

New York London
Paris San Diego San Francisco São Paulo Sydney Tokyo Toronto

ACADEMIC PRESS, INC.
111 Fifth Avenue, New York, New York 10003

United Kingdom Edition published by
ACADEMIC PRESS, INC. (LONDON) LTD.
24/28 Oval Road, London NW1 7DX

Library of Congress Cataloging in Publication Data
Main entry under title:

Acoustic communication in birds.

(Communication and behavior)
Vol. 1: Edited by Donald E. Kroodsma, Edward H. Miller, and Henri Ouellet.
Includes bibliographical references and indexes.
1. Bird-song. 2. Animal communication.
I. Kroodsma, Donald E. II. Miller, Edward H.
III. Ouellet, Henri. IV. Series.
QL698.5.A26 598.2'59 82-6730
ISBN 0-12-426801-3 (v. 1)

PRINTED IN THE UNITED STATES OF AMERICA

82 83 84 85 9 8 7 6 5 4 3 2 1

The contributors dedicate these volumes to Peter Marler. Peter's research and writings have placed the study of animal communication in the fore of modern evolutionary and ecological thought in ethology, and his impact is apparent from the sheer number and diversity of citations he receives in these two volumes. Research into almost any facet of bird acoustics puts one into contact with his work, on topics such as learning, geographic variation, species-specificity, "design features," individual variation, and grading. In addition, Peter has touched many workers individually, as graduate students, postdoctoral researchers, and peers, whereby his influence upon research and thinking on animal communication has carried yet further.

We are sure that this treatise reflects Peter Marler's influence upon the field of avian acoustics, and hope that it also reflects our esteem and respect for him as a scientist, teacher, and colleague.

Contents

6 Grading, Discreteness, Redundancy, and Motivation–Structural Rules

EUGENE S. MORTON

7 The Coding of Species-Specific Characteristics in Bird Sounds

PETER H. BECKER

8 Character and Variance Shift in Acoustic Signals of Birds

EDWARD H. MILLER

9 The Evolution of Bird Sounds in Relation to Mating and Spacing Behavior

CLIVE K. CATCHPOLE

Contributors

Numbers in parentheses indicate the pages on which the authors' contributions begin.

Arthur P. Arnold (75), Department of Psychology, and Laboratory of Neuroendrocrinology, Brain Research Institute, University of California, Los Angeles, California 90024

Peter H. Becker (213), Institut für Vogelforschung, Vogelwarte Helgoland, D 2940 Wilhelmshaven 15, West Germany

John H. Brackenbury (53), Department of Biology, University of Salford, Salford M5 4WT, England

Clive K. Catchpole (297), Department of Zoology, Bedford College, University of London, London NW1 4NS, England

Robert J. Dooling (95), Department of Psychology, University of Maryland, College Park, Maryland 20742

Edward H. Miller (253), Vertebrate Zoology Division, British Columbia Provincial Museum, and Biology Department, University of Victoria, Victoria, British Columbia, Canada

Eugene S. Morton (183), National Zoological Park, Smithsonian Institution, Washington, D.C. 20008

Douglas G. Richards (131), Kewalo Basin Marine Mammal Laboratory, University of Hawaii, Honolulu, Hawaii 96814

David C. Wickstrom (1), Library of Natural Sounds, Laboratory of Ornithology, Cornell University, Ithaca, New York 14850

R. Haven Wiley (131), Department of Zoology, University of North Carolina, Chapel Hill, North Carolina 27514

Foreword

It is entirely appropriate that these volumes of collected essays on aspects of the study of bird vocalizations should be dedicated to Peter Marler. While the scientific study of bird vocalizations was initiated by W. H. Thorpe, its subsequent expansion and elaboration was due primarily to Peter Marler.

W. H. Thorpe was initially an entomologist. Finding that some aspects of insect behavior were more labile than had formerly been supposed, he came to feel that the most pressing problems in the study of animal behavior concerned the interface between "instinct," as it was then called, and learning. As an amateur ornithologist, he realized that birds provided exceptionally suitable material for this work. Birds have a repertoire of relatively stereotyped movement patterns and yet at the same time exhibit marked learning ability. Accordingly, in 1950 he set up "An Ornithological Field Station" (now called the Sub Department of Animal Behaviour) at Madingley, Cambridge. Knowing that Chaffinch (*Fringilla coelebs*) song was subject to at least some individual variation and some flexibility (e.g., Poulsen, 1951), he decided to study the ontogeny of Chaffinch vocalization, and initiated a program of hand-rearing Chaffinches and studying their song development after varying exposure to the species' song (Thorpe, 1963).

Peter Marler was the first graduate student at Madingley. When he came, he already had a Ph.D. in botany and a tenured job. It must have been a decision of considerable courage to give up that job in order to take a Ph.D. in a subject that really interested him. His Ph.D. was a field study of the Chaffinch: he wanted to work in the field and the choice of the Chaffinch as subject meshed well with Thorpe's own work. At that time I was studying the courtship of captive Chaffinches and I am sure that Peter will remember the long discussions, and sometimes disputes, that we had over whether this or that posture should be called the lopsided wings-drooped posture or something a bit snappier—issues that seemed terribly important to us then!

Peter stayed on as a postdoctoral worker and took on the song learning work. His training as an all-round biologist stood him in good stead, and at that time he became interested not only in the ontogenetic problem that Thorpe was studying—how does Chaffinch song develop—but also in the other three questions with which ethologists are concerned—those of causation, function, and evolution. When he subsequently moved to Berkeley, he took the song problem with him. Changing his British speech for what seemed to some of us a rather extreme American accent, and the Chaffinch for White-crowned Sparrows (*Zonotrichia leucophrys*) and juncos, he started his well-known studies of ontogeny. At around that time the study of bird song really took off and it can now be regarded as having produced material of crucial importance for all four major questions of ethology.

Taking the ontogenetic question first, three fundamentally important principles have been established largely or entirely through work on bird song. The principle of sensitive periods in development, though apparent from earlier work on imprinting and other phenomena, owes a great deal to studies of bird song. The fact that what an animal learns is constrained in part by its species, though again an idea coming also from other sources of evidence, was established most firmly in the 1950s and early 1960s by the work on bird song. And third, the view that the elaboration and perfection of the song pattern depend on comparison between the vocal output and a previously established template casts a new light on many aspects of ontogeny. In all of these issues Peter Marler and his colleagues played a leading role. Understanding the processes of song development was facilitated by comparisons between studies of different species (e.g., Immelmann, 1969; Konishi, 1964; Nicolai, 1956), and Marler himself used a comparative approach to good effect, relating plasticity to the ecology of the species (Marler, 1967).

Marler's field studies of the Chaffinch had inevitably alerted him to problems concerned with the causation of bird song, and his thesis contains a great deal of observational material on the factors determining when a Chaffinch sings. He soon became interested also in the patterning of bird song, and was one of the first people to study the detailed sequencing of different song types in an individual's vocal output (e.g., Isaac and Marler, 1963). However, a more detailed study of the neural mechanisms underlying bird song arose from one of the findings of the studies of song ontogeny. Domestic Fowl (*Gallus domesticus*) and Ringed Turtle-Doves (*Streptopelia risoria*) were found to be capable of developing all the normal species' vocal signals, even though deafening took place soon after hatching. On the other hand, Konishi and Nottebohm (e.g., 1969) found that, in a number of species of songbirds, deafening before full song had developed resulted in consistent abnormalities in their song, and often in a regression to a rather amorphous type of vocal output. However, birds that had already learned to sing could continue singing after deafening. Since it was known that the control of human speech is influenced by the speaker's perception of himself speaking, the finding that deaf birds could continue to sing normally was surprising. One possibility was that the deafened bird was using feedback from the muscles. Nottebohm therefore investigated the effects of severing the hypoglossal nerves in the syrinx. This has led to the discovery of laterality in the motor control of bird vocalizations, and what is more of a partially reversible laterality. Furthermore, it has led to a detailed investigation of the brain mechanisms underlying bird song, and has provided us with quite new data on the role of hormones in the development of neural mechanisms (e.g., Nottebohm, 1980).

Marler's training as an all-round naturalist led him early on to ask functional questions. He was concerned with the functions of different calls in the Chaffinch's repertoire. What information did each carry? This led him into an attempt to categorize the nature of the information carried in animal signals (Marler, 1961). While not everyone will agree with the view that natural selection always acts to enhance the effectiveness of signals in transmitting information about the signaler, Marler's formulations greatly facilitated the study of communication.

He was also concerned with the diversity of avian vocalizations. Noticing that the alarm calls of different species tended to resemble each other, while the songs were very different, he speculated about the selective factors controlling the form of avian vocalizations—selective factors that included the optimal degree of audibility in the habitat in question as well as the particular response to be elicited in the responding individual (e.g., Marler, 1955). In addition, in comparing the ontogeny of different avian species he was forced to ask why learning plays a much greater part in some species than others, why the sensitive period occurs at different ages in different species, why learning is constrained in one way in some species and in another way in others, and so on. We do not yet know the answers to all these questions, but it was Marler's pioneering work in the early 1950s that posed them.

Finally, and inevitably, functional questions led to questions about the course of evolution of avian vocalizations. Peter Marler was concerned with this problem at a very early stage: as a student, he went on an expedition to the Canary Islands and became concerned with the differences between the songs of the species found there and their mainland counterparts (Marler and Boatman, 1951). His work on bird ontogeny led him into questions of the relations between song and call notes, and the functional questions he raised were inevitably linked with evolutionary questions about how song evolved.

While Peter Marler's work on avian vocalizations has led to progress in answering all four of the major questions in which ethologists are interested, that is not all. Perhaps influenced by Thorpe, who was a pioneer in emphasizing the importance of perceptual processes at a time when most ethologists were thinking in relatively mechanistic terms and were not yet sensitive to his suggestions, Peter Marler became interested at an early stage in the relations between the physical structure of bird vocalization and their quality as perceived by the recipient (e.g., Marler, 1969). And while outside the scope of this volume, Marler's more recent work on primate vocalizations, and his emphasis on common features between avian communication and human language (Marler, 1970) has done much to stimulate research.

I would suggest that the study of bird song is the example *par excellence* of the ethological approach. It involves the study of a naturally occurring pattern of behavior against a background of the natural history of the species concerned, but employs an experimental methodology. It involves questions about the causation, ontogeny, function, and evolution of the pattern in question, questions that are at the same time independent and interfertile. It is probably true to say that the study of bird song has done as much for the advancement of ethology as the study of any other specific aspect of behavior. In this, Peter Marler has played a major role.

Robert A. Hinde

REFERENCES

Immelmann, K. (1969). Song development in the Zebra Finch and other estrilidid finches. *In* "Bird Vocalizations. Their Relation to Current Problems in Biology and Psychology" (R. A. Hinde, ed.), pp. 61–74. Cambridge Univ. Press, London and New York.

Isaac, D., and Marler, P. (1963). Ordering of sequences of singing behaviour of Mistle Thrushes in relation to timing. *Anim. Behav.* **11,** 179–188.

Konishi, M. (1963). The role of auditory feedback in the vocal behavior of the Domestic Fowl. *Z. Tierpsychol.* **20,** 349–367.

Konishi, M. (1964). Effects of deafening on song development in two species of juncos. *Condor* **66,** 85–102.

Konishi, M., and Nottebohm, F. (1969). Experimental studies in the ontogeny of avian vocalizations. *In* "Bird Vocalizations. Their Relation to Current Problems in Biology and Psychology" (R. A. Hinde, ed.), pp. 29–48. Cambridge Univ. Press, London and New York.

Marler, P. (1955). Characteristics of some animal calls. *Nature (London)* **176,** 6.

Marler, P. (1961). The logical analysis of animal communication. *J. Theoret. Biol.* **1,** 295–317.

Marler, P. (1967). Comparative study of song development in sparrows. *Proc. 14th Int. Ornith. Cong. Oxford,* pp. 213–244, Blackwell, Oxford.

Marler, P. (1969). Tonal quality of bird sounds. *In* "Bird Vocalizations. Their Relation to Current Problems in Biology and Psychology" (R. A. Hinde, ed.), pp. 5–18. Cambridge Univ. Press, London and New York.

Marler, P. (1970). Birdsong and speech development: could there be parallels? *Am. Sci.* **58** (6), 669–673.

Marler, P., and Boatman, D. J. (1951). Observations on the birds of Pico, Azores. *Ibis* **93,** 90–99.

Nicolai, J. (1956). Zur Biologie und Ethologie des Gimpels (*Pyrrhula pyrrhula* L.). *Z. Tierpsychol.* **13,** 93–132.

Nottebohm, F. (1980). Brain pathways for vocal learning in birds. A review of the first 10 years. *Prog. in Psychobiol. and Physiol. Psychol.* **9,** 85–124.

Poulsen, H. (1951). Inheritance and learning in the song of the Chaffinch (*Fringilla coelebs*). *Behaviour* **3,** 216–228.

Thorpe, W. H. (1963). "Learning and Instinct in Animals." Methuen, London.

Preface

We began corresponding about co-editing a book on bird sounds in 1978, excited by the enormous increase in evolutionary understanding and interpretation of communication systems since Robert Hinde's edited volume, *Bird Vocalizations,* appeared in 1969. The rapid mushrooming and splintering of ideas and observations on animal communication were daunting, but we nevertheless shared the belief that a representative and useful collection of writings on evolution and ecology of bird acoustics could be assembled. Our original intent was to compile both taxonomic and conceptual reviews, but the meager knowledge of acoustic signals of many important avian taxa made the first of these impossible. Consequently, we solicited contributions from active researchers in bird acoustics, for chosen areas of evolutionary and behavioral ecology, and sought to complement them with reviews of sound recording techniques, and sound production, reception, and processing.

Volume 1 begins with several of these background chapters. The first discusses sound recording, makes certain recommendations, and points out common errors. The others outline some of the complex events and processes between sound production and behavioral response to sound. The remaining chapters stand apart from the first ones, a gap that accurately reflects our poor understanding of the processes which ultimately link individual physiological responses to population-genetical changes, and to larger-scale evolutionary trends. Bridging this gap will require diverse research endeavors and theorizing, some of which are touched on in Chapters 5–9 of Volume 1 and Chapters 1–9 of Volume 2.

The chapters are varied. They range from lengthy, well balanced, detailed syntheses of subjects such as coding of species-specificity (Becker), individuality (Falls), and environmental acoustics (Wiley and Richards), etc., to briefer more partisan explications of ideas and observations on dialects (Baker), subsong (Marler and Peters), sexual selection (Catchpole), ''motivation–structural rules'' (Morton), etc. Others present novel views and reviews on doggedly troublesome concepts such as duetting (Farabaugh), vocal mimicry (Baylis), and geographic variation (Mundinger). We have contributed chapters on ontogeny, the evolution of complex vocalizations, and character shift, and Beer wrote a general chapter on conceptual issues relevant to bird acoustics. Despite the inevitable deficiencies and uneven style and coverage in a volume of this sort, we feel that many important, current topics are refreshingly well reviewed, and that these volumes will serve as a stimulating and valuable reference for students of communication and bioacoustics. We thank the authors for their care and labor; Henri Ouellet, for assuming the unrewarding role of taxonomic editor; and Robert Hinde, for writing the Foreword.

Donald E. Kroodsma
Edward H. Miller

Note on Taxonomy

The editors, acknowledging the current unstable state of avian taxonomy and nomenclature, invited me to read the manuscript in order to ensure uniformity in taxonomic and nomenclatural usage. This proved to be particularly appropriate because none of the contributors is a taxonomist and because the sources used in this work are very diverse and cover an extensive time span.

Scientific names were nearly all standardized here after the *Reference List of the Birds of the World* (Morony *et al.*, 1975); this reference is readily available, even to the nonornithologists, includes nearly all the species currently recognized, and is the most up-to-date reference of its kind. Domestic or laboratory birds have usually been referred to their wild counterpart with the exception of "the chicken," which has been designated here as "Domestic Fowl" and "*Gallus domesticus*" because this work addresses itself to a broader audience than ornithologists. However, the Canary remains *Serinus canaria* and the Turkey *Meleagris gallopavo,* but the quail has become the Common Quail *Coturnix coturnix* and the Ring Dove the Ringed Turtle-Dove *Streptopelia risoria*. Otherwise Morony *et al.* (1975) have been followed closely, although I have been inclined in several instances to diverge from it.

For English names a variety of sources has been used. The A.O.U. Checklist (1957) and its supplements (1973, 1976) were used as the standard reference for all Holarctic and Neotropical species appearing in the checklist. I have referred to Voous (1973, 1977) for Palearctic species, to Hall and Moreau (1970) and Snow (1978) for Africa, to Ridgeley (1976) for Central America, and to Meyer de Schauensee (1970) for South America. In the few instances where Asiatic and Australian species were mentioned, I have referred to a variety of sources and had to use my best judgment.

Henri Ouellet

REFERENCES

American Ornithologists' Union. (1957). "Check-list of North American Birds," 5th ed. Lord Baltimore Press, Baltimore.

American Ornithologists' Union. (1973). Thirty-second supplement to the American Ornithologists' Union check-list of North American birds. *Auk* **90,** 411–419.

American Ornithologists' Union. (1976). Thirty-third supplement to the American Ornithologists' Union check-list of North American birds. *Auk* **93,** 875–879.

Hall, B. P., and Moreau, R. E. (1970). "An Atlas of Speciation in African Passerine Birds." Trustees of the British Museum (Natural History), London.

Meyer de Schauensee, R. (1970). "A Guide to the Birds of South America." Livingston, Wynnewood, Pennsylvania.

Morony, J. J., Jr., Bock, W. J., and Farrand, J., Jr. (1975). "Reference List of the Birds of the World." American Museum of Natural History, New York.

Ridgeley, R. S. (1976). "A Guide to the Birds of Panama." Princeton Univ. Press, Princeton, New Jersey.

Snow, D. W. (1978). "An Atlas of Speciation in African Non-Passerine Birds." Trustees of the British Museum (Natural History), London.

Voous, K. H. (1973, 1977). List of Recent Holarctic bird species. *Ibis* **115,** 612–638; **119,** 223–250, 376–406.

Introduction

Natural historians study four major aspects of animal behavior: phylogeny, causation, ontogeny, and adaptation (Hailman, 1977). These approaches have imparted the evolutionary flavor to ethology since its inception, so that ethology has come to be viewed as a truly biological discipline. Contributions to this self-conception have been diverse and numerous, from Darwin's writings on expressive behavior, to current mathematical renderings of optimal foraging. With the exception of phylogeny, the main concerns of ethologists have likewise been central to research in avian acoustics, as is evident from these volumes. In the following comments, we discuss some issues in the study of bird sounds, with particular reference to evolutionary considerations.

I. PHYLOGENY

The promise of ethology to systematics and phylogenetics remains largely unfulfilled, despite the long-lived comparative tradition in ethology (Lanyon, 1969; Mundinger, 1979). The tradition arose and is sustained mainly by studies on stereotyped motor patterns in visual displays of birds and fish (e.g., Lorenz, 1941; van Tets, 1965). It has proved less successful in research on taxa like mammals, in which senses like olfaction are so important, and in which movements and postures are less readily classified and quantified (e.g., Eisenberg, 1967; Golani, 1976). Comparative studies on bird sounds have not led to significant systematic advances for very different reasons, related to sound variability. There are several reasons for this. First, sounds can be emitted easily and cheaply, and thus can often assume numerous forms with little selective penalty (see comments below on nonadaptive evolution). In addition, sounds used over short distances often encode minor variations in the emitter's state, and need not identify the signaling species, so are often highly graded. Variability in sounds is increased further by vocal learning in many or most species, and by short- and long-term neuroendocrine effects in individuals. Finally, the physical simplicity of sound, and the extent to which anatomical and neuromuscular changes affect sound properties, can result in rapid divergence between homologous sounds of related taxa. The high variability consequent on these factors is daunting to taxonomic and phylogenetic research, and makes quantitative analyses very difficult. All in all, bird sounds make difficult material for systematics because of their numerous variations resulting from emitters' developmental, physical, genetic, and physiological characteristics.

Taxa (or call types) in which vocal learning is slight and in which species-specific characteristics are unimportant (as in short-range calls) offer the best material for systematics (Mundinger, 1979). Thus, contact calls are remarkably conservative in Anatidae, and their detailed study could resolve certain affinities (Thielcke, 1970). Other call characteristics which should be evolutionarily conservative are those which conform with "motivation–structural rules" (see below).

The utility of acoustic characteristics in systematics depends on the recognition of shared derived homologies (Eldredge and Cracraft, 1980). Only through study of their distribution in taxa of interest can monophyletic groups be defined. For example, consider the Calidridinae, a subfamily of Scolopacidae which includes 24 closely related arctic and subarctic species. Unpaired males in these species employ simple and conspicuous aerial displays to simultaneously attract mates and repel competing males. Certain species, like Dunlin (*Calidris alpina*) and Baird's Sandpiper (*Calidris bairdii*), emit long series of rhythmically repeated buzzy (pulsed) calls. Others, like the Least Sandpiper (*Calidris minutilla*) and Stilt Sandpiper (*Micropalama himantopus*) emit remarkably similar series of calls which are presumably homologous, but the calls are never pulsed, and contain rhythmic frequency modulation at most (Figs. 1 and 2). The nonpulsed form is also present in many related, non-calidridine taxa, including Willets (*Catoptrophorus semipalmatus*) and curlews (*Numenius* species), so is presumably ancestral within the Calidridinae (Miller, 1982). Based on this characteristic then, *C. alpina* and *C. bairdii* share a common ancestor with one another more recently than they do with *C. minutilla* or *Micropalama*. By such reasoning, and using various characteristics, it is possible to construct a testable phylogeny. To date, only Mundinger (1979) has employed a cladistic approach in studying affinities of bird species based on acoustic characteristics.

II. CAUSATION

Several chapters in these two volumes deal with what Bates (1960) has termed "skin-in" biology. They concern sound production, neural control of song, and auditory perception (Chapters 2, 3, and 4 of Vol. 1). The latter area has received most attention from evolutionists, because of the prospect of detecting close adaptation of hearing abilities to species-specific sound characteristics. Such

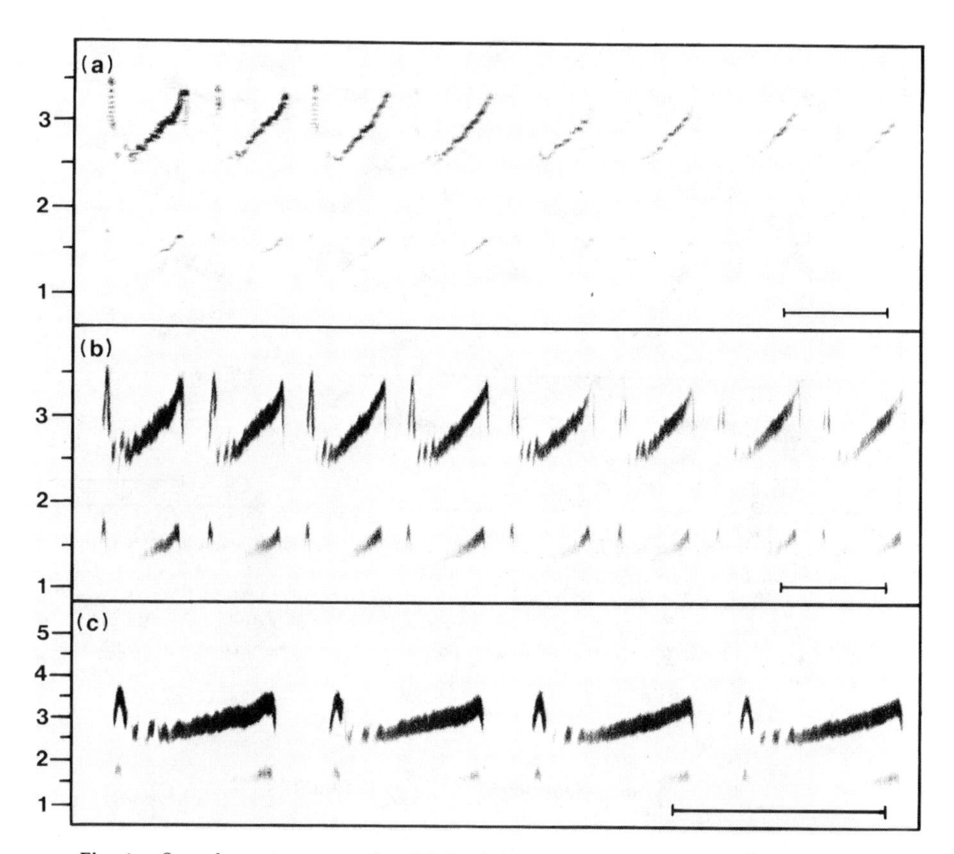

Fig. 1. Sound spectrograms of aerial display calls by a male Least Sandpiper (*Calidris minutilla*), showing no rhythmic FM and lack of pulsing. Part of a long rhythmic sequence is illustrated. Panel c corresponds to the first four calls of panels a and b, which are identical except for analyzing filter bandwidth (45 Hz for a, 300 Hz for b and c). The frequency scales are in kiloHertz; the time marker in the bottom right corner of each panel is 250 msec.

adaptation is not restricted to invertebrates and lower vertebrates, but is also known for certain mammals (Suga, 1978). Evidence for similar specificity in birds is weak, and in general species-typical frequency spectra correspond only approximately to hearing curves, while the latter differ little across species.

Neurophysiology lends itself less easily to evolutionary interpretation, because linkages of cause and effect are so difficult to trace. Some evidence points to unsuspectedly close adaptation even now, however. For example, male Long-billed Marsh Wrens (*Cistothorus palustris*) in nature learn about 50 and 150 songs in New York and San Francisco populations, respectively, and correlated

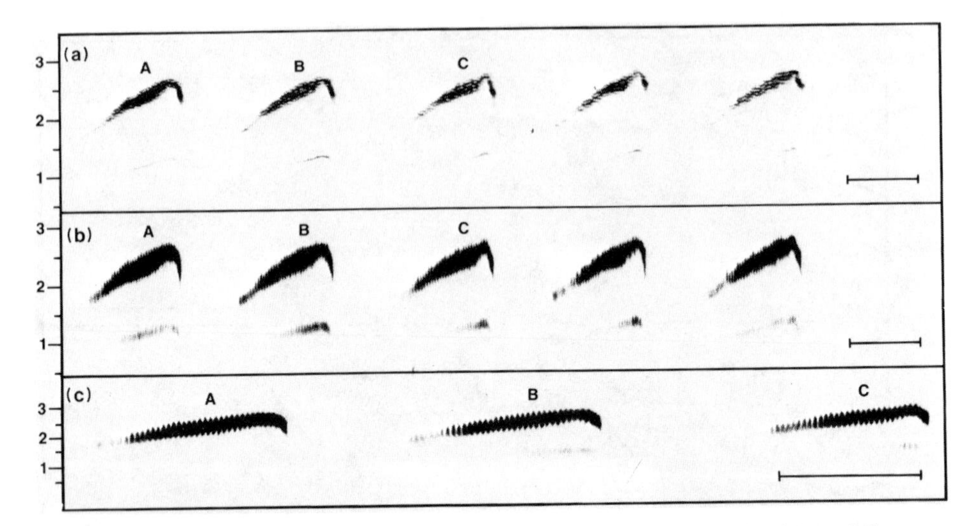

Fig. 2. Sound spectrograms of aerial display calls by a male Stilt Sandpiper (*Micropalama himantopus*), showing rhythmic FM and lack of pulsing. Part of a long rhythmic sequence is illustrated. Panel c corresponds to the first three calls of panels a and b, which are identical except for analyzing filter bandwidth (45 Hz for a, 300 Hz for b and c). The frequency scales are in kiloHertz; the time marker in the bottom right corner of each panel is 500 msec.

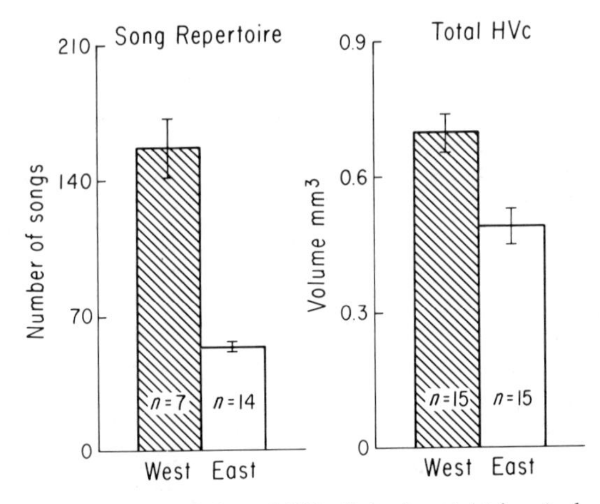

Fig. 3. Song repertoire size and total HVc (left plus right hemispheres—see Vol. 1, Chapter 3, for further discussion of these brain nuclei) for Long-billed Marsh Wrens from San Francisco, California (West) and Hudson River, New York (East) populations. Song repertoire sizes and the brain space devoted to singing are larger in Western populations (R. Canady, D. E. Kroodsma, F. Nottebohm, unpublished data).

with this 3:1 ratio in repertoire size is an approximate 3:2 ratio in the volume of several song control nuclei in the forebrain (Fig. 3). Laboratory experiments indicate that San Francisco males can learn more songs than New York males, a difference which is probably related to higher population density and overall intensity of competition for territories and mates in western populations (R. Canady, D. E. Kroodsma, and F. Nottebohm, unpublished data).

III. ONTOGENY

Most aspects of the ontogeny of acoustic behavior in birds are poorly known. Research has emphasized song learning in oscines, and little effort has been put into other groups (but see Andrew, 1969; Cosens, 1981; Wilkinson and Huxley, 1978). One reason for this bias is the relatively stereotyped, discrete nature of most oscine song, and the variable, highly graded nature of sounds in most other taxa (e.g., Huxley and Wilkinson, 1977; Mace, 1981; Mairy, 1979a,b). Few trends of broad evolutionary significance have emerged from studies of ontogeny yet, despite the documentation of widespread characteristics. For example, most young songbirds need models for normal song acquisition (see the Appendix of Vol. 2), and are maximally sensitive to appropriate models only during discrete time periods at a certain age. But why is song learning so significant for some sound types and not for others? Why is it so pronounced in oscines, and not in other groups? What roles have phylogeny and adaptation to local conditions played in the origin and maintenance of open developmental programs of song acquisition?

Vocal ontogenies have undoubtedly coevolved with other life history strategies, yet searches for correlations have yielded few generalizations. Further comparative work is needed, not only among closely related species, but also among populations of the same species and among individuals of the same population. The timing of sensitive periods for song learning may vary between migratory and nonmigratory populations, or between populations at different latitudes. Juveniles within a population may face different social and physical environments, and such environmental factors may influence the timing of vocal learning. Thus, Long-billed Marsh Wrens hatching early in the season experience longer days and hear more adult song than those hatching later in the season; birds hatching later can learn songs better their next spring, indicating that these environmental factors do have an important impact on vocal development (Kroodsma and Pickert, 1981). Much research on environmental influences and neuroendocrine bases of song development is needed before the evolutionary significance of different vocal ontogenies can be appreciated.

IV. ADAPTATION

When bird sounds are viewed as phenotypic components, their assessment as adaptations is sure to follow. There has been much recent study of adaptiveness of sound structure related to transmission in natural environments. This work has resulted in important findings which lead to prediction of optimal sound characteristics. For example, if maximal transmission distance is important, sounds should have characteristics which are resistant to attenuation and degradation. This prediction has been supported for long-distance signals even within species (Gish and Morton, 1981; Wasserman, 1979). More comprehensive studies suggest some problems with the assumption, though. Thus, Lemon *et al.* (1981) found that frequency characteristics of songs of 19 species of Parulidae were significantly related to singing height, in a manner contrary to prediction (Fig. 4). They comment (p. 1174) that perhaps "warblers are not trying to maximize the distance their songs carry but rather are optimizing the distance, sometimes greater, sometimes lesser, so as to communicate mainly with those individuals who are most significant to them biologically" (see also Krebs and Davies, 1981). Their suggestion can be generalized for different kinds of sounds, and not just long-distance ones: natural selection should promote the selective transmission of sound characteristics to intended receivers under average conditions; facultative responses may improve transmission, for example, by singing more

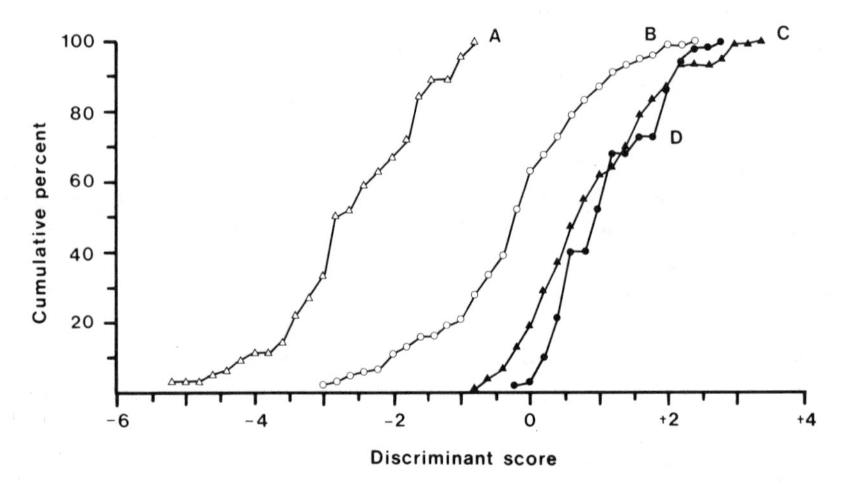

Fig. 4. Cumulative frequency plots of discriminant scores (from single function discriminant analysis) for songs of groups of warbler species (Parulidae) defined by singing height. For example, group A consists of four species with the highest mean singing height, group B contains six species, group C four species, and group D five species with lowest mean singing height. [After Fig. 1 of Lemon *et al.* (1981), using data provided by R. E. Lemon (*in litt.*).]

loudly or persistently when a receiver is known to be distant, or under noisy conditions, etc. Few situations must exist where maximal and optimal transmission distances are the same. In any case, since most displays have multiple functions, average adaptations and optimal solutions are probably the rule (Nuechterlein, 1981).

Another possibility to consider is that many or most spatio-temporal changes in sounds have no selective or adaptive significance (e.g., Wiens, 1982). P. J. B. Slater and colleagues have extensively documented factors influencing song features in the Chaffinch (*Fringilla coelebs*), and point out that immigration, learning errors, nonlearning of rare song types, and other factors can all lead to nonadaptive evolutionary change (e.g., Ince *et al.*, 1980; Slater and Ince, 1979). Various possible explanations for the origin of dialects in Japanese monkeys (*Macaca fuscata*) have been reviewed by Green (1975). He points out that several causes may contribute to dialectical differentiation, some concerned with the initial establishment of differences, some with their spread within local populations, and some with their maintenance there. Thus, "a behavioral founder-effect . . . with socially facilitated mimicry, enhanced by cultural propagation with feeding as a reinforcer" could account for his observations (Green, 1975, p. 308). The founder event is probably never totally random, but rather "the rise of the tonal theme at the three [study] sites indicates that the broad acoustic structure is circumscribed by the genetic constraints linking tonal sounds to affinitive circumstances" (p. 309); fine differences are probably under less stringent control. Comparable comments can be applied to birds.

Predictions about the adaptiveness of sound characteristics depend crucially upon the paradigm within which one works. A model of "honest" nonmanipulative communication predicts a straightforward match between optimal transmission distance and transmissibility of sound characteristics; other models yield different predictions (Dawkins and Krebs, 1978; Gish and Morton, 1981; Krebs and Davies, 1981; Richards and Wiley, 1980). Thus, if it is advantageous to inform a neighbor of one's distance from him, a song should exhibit predictable differential attenuation or degradation of its characteristics. If it is advantageous to conceal this information from a neighbor, characteristics should show unpredictable patterns of attenuation and degradation and should have highly transmissible features.

The selective value to communicating derives from the average effect induced or maintained in receivers. The evaluation of "optimality" must extend beyond the physical features of signals before and after average transmission; it must also include the relationship of signal structure to physiological/motivational states induced in receivers (see Collias, 1960; Eisenberg, 1981; Morton, 1977). General principles that apply to this realm are not clearly documented or understood, but it seems likely that the evolutionary potential of sound structure is limited for neurophysiological reasons, and other organismal properties are probably linked to sound characteristics, too, but for yet unknown reasons. For example, duration plus three frequency measures of alarm calls in antelope ground squirrels (*Ammospermophilus*) are highly correlated with measures of the rostrum, and two of these are further correlated with habitat (Bolles, 1980).

The physical structure of sounds is under intense study to reveal adaptations to the physical environment, and such investigations will increasingly extend to the physiological realm. Despite these important trends, and the critical need for good detailed descriptions, it must be noted that the significance of different sounds to receivers can be most directly appraised without any reference to

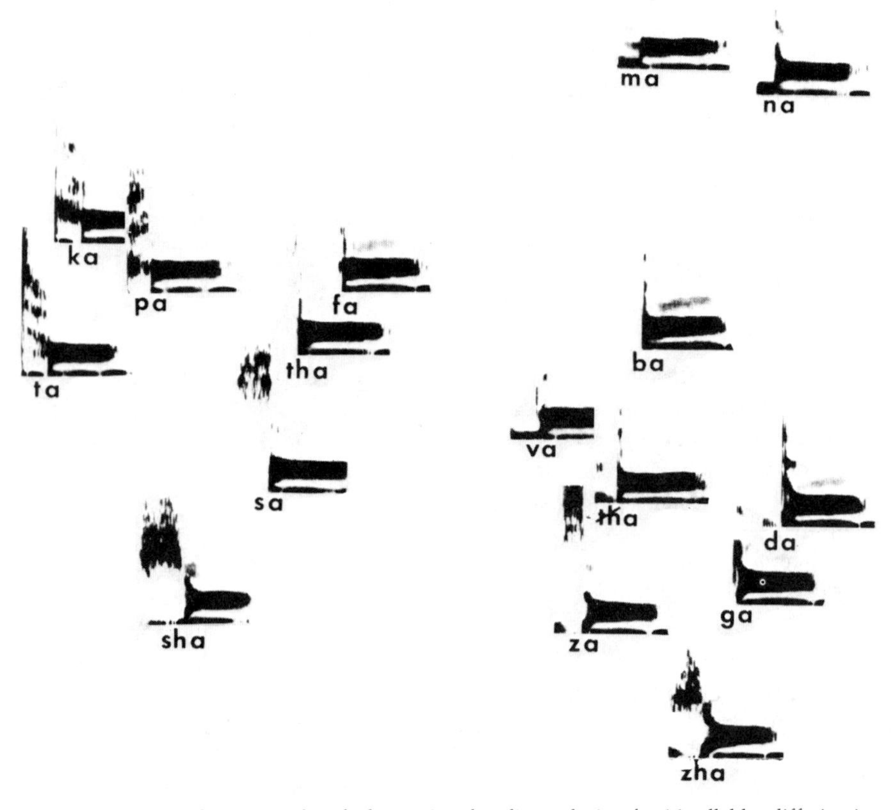

Fig. 5. Two-dimensional multidimensional scaling solution for 16 syllables differing in the initial consonant. The proximity of the syllables in the diagram reflects how easily they were confused with one another by subjects: large distances indicate low confusability, and small distances indicate high confusability. Morphological characteristics of the sounds suggest some structural and perceptual congruence. Thus, ka, pa, and ta are voiceless stops; fa, tha, sa, and sha are voiceless fricatives, etc. [From Shepard (1980, Fig. 2); reproduced with permission of the American Association for the Advancement of Science, and R. N. Shepard.]

sounds' physical attributes. Thus, by pairwise testing of different sounds' confusability and similarity to subjects, the proximity of sound signals to one another in subjects' perceptual space can be estimated (Fig. 5; Shepard, 1980). Congruence between physical and perceptual proximity can be investigated subsequently.

V. CONCLUDING COMMENTS

There is a growing trend for research on bird acoustics to become more rigorous in scope, design, execution, and analysis. This is desirable and healthy, but purely descriptive studies and documentary audio recording also need to be sharply increased. Displays of most extant species have not even been taped or filmed, much less studied in detail. There is an urgent need for extensive audio recording and filming of species which are declining in abundance or range, before they become extinct. This need is acute for faunas in many areas of the world, especially those of the vast but rapidly disappearing tropical forests (Stuessy and Thomson, 1981). Without such material, we will be forever unable to know, appreciate, or explore the remarkable and diverse acoustic behavior of birds. Marshall (1978, p. 1) poignantly expresses this view in his monograph on small Asian night birds:

> Natural forests are hard to find and difficult to reach, being mostly contracted to high altitudes. They are disappearing at an appalling rate, invariably by clear-felling, to be replaced by wretched crops for a couple of seasons, then abandoned to desolate *Imperator, Eupatorium,* or other bushes. My study devolved into a crash program just to hear and tape-record the owls before they become extinct. Frills such as play-back experiments gave way to anguished efforts at identification.

REFERENCES

Andrew, R. J. (1969). The effects of testosterone on avian vocalizations. *In* "Bird Vocalizations. Their Relation to Current Problems in Biology and Psychology" (R. A. Hinde, ed.), pp. 97–130. Cambridge Univ. Press, London and New York.

Bates, M. (1960). "The Forest and the Sea; a Look at the Economy of Nature and the Ecology of Man." Random House, New York.

Bolles, K. (1980). "Variation and Alarm Call Evolution in Antelope Squirrels, *Ammospermophilus* (Rodentia: Sciuridae)." Ph.D. thesis, Univ. of California, Los Angeles, California.

Collias, N. E. (1960). An ecological and functional classification of animal sounds. *In* "Animal Sounds and Communication" (W. E. Lanyon and W. N. Tavolga, eds.), pp. 368–391. Amer. Inst. Biol. Sci., Washington, D.C.

Cosens, S. E. (1981). Development of vocalizations in the American Coot. *Can. J. Zool.* **59,** 1921–1928.

Dawkins, R., and Krebs, J. R. (1978). Animal signals: information or manipulation? *In* "Behavioural Ecology. An Evolutionary Approach'' (J. R. Krebs and N. B. Davies, eds.), pp. 282–309. Sinauer Assoc., Sunderland, Massachusetts.

Eisenberg, J. F. (1967). A comparative study in rodent ethology with emphasis on evolution of social behavior, I. *Proc. U.S. Nat. Mus.* **122,** 1–51.

Eisenberg, J. F. (1981). ''The Mammalian Radiations. An Analysis of Trends in Evolution, Adaptation, and Behavior.'' Univ. of Chicago Press, Chicago, Illinois.

Eldredge, N., and Cracraft, J. (1980). ''Phylogenetic Patterns and the Evolutionary Process.'' Columbia Univ. Press, New York.

Gish, S. L., and Morton, E. S. (1981). Structural adaptations to local habitat acoustics in Carolina Wren songs. *Z. Tierpsychol.* **56,** 74–84.

Golani, I. (1976). Homeostatic motor processes in mammalian interactions: a choreography of display. *In* ''Perspectives in Ethology'' (P. P. G. Bateson and P. Klopfer, eds.), Vol. 2, pp. 69–134. Plenum, New York.

Green, S. (1975). Dialects in Japanese monkeys: vocal learning and cultural transmission of locale-specific vocal behavior? *Z. Tierpsychol.* **38,** 304–314.

Hailman, J. P. (1977). ''Optical Signals, Animal Communication and Light.'' Indiana Univ. Press, Bloomington, Indiana.

Huxley, C. R., and Wilkinson, R. (1977). Vocalizations of the Aldabra White-throated Rail *Dryolimnas cuvieri aldabranus. Proc. R. Soc. London Ser. B* **197,** 315–331.

Ince, S. A., Slater, P. J. B., and Weismann, C. (1980). Changes with time in the songs of a population of Chaffinches. *Condor* **82,** 285–290.

Krebs, J. R., and Davies, N. B. (1981). ''An Introduction to Behavioural Ecology.'' Blackwell, Oxford.

Kroodsma, D. E., and Pickert, R. (1980). Environmentally dependent sensitive periods for avian vocal learning. *Nature (London)* **288,** 477–479.

Lanyon, W. E. (1969). Vocal characters and avian systematics. *In* ''Bird Vocalizations. Their Relation to Current Problems in Biology and Psychology'' (R. A. Hinde, ed.), pp. 291–310. Cambridge Univ. Press, London, England.

Lemon, R. E., Struger, J., Lechowicz, M. J., and Norman, R. F. (1981). Song features and singing heights of American warblers: maximization or optimization of distance? *J. Acoust. Soc. Am.* **69,** 1169–1176.

Lorenz, K. (1941). Vergleichende Bewegungsstudien an Anatinen. *J. Orn.* **89** (*Suppl.*), 194–294.

Mace, T. R. (1981). ''Causation, Function, and Variation of the Vocalizations of the Northern Jacana, *Jacana spinosa*.'' Ph.D. thesis, Univ. of Montana, Missoula, Montana.

Mairy, F. (1979a). Le roucoulement de la tourterelle rieuse domestique, *Streptopelia risoria* (L.) I. Variation morphologique de sa structure acoustique. *Bull. Soc. Roy. Sci. Liège* **9–10,** 355–377.

Mairy, F. (1979b). Le roucoulement de la tourterelle rieuse domestique, *Streptopelia risoria* (L.) II. Aspects causaux et sémantiques de la variation de sa morphologie acoustique. *Bull. Soc. Roy. Sci. Liège* **9–10,** 378–390.

Marshall, J. T. (1978). Systematics of smaller Asian night birds based on voice. *Amer. Orn. Union, Orn. Monogr.* **25,** 1–58.

Morton, E. S. (1977). On the occurrence and significance of motivation-structural rules in some bird and mammal sounds. *Amer. Natur.* **111,** 855–869.

Miller, E. H. (1982). Aerial displays of calidridine sandpipers, their structure and systematic significance. Unpublished data.

Mundinger, P. (1979). Call learning in the Carduelinae: ethological and systematic considerations. *Syst. Zool.* **28,** 270–283.

Nuechterlein, G. L. (1981). Variations and multiple functions of the advertising display of Western Grebes. *Behaviour* **76,** 289–317.

Richards, D. G., and Wiley, R. H. (1980). Reverberations and amplitude fluctuations in the propagation of sound in a forest: implications for animal communication. *Amer. Natur.* **115,** 381–399.

Shepard, R. N. (1980). Multidimensional scaling, tree-fitting, and clustering. *Science* **210,** 390–398.

Slater, P. J. B., and Ince, S. A. (1979). Cultural evolution in Chaffinch song. *Behaviour* **71,** 146–166.

Stuessy, T. F., and Thomson, K. S. (eds.) (1981). "Trends, Priorities and Needs in Systematic Biology." Report to the Systematic Biology Program, National Science Foundation. Association of Systematics Collections, Lawrence, Kansas.

Suga, N. (1978). Specialization of the auditory system for reception and processing of species-specific sounds. *Fed. Proc., Fed. Am. Soc. Exp. Biol.* **37,** 2342–2354.

Thielcke, G. (1970). Die sozialen Funktionen der Vogelstimmen. *Vogelwarte* **25,** 204–229.

Van Tets, G. F. (1965). A comparative study of some social communication patterns in the Pelecaniformes. *Amer. Orn. Union, Orn. Monogr.* **2,** 1–88.

Wasserman, F. E. (1979). The relationship between habitat and song in the White-throated Sparrow. *Condor* **81,** 424–426.

Wiens, J. A. (1982). Song pattern variation in the Sage Sparrow (*Amphispiza belli*): dialects or epiphenomena? *Auk* **99,** 208–229.

Wilkinson, R., and Huxley, C. R. (1978). Vocalizations of chicks and juveniles and the development of adult calls in the Aldabra White-throated Rail *Dryolimnas cuvieri aldabranus* (Aves: Rallidae). *J. Zool.,* **186,** 487–505.

1

Factors to Consider in Recording Avian Sounds

DAVID C. WICKSTROM

ACOUSTIC COMMUNICATION IN BIRDS
VOLUME 1

ISBN 0-12-426801-3

I. INTRODUCTION

The recording of bird sound was once a scientific achievement in itself. Today, using relatively inexpensive equipment, anyone can make subjectively good recordings. While this may be a pleasurable pastime, it can scarcely be called a scientific endeavor. Documentation and calibration are two key words for the serious recordist. The scientific utility of a given recording is diminished greatly if good data are not kept on what is being recorded. Similarly, if the characteristics of the recording system are not documented, the precision of conclusions drawn from analysis of the recordings is severely degraded.

An ideal recording system would reproduce in the minutest detail an exact electrical analog of the sound present at the time of recording. Unfortunately, this ideal can only be approximated.

Most recording equipment design is predicated on its use for human speech or music. These assumptions are not necessarily valid for avian recording. If recordists are not aware of the limitations of their recording systems, they can unknowingly precondition their data.

In the case of field recording where the expense of getting to the location can be substantial, it is illogical to use anything but the best equipment that can be accommodated. Even if the recording is for a specific research project, it will, ideally, be deposited with an archive. When a recording becomes a library resource, the better the quality and documentation, the wider its usefulness.

From the moment sound emanates from a bird, it becomes contaminated. Contamination comes from the environment and the recording equipment. The finished recording consists not only of the bird sound, but of all the sounds present in the environment at the time of the recording as well as artifacts generated by the recording equipment.

A primary objective in recording is to optimize the amount of desired sound with respect to the amount of extraneous information, in other words, to maximize the signal-to-noise ratio. To accomplish this requires an understanding of the behavior of sound in air (see Chapter 5, this volume), the behavior of the subject, and the characteristics of the recording system. No matter what technological wonders are available for postrecording processing, the easiest time to maximize the signal-to-noise ratio is when the recording is made.

There is no piece of electronic apparatus that does not, in some way, change the signal passing through it. In most components, the action of the device is largely beneficial (the amplifier increases the signal), but there is some cost (addition of noise and distortion). It is important to know what types of distortion a system introduces since the usefulness of the final recording is affected. It is noteworthy that the existing standards for the measurement of distortion do not correlate well with perceived quality.

In this chapter I will first review some terms, and then break the recording system into its component parts: microphones, recorders, tape, and signal-processing equipment and interface components. The final part of the chapter will discuss the combination of components into a recording system.

II. TERMINOLOGY

Decibels: One of the most common terms used when working with sound is the decibel (dB). An understanding of the term is crucially important when working with sound. Two things to keep in mind are that the term dB (1) simply means that a logarithmic power ratio is being used and (2) unless referenced to something, does not represent an absolute measurement. For power the formula is:

$$\text{dB} = 10 \log \frac{P_1}{P_2}$$

For voltage:

$$\text{dB} = 20 \log \frac{E_1}{E_2}$$

where E is voltage (see Davis and Davis, 1975). Since it is possible to find sounds occurring over a range of 180 dB (although 130 dB is more usual), and most recording systems have a dynamic range of around 50 dB, one can see the necessity of making some informed decisions.

Various standards exist specifying the methods to be used when evaluating audio and sound equipment. Some of the organizations involved are the American National Standards Institute (ANSI), the International Organization for Standardization (ISO), the International Electrotechnical Commission (IEC), the Society of Motion Picture and Television Engineers (SMPTE), the Institute of High Fidelity (IHF), and the Audio Engineering Society (AES). Other than these organizations, manufacturers of audio test equipment are excellent sources of information and application notes. The following descriptions are purposefully general as space does not permit delving into the various standards. When comparing equipment be sure that the same test methods were used.

Frequency Response: This is the degree of amplitude change with respect to frequency. This specification is usually presented graphically with amplitude on the y axis and frequency on the x. The ideal of no change in amplitude with frequency would be presented as a straight line, i.e., a "flat" response. In written form, upper and lower frequency limits are given along with the limits of amplitude variation. It is important to have a *system* response as wide as the frequency range of the subject.

Signal-to-Noise Ratio: Expressed in decibels, the signal-to-noise ratio is a measure of the distance between a reference level (usually the maximum capability) and the noise floor of the device. In many cases the specification is "weighted" according to a specific frequency curve. Ideally, an unweighted specification is also given. The unweighted value is the more useful number for analysis as it is not a frequency-sensitive specification.

Dynamic Range: Related to the signal-to-noise ratio, the dynamic range gives the range of amplitude variation that a device can accommodate at a given setting.

Noise: Sometimes a specification simply labeled "noise" is given. This is a measurement of the device's inherent noise level. The reference point and the band width of this measurement should be given somewhere in the specifications (e.g., all measurements referenced to 0.775 V).

Phase Shift: This specification is seldom seen. If predictable phase response is a requirement, the system should be checked by a competent technician.

Drift, Wow, Flutter, Scrape Flutter: These specifications refer to tape recorders. Drift is usually considered to be a speed variation of up to 0.1 Hz.; Wow from 0.1 to 10 Hz; Flutter from 10 to 300 Hz.; and Scrape flutter from 3 to 5 kHz (McClurg, 1976). The specification most often given is the combined Wow and Flutter weighted by some standard curve.

Total Harmonic Distortion (THD): This specification gives the level of harmonic content at the output of a device, relative to the level of the fundamental. It is most often expressed as a percentage but it can also be expressed in dB. For example, harmonic content 40 dB below the fundamental is the same as 1% THD.

Intermodulation Distortion (IM): The purpose of this test family is to ascertain how much interaction there will be between two tones passing through a device. IM is expressed as a percentage. The appropriate standard should be consulted for the specifics of the test.

It is worth noting that the two distortion specifications refer to a device being tested under static operating conditions. At the present time, dynamic tests are evolving but are yet to be standardized.

III. MICROPHONES

The first component in the system that sound encounters is the microphone, which has the crucial task of converting some aspect of acoustical energy to an electrical analog. The description of a microphone includes the type of converter, its polar pattern, frequency response, and efficiency.

A. Types of Converters and Their Characteristics

1. *The Dynamic Microphone*

The dynamic microphone consists of a diaphragm which is attached to a tube. Around the tube is wound a coil of wire. The coil is positioned in a magnetic field in such a way that, when sound moves the diaphragm, the coil moves through the magnetic flux, inducing a current in the coil.

In simplest form, the dynamic microphone will exhibit a broad midrange peak in its frequency response (Olson, 1957). Flatter amplitude response is achieved by distributing the mass of the diaphragm and by using filter networks (usually acoustic, but sometimes electrical or combinations). The specific virtues of dynamic microphones are their durability, environmental immunity, and electrical simplicity. The inherent noise of the dynamic is limited to that generated by thermal agitation of the coil and diaphragm and is therefore very low.

On the negative side, the dynamic, with its large diaphragm and voice coil, is susceptible to mechanical noise. The average output level of the dynamic is lower than a condenser. If the following electronics are lacking in gain or are noisy, the system noise may be quite high. The high mass of the diaphragm assembly makes it difficult for the dynamic to handle fast rise time signals. The dynamic is not a good choice for extremely low frequency response and its stability is affected by temperature.

2. *Condenser Microphones*

Condenser or capacitor microphones use varying capacitance to generate the electrical analog. The diaphragm is made of very thin conductive material stretched over a support. Close behind the diaphragm is a conductive plate. The two surfaces form a capacitor. As the sound moves the diaphragm, the capacitance varies. This variation of capacitance is used in one of two ways to create the audio output. The most common method is to polarize the element using either an externally supplied dc voltage, or an electret material (a plastic that has a virtually permanent charge). The capacitance change causes a proportional change in voltage that, after an impedance change, becomes the audio output. The other method places the capacitor microphone element in the resonant circuit of an oscillator. The oscillator changes frequency with the change in capacitance, creating an FM output. The FM is demodulated to produce the audio output.

The condenser microphone once confined to studio use is a type that no longer lends itself to generalizations. With the advent of solid-state electronics and electret elements, its usefulness has been greatly expanded. The condenser can offer wide, dependable frequency response, high output, long-term stability, and relative insensitivity to mechanical noise. While not always the case, the design can offer predictable phase response and fast rise times.

The classic difficulties with condensers involve the need for associated electronics. With one version, the requirement is for a polarizing voltage and an impedance converter. The electret eliminates the need for external polarization, but still requires the impedance converter. The impedance converter is commonly called either an amplifier or preamplifier, although I know of no condenser microphone with gain in its electronics. The FM microphone requires an oscillator and demodulator. Assuming the electronics of a condenser microphone to be trouble free, power is still required to operate them.

The light diaphragm, desirable from the sonic standpoint, is easily contaminated. The greatest enemy of the condenser is humidity. If moisture finds its way into the element, it will raise the noise of the microphone, and in extreme cases cause conduction through the element, resulting in large amounts of low-frequency noise at the output. With the electret, the charge on the element will be diminished, reducing the sensitivity of the microphone. In addition, the electret's sensitivity is permanently reduced by elevating its temperature above that for which it was designed (usually above 100°–120°F).

3. *The Piezoelectric Microphone*

The piezoelectric microphone, otherwise called crystal or ceramic, uses the piezoelectric effect to produce its output. A piezoelectric material is by definition any material that produces an electrical voltage when deformed. A diaphragm is either attached to this material or the material's surface itself becomes the diaphragm. While attractively simple, this microphone type has not kept pace with others in quality, and is found only in specialized applications.

4. *The Ribbon Microphone*

The ribbon microphone uses a very light metallic ribbon suspended in a strong magnetic field. As the ribbon moves, a voltage is developed proportional to the particle velocity of the sound wave.

B. Polar Patterns

The polar pattern of a microphone is a graphic representation of the sensitivity of the microphone with respect to the angle of incidence of the sound. There are three basic patterns: omnidirectional, bidirectional or figure eight, and unidirectional/cardioid. Various prefixes, such as super or hyper, are added to better define the degree and type of pattern. Representative polar patterns are shown in Fig. 1. It is easy to forget that a polar plot represents a three-dimensional pattern, as shown in Fig. 2. (This figure assumes symmetry, which is not always the strict case.)

In their elemental form, all microphones except the ribbon have an omnidirectional pickup pattern. All except the ribbon respond to variation in air pressure. In the case of a single diaphragm microphone, acoustical networks are used to create the desired pattern. The purpose of the delay network is to cause the sound

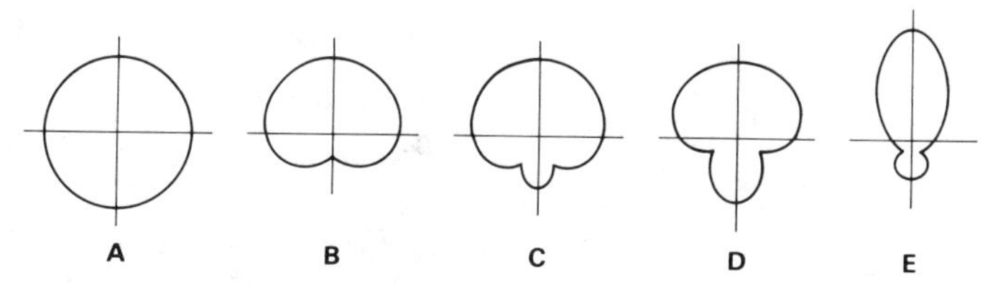

Fig. 1. Microphone polar patterns. (A) Omnidirectional. (B) Cardioid. (C) Hypercardioid. (D) Supercardioid. (E) Shotgun.

coming from the undesired direction to arrive at both sides of the diaphragm at the same instant, resulting in a net pressure differential of zero. The difficulty of creating a microphone-sized acoustical delay equally effective at all frequencies is one reason for the frequency dependence of polar patterns. As the frequency rises, the wavelength diminishes to a point where the physical size of the microphone exerts control over the pattern. With the addition of directional sensitivity, the microphone no longer responds to the pressure but to the pressure gradient.

C. Microphones for Special Applications

1. *Measurement Microphones*

Most microphones sold are manufactured for use in standard audio applications. There are a few manufacturers that offer or specialize in measurement grade microphones. The best known of these is Bruel and Kjaer. The primary purpose of these microphones is to make calibrated acoustic measurements. Extensive documentation is offered as well as an excellent series of application notes. If reliable, repeatable readings are required, the usefulness of this type of equipment should be explored.

2. *Hydrophones*

A hydrophone is a microphone designed for use underwater. Under certain conditions they can be used in air. Specific literature should be consulted.

3. *Wireless Microphone*

A wireless microphone substitutes a radio link for the microphone cable. Commercial units exist with excellent specifications. It should be noted that some incorporate noise-reduction systems. The manufacturer should be consulted for specific applications.

D. Microphone Specifications

It is necessary to view published specifications with a degree of skepticism. They apply only to new units and even then the production variation can be quite wide. If there is a specific parameter that are important to a recording, the

Fig. 2. Three-dimensional microphone polar patterns. (A) Cardioid. (B) Omnidirectional. (C) Figure eight. (D) Shotgun. (E) Supercardioid (see pp. 9–12). (Courtesy Sennheiser Corp.)

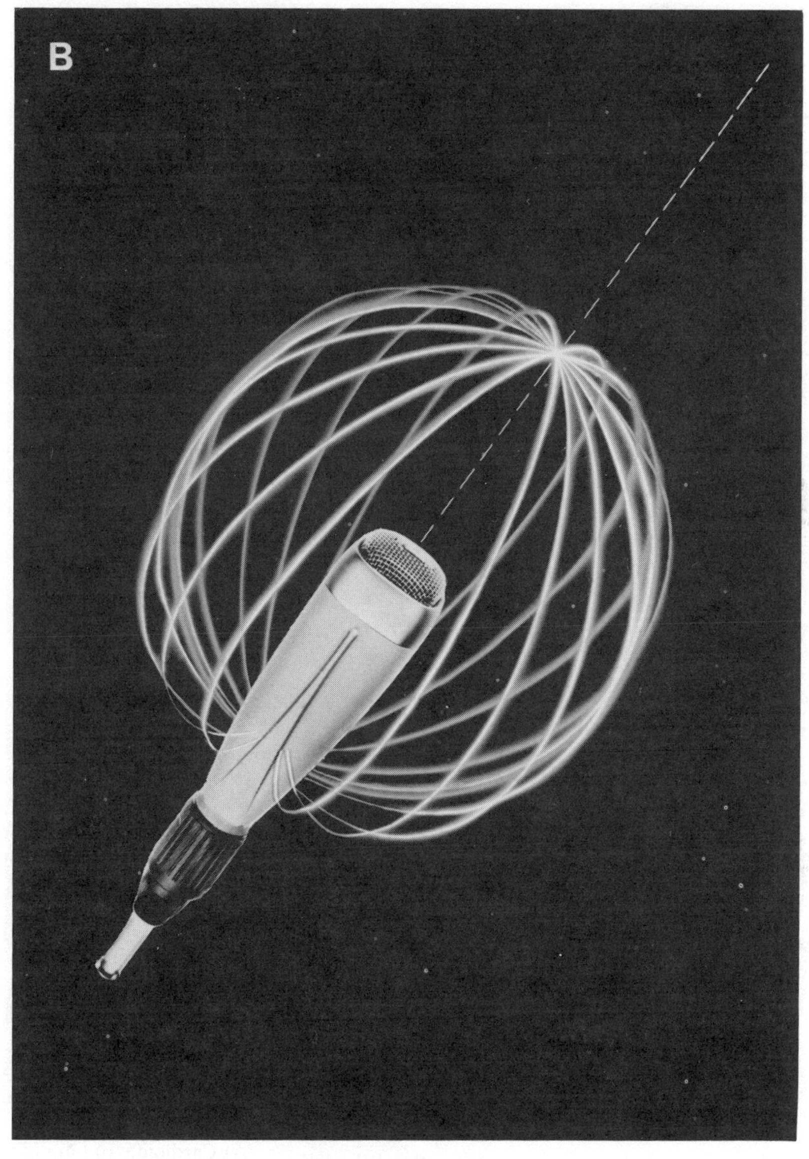

Fig. 2 (*continued*)

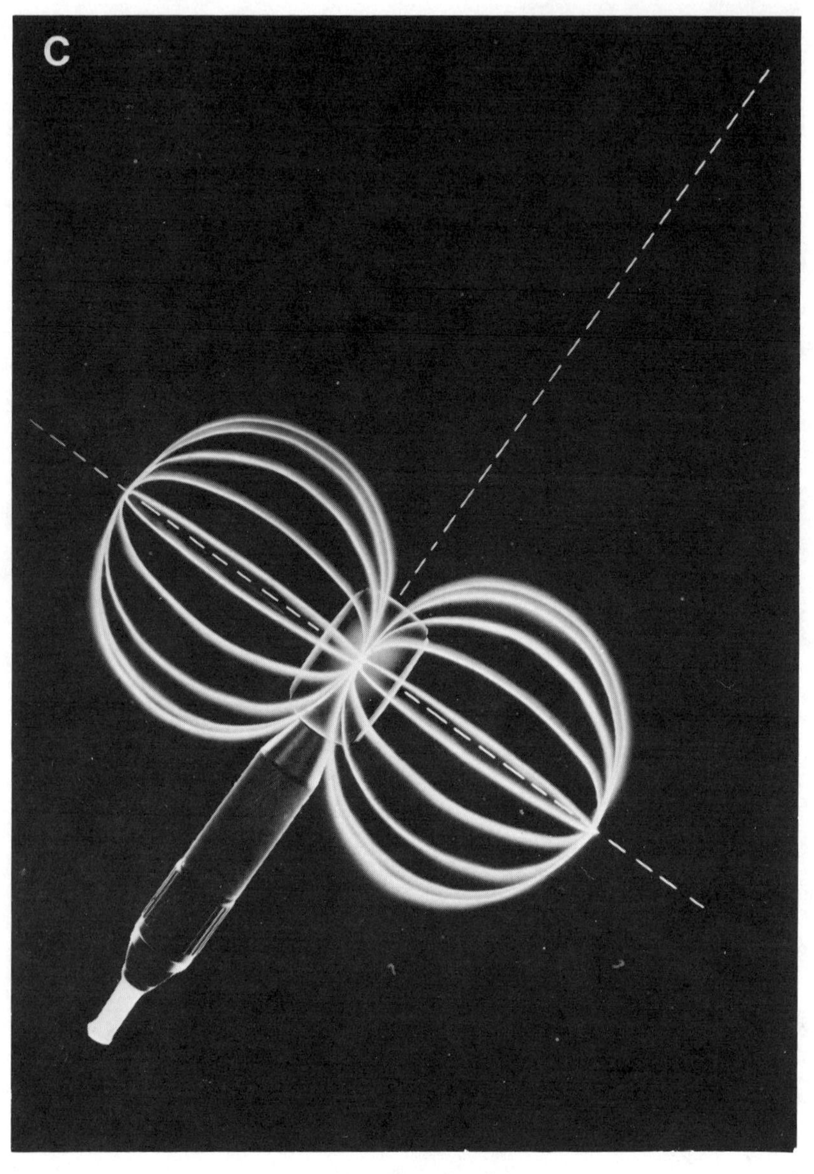

Fig. 2 (*continued*)

Fig. 2 (*continued*)

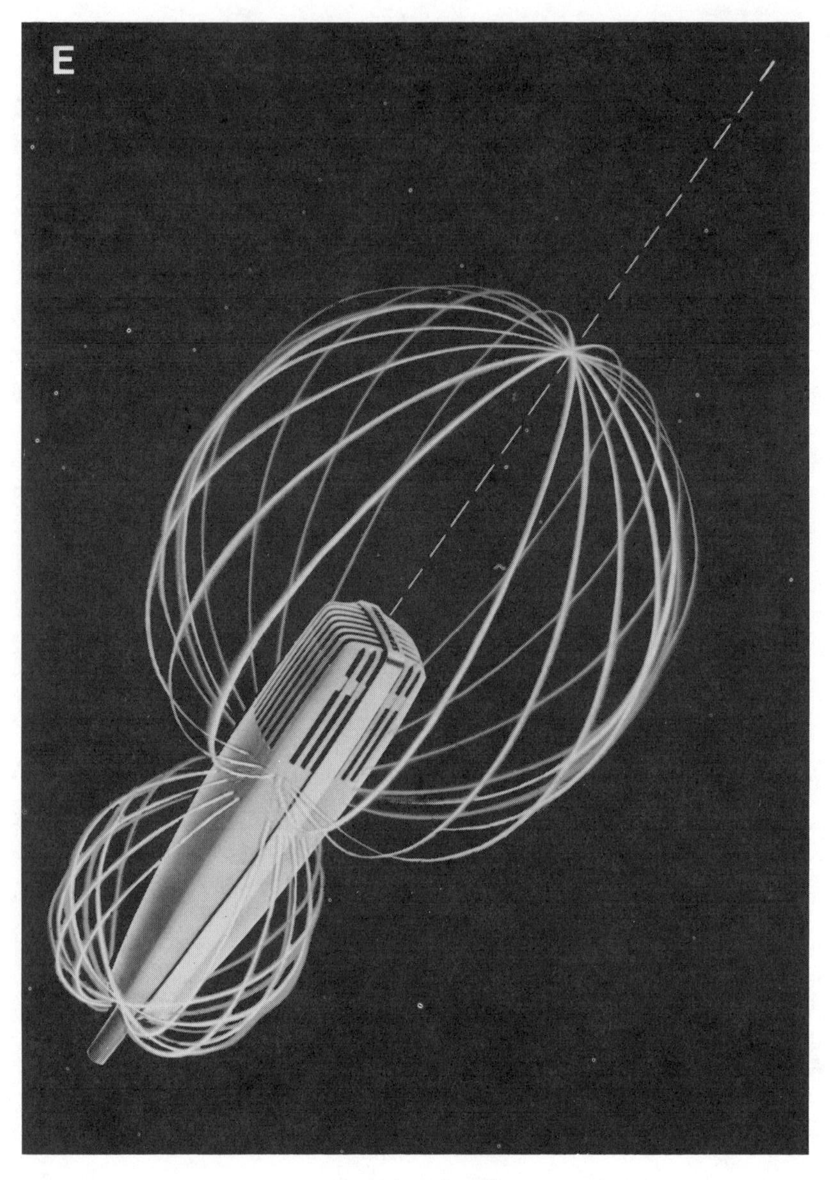

Fig. 2 (*continued*)

microphone should be tested. If stability and consistency of the system's performance are important to a recording project, means of calibrating the system in the field must be obtained.

1. *Output Level and Sensitivity*

For avian recording, the efficiency of a microphone is of major importance, since the sound level is in many cases extremely low. Even the best electronics contribute some noise. All things being equal, however, the microphone with the higher output level is the more desirable. Two terms used to describe a microphone's efficiency in converting sound to electricity are output level and sensitivity. When comparing microphones, be sure to read the specification carefully and ascertain what reference is being used. There are two reference levels in common use. One is a level of 74 dB sound pressure level (SPL) at the microphone. The other is 1 Pascal or 94 dB SPL. Before the specifications can be compared, one must know to what the reading is referenced and whether it is an open circuit or a power reading.

If the specifications describe a microphone operating into a matching load, the output will be 6 dB lower than in the unloaded configuration, which is the way most microphone input circuits are designed. It is possible to reconcile the various rating methods so the output levels can be compared (see Section VII,C).

2. *Noise*

Noise is specified only for microphones that have associated electronics. A noise specification of 17 dB means that the noise at its output is the same as if the microphone had no inherent noise, and was in a sound field of 17 dB SPL. Again, this is usually a weighted specification and the appropriate standard should be consulted.

The noise of a microphone is sometimes given as a signal-to-noise ratio. In most cases, the reference point is 94 dB SPL. This is all well and good except that in field recording it is not uncommon to have an SPL at the microphone of 50 dB. What was an impressive signal-to-noise ratio of 78 dB in reference to 94 dB SPL becomes 34 dB in reference to the 50 dB SPL level. In many cases the noise of a condenser microphone is determined by replacing the capsule with a capacitor of the same value. This does give the noise of the electronics, but it assumes a perfectly dry element, which is seldom found in the field.

3. *Impedance*

The output impedance of a microphone is specified in ohms. For most microphones that are called low impedance, it is on the order of 200 ohms. The importance of this specification is that, to achieve the best signal-to-noise ratio, the impedance of the microphone should be appropriate to the input of the recorder.

4. *Maximum Level*

The maximum sound level a microphone can handle is specified in decibels SPL for a given level of distortion. If the microphone is used close to the source it would be a good idea to consult a reference such as Beranek (1971) on the behavior of sound and microphones in the near field.

5. *Polarity*

The polarity of a microphone is usually defined as the number of the in-phase connector pin. This means a positive-going pressure at the microphone produces a positive-going voltage on the specified pin.

6. *Powering*

A condenser microphone needs some source of electricity to operate. This can be supplied internally by a battery enclosed in the microphone housing or externally by a power source. External power sources can use mains current, batteries, or an associated piece of equipment as the source of electricity.

If the power source is external to the microphone, there must be a way of delivering electricity to the microphone. This can be done with additional conductors in the microphone cable or by combining the power with the audio, using one of several different methods (Phantom, T, etc.). Each method has its relative merits. When buying a condenser microphone make sure that its powering requirements have been accounted for.

E. Accessories

The selection of a microphone involves more than just the initial selection. The microphone chosen must be easy to handle and protected from potential damage. By looking through the catalogs, one can find various useful accessories for mounting microphones.

1. *Windscreens*

A major problem in field recording is eliminating the noise and subsequent overloading caused by wind striking the diaphragm. Wind manifests itself as low-frequency noise that may not be evident on the meter or monitor. The effectiveness of windscreens is seldom given, and in most cases, the recordist must test their effectiveness. The larger the windscreen, and the higher its acoustic resistivity, the better it will reduce wind noise. A windscreen is always a compromise between the degree of shielding offered and the degree to which it changes the response of the microphone.

Most windscreens sold today are made of a plastic foam, either permanently attached to the microphone or designed to be slipped on. Two common pitfalls with this type of windscreen are the use of a foam not designed for the purpose, and the failure to cover the back vents of a directional microphone. Some microphones have a built-in windscreen for use as a blast filter. Their primary purpose is for the close miking of vocalists, but they are not always effective as field microphones. In extreme circumstances, two windscreens can be combined, the resulting loss of response being the lesser of the evils.

2. *Filters*

If a high-pass filter is included in a microphone, it can be an aid in reducing wind and handling noise. Its amplitude and phase response should be such that it does not interfere with the sound being recorded and, as in the case of any signal-modifying device, its use should be noted.

3. *Shock Mounts*

Noise generated by the movement of the microphone can come from a number of sources. It can result from the microphone shaking, the cable conducting noise to the microphone housing, and either the microphone or the cable brushing against objects. Various shock mounts, cables, and handles are sold to aid in reducing handling noise, and some succeed. It is also possible to buy microphones with internal shock mount systems. Any shock mount will be sensitive to the weight of the microphone, the frequency of vibration, the rate of acceleration, and the type of attachment point. One should try the mount, making sure it is workable as a piece of field gear, and that it actually attenuates handling noise. If a shock mount seems to be doing little or no good, make sure the cable (or anything else) is not short-circuiting the mount. Many manufacturers sell shock cables and proper mechanical terminations along with their mounts.

F. Collecting Specific Sounds in the Field

1. *The Parabola*

A parabola focuses onto a single point incoming sound waves that are parallel to its axis. Its effectiveness is determined by the diameter of the reflector in relation to the wavelength of the sound. Its gain and directivity increase proportionately with decreasing wavelength. For wavelengths larger than the parabola's diameter, the response is predominantly that of the microphone itself. For a parabola to be minimally effective at 100 Hz, the diameter must be a little over

11 ft. The practical solution is to use as large a reflector as possible in conjunction with an omnidirectional microphone. The resulting combination is omnidirectional up to the point the diameter becomes significant with respect to wavelength and has increasing directivity and gain as the frequency increases. Approximate wavelengths of sound in air corresponding to common parabola diameters are: 36 inches = 370 Hz; 24 inches = 565 Hz; 18 inches = 750 Hz; 13 inches = 1040 Hz.

Placement of the microphone in the parabola is governed by the designer's choice of focal point. A parabola can be constructed with varying degrees of curvature, each equally effective as a reflector. The deeper the curvature, the closer to the back of the dish is the focal point. If the focal point is placed inside the plane of the edge of the reflector, it is shielded from the wind, but since the reflector is also a resonant cavity, the microphone picks up the resonance. Noise from the movement of the parabola (either from wind or handling) will be louder the closer the microphone is to the parabola. If the focal point is outside the plane of the edge of the reflector, these effects are minimized but the benefit of shielding from the wind is lost. Commercial realizations exist with the focal point at any of these locations and a number exist with the focus at the approximate edge. The microphone is not always placed at the precise focal point. This is done both to broaden the beam width at the extremely high frequencies and to involve more of the microphone's diaphragm. Note that not all microphones work efficiently in a parabola and a gain test should be a part of choosing a microphone.

2. *Shotgun Microphones*

The shotgun microphone is a cardioid microphone fitted with an interference tube. The interference tube causes phase cancellation of sound arriving off-axis. A phrase often heard describing a shotgun microphone is that the microphone has "greater reach." This does not mean that the microphone has some ability to collect more energy from the source. As the distance from the source increases, the energy per unit area decreases. The shotgun microphone can respond only to the energy reaching its diaphragm. The advantage lies in its ability to reject more of the off-axis sound.

Consider that the microphone is surrounded by a sphere of noise. Interpose in this sphere of noise the three-dimensional polar pattern of Fig. 2. It can be seen that the more narrow the pattern, the less noise will be accepted. The wider the microphone's pattern, beyond that needed to receive the desired sound, the more noise is accepted. If the noise is not equally distributed spatially, then the microphone's specific angular rejection characteristics become important. Since it is not always possible to aim a microphone precisely, some manufacturers construct their shotgun microphones in a way that reduces the high-frequency beaming. This is similar in effect to the slight defocusing of a parabola.

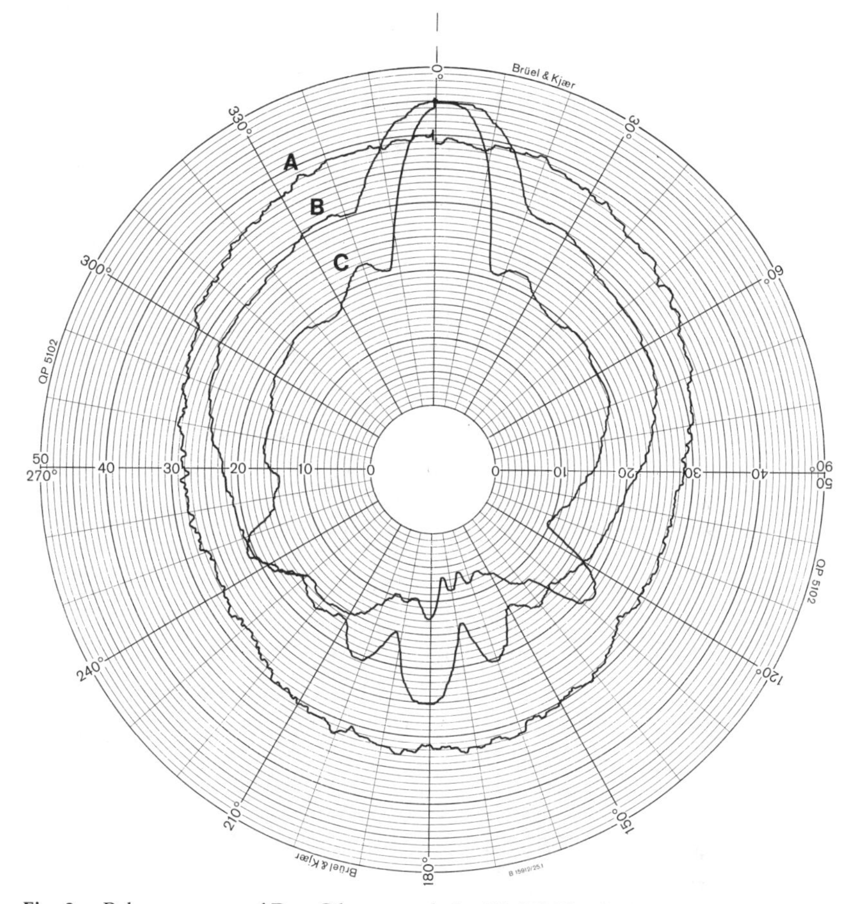

Fig. 3. Polar response of Dan Gibson parabola. (A) 500 Hz. (B) 2500 Hz. (C) 5000 Hz.

G. Some Illustrative Tests

1. *Representative Equipment*

To examine how these devices function, I tested three different parabolas as well as one omnidirectional microphone and one shotgun microphone. These tests were performed under controlled conditions and, while comparable, they are not absolute.

The Dan Gibson parabolic microphone is an assembly incorporating an 18-inch reflector, dynamic microphone, electrical equalizer, and monitor amplifier. Its polar response is shown in Fig. 3, its axial frequency response in Fig. 4. Since the response curve includes the response of the filter network, the filter's re-

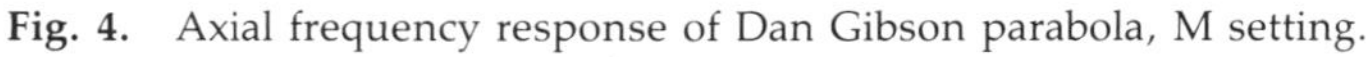

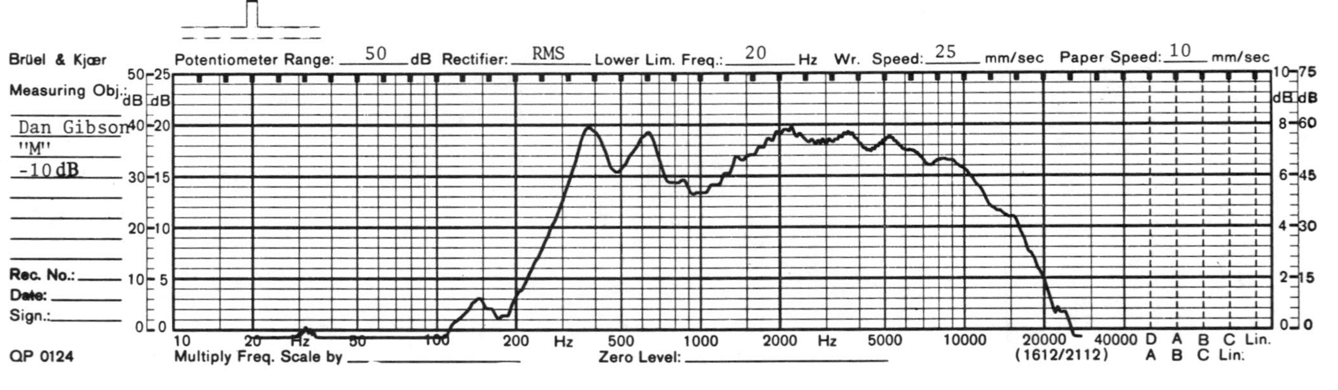

Fig. 4. Axial frequency response of Dan Gibson parabola, M setting.

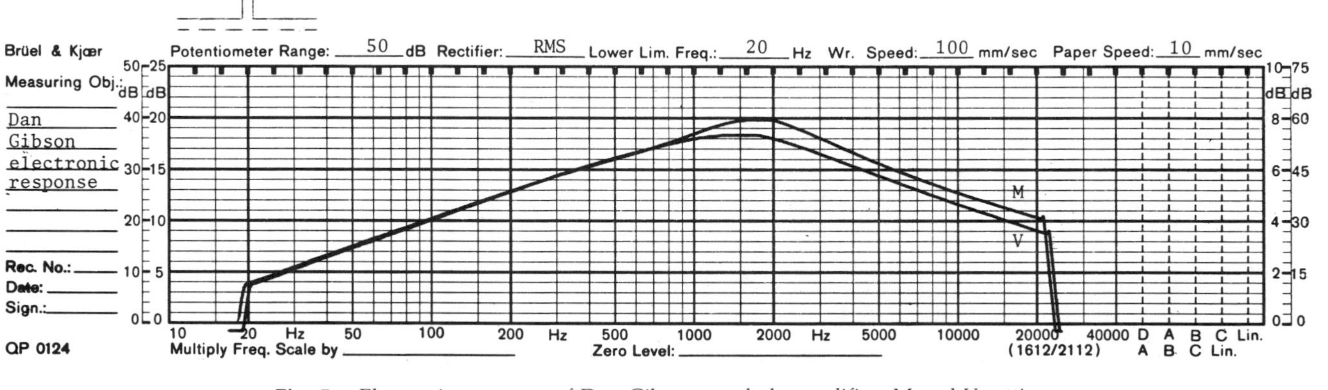

Fig. 5. Electronic response of Dan Gibson parabola amplifier, M and V settings.

sponse alone is shown in Fig. 5. The assembly offers the advantage of a translucent dish to allow for visual aiming. It does not offer the option of interchangeable microphones, and it requires batteries for operation.

The Sony PBR-330 has a diameter of approximately 13 inches. It is offered as an accessory item, and is designed to accommodate a wide range of microphones. The parabola was fitted with a Sennheiser MKH-104 omnidirectional condenser microphone. The polar pattern and axial frequency response are shown in Figs. 6 and 7. At the same time, the microphone's axial response was tested without the reflector (Fig. 8). Band-limited pink noise was fed through the source (200–8000 Hz, 18 dB/octave) and the output level of the microphone was noted. The difference between the microphone alone and with the reflector is

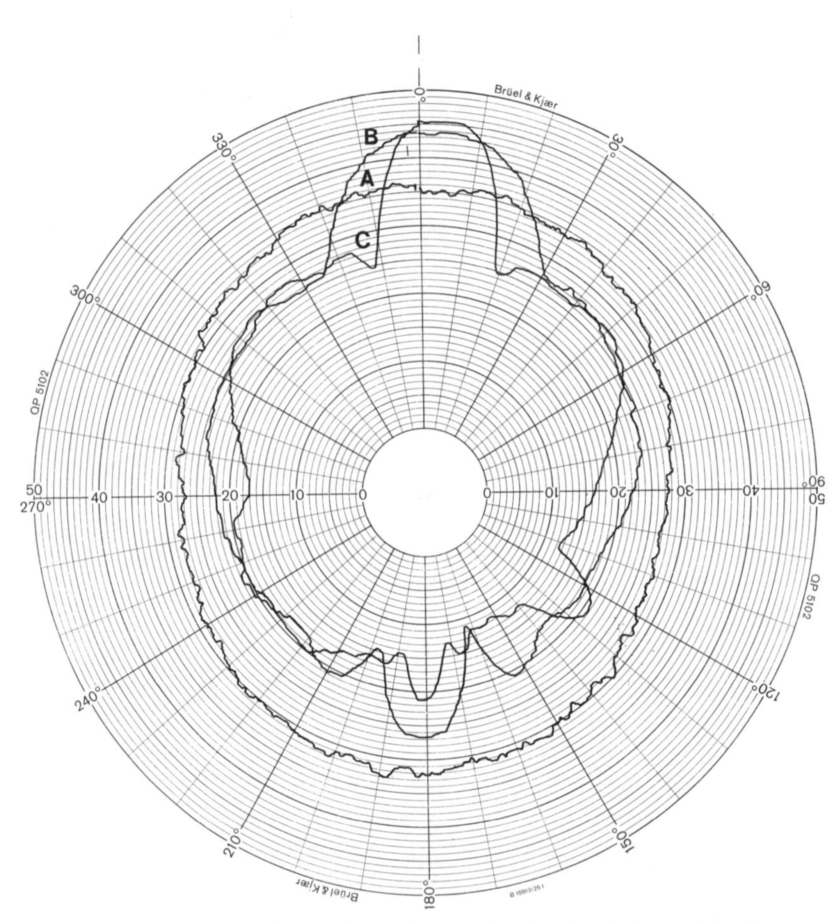

Fig. 6. Polar response of Sony PBR-330 parabola with Sennheiser MKH-104 microphone. (A) 500 Hz. (B) 2500 Hz. (C) 5000 Hz.

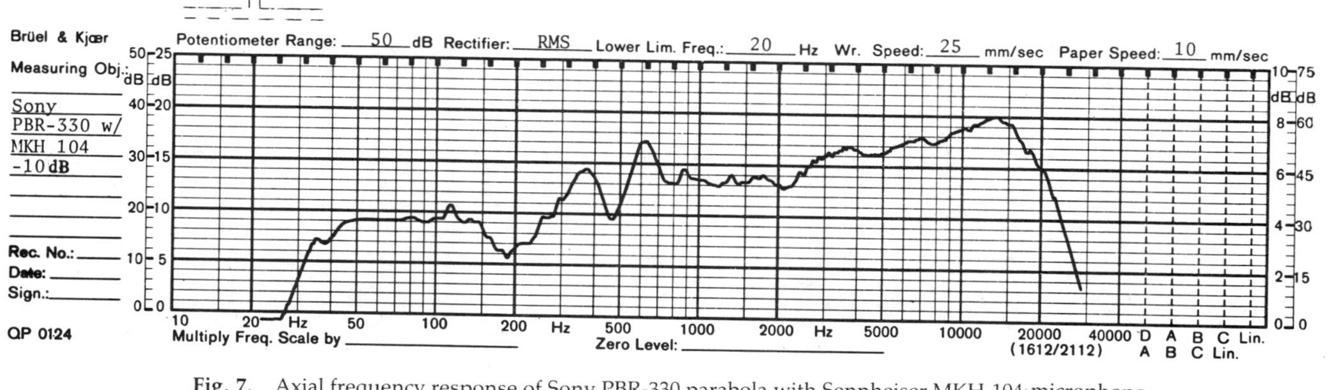

Fig. 7. Axial frequency response of Sony PBR-330 parabola with Sennheiser MKH-104 microphone.

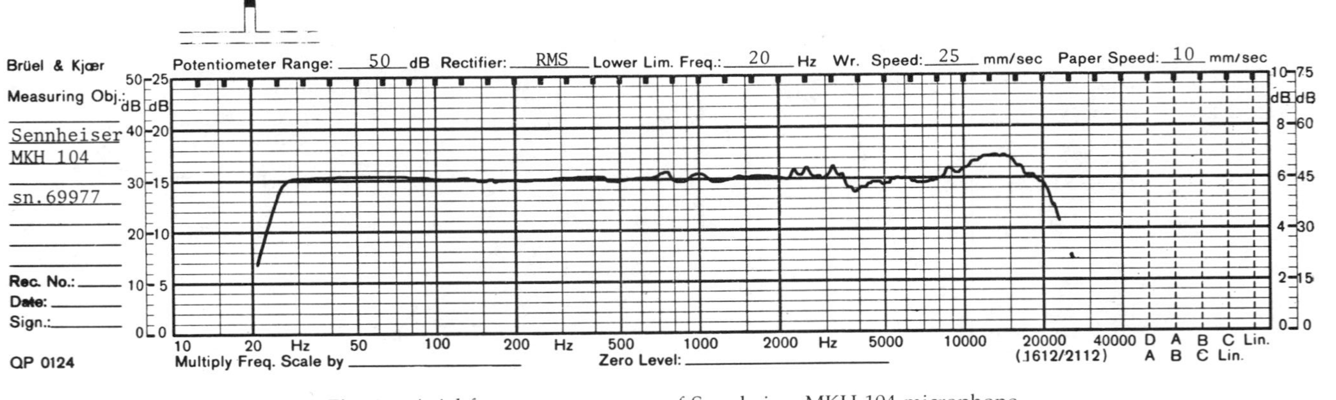

Fig. 8. Axial frequency response of Sennheiser MKH-104 microphone.

+14.9 dB. The Sony offers the advantage of a plastic reflector, and also gives the user the option of selecting microphones appropriate to particular applications.

In that both of the parabolas discussed are relatively small for many avian sounds, the same tests were conducted using a 36-inch aluminum parabola fitted with the Sennheiser MKH-104. The results are shown in Figs. 9 and 10. While its additional size is a hindrance in the field, its better gain and directivity indicate it should be considered. The difference between the microphone's output by itself and mounted in the reflector is +18.9 dB.

For purposes of comparison, the response of a Sennheiser MKH-805 shotgun microphone is shown in Figs. 11 and 12.

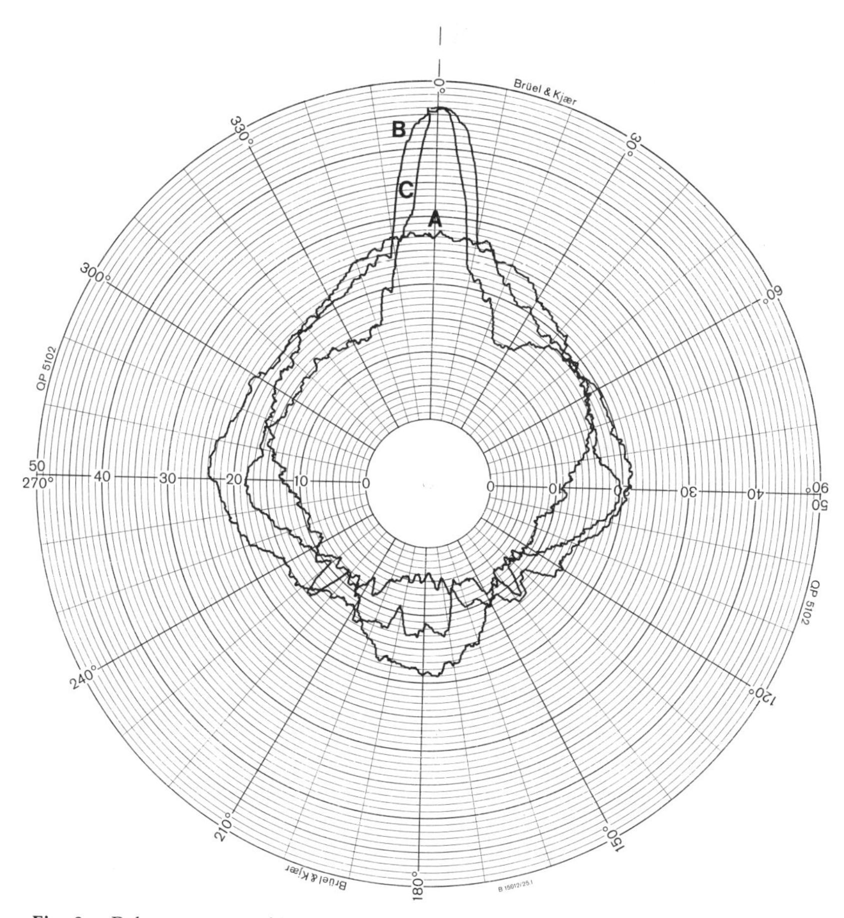

Fig. 9. Polar response of 36-inch parabola with Sennheiser MKH-104 microphone. (A) 500 Hz. (B) 2500 Hz. (C) 5000 Hz.

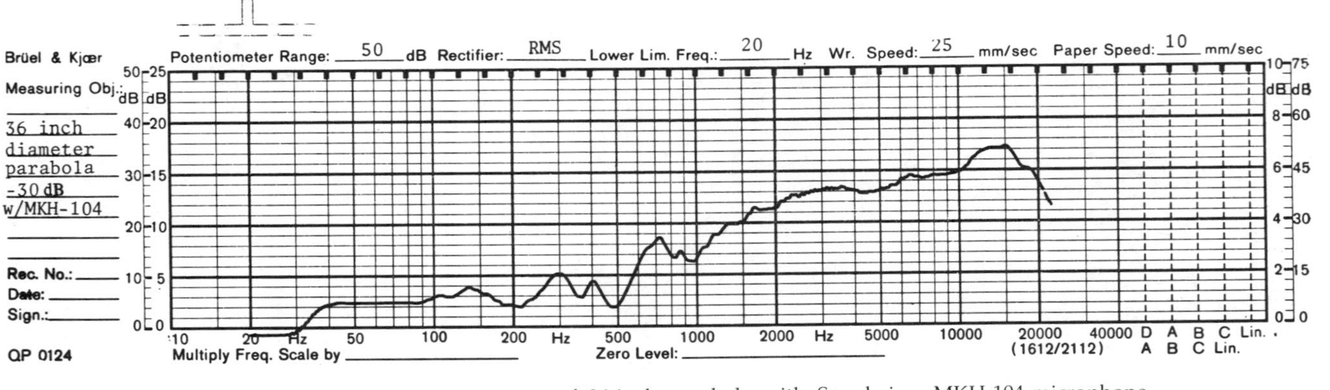

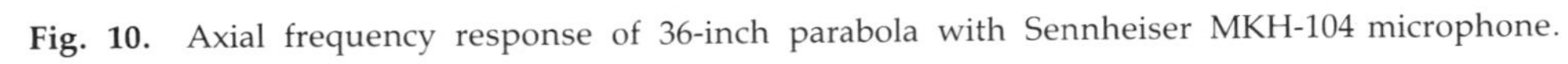

Fig. 10. Axial frequency response of 36-inch parabola with Sennheiser MKH-104 microphone.

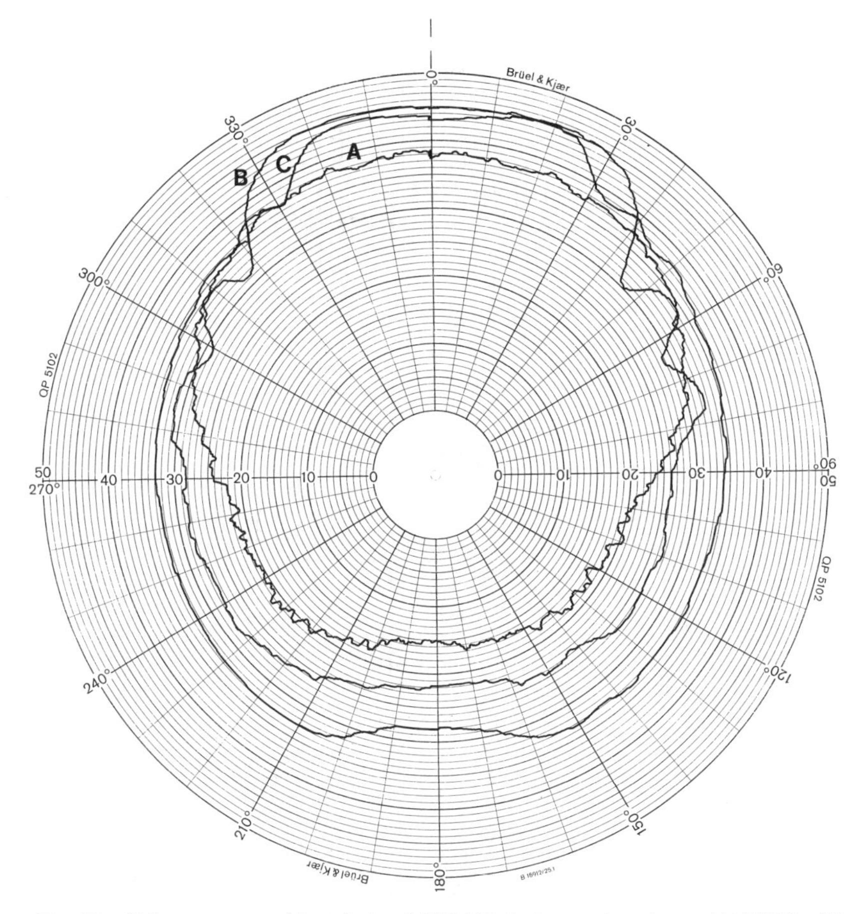

Fig. 11. Polar response of Sennheiser MKH-805 shotgun microphone. (A) 500 Hz. (B) 2500 Hz. (C) 5000 Hz.

2. Positioning a Microphone in a Reflective Environment

To illustrate one of the problems in microphone positioning, consider Fig. 13. It can be seen that in one position there are many potential paths by which the sound can reach the microphone. In the other, where the microphone is placed at a boundary, there is only one path (until very short wavelengths are involved). An experiment was set up with the microphones positioned as indicated. The resultant response curves are shown in Fig. 14. It can be inferred that, in a reflective environment, microphone placement is important if anomalies in the recording are to be avoided.

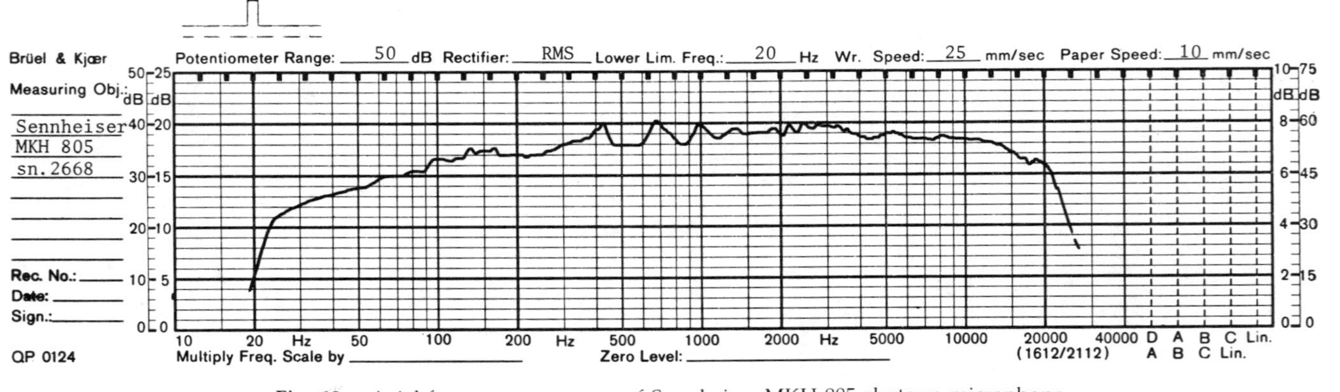

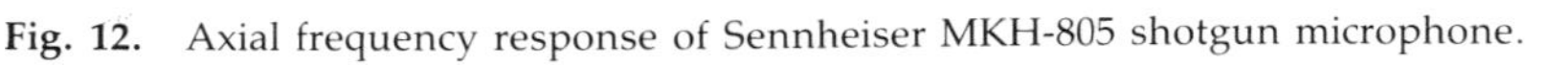
Fig. 12. Axial frequency response of Sennheiser MKH-805 shotgun microphone.

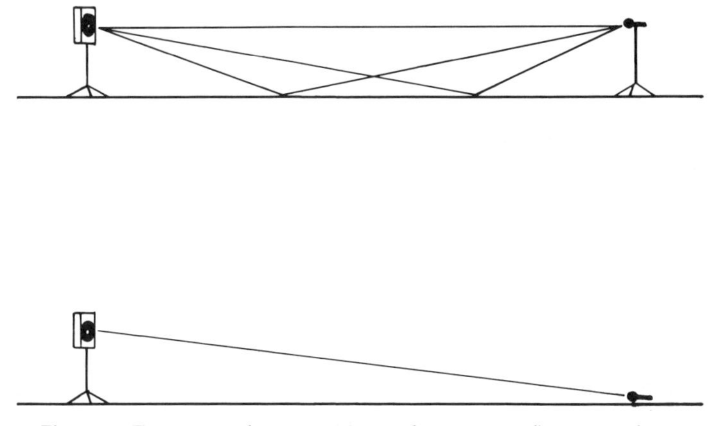

Fig. 13. Two microphone positions relative to a reflective surface.

H. Additional Considerations in Evaluating Microphones

1. Suitability of a Microphone for an Application

When evaluating a microphone, its physical suitability for the intended use must be considered as much as its electroacoustic capabilities. When procuring equipment of any kind, it is important to consider what it was designed to do. This may be stating the obvious, but if a piece of equipment is designed for a specific purpose, it is likely to be best suited to that application. Since equipment is seldom designed for use in avian recording, one is required to choose devices that have applications analogous to avian recording. By comparing the design goals with your application, you can get some indication of suitability.

In most bird sound recordings, I would say that the signal-to-noise ratio is fixed at the time it is made. In other words, the environmental noise recorded exceeds the electronic noise of the system. To improve the signal-to-noise ratio requires that more signal reach the microphone element. Four ways to accomplish this are: (1) get the source and microphone closer; (2) capture more of the energy; (3) reject unwanted sound; (4) some combination of these. Getting the microphone closer to the source is an attractive option. While it requires a lot of wire, proper impedance microphones, and much patience, it is a technique that should be used more than it is. The need to capture more energy indicates the need for a parabolic reflector. Rejecting unwanted noise is the purpose of all directional microphones, especially the shotgun microphone.

2. Checking a Microphone in the Field

If one is trying to make recordings where absolute SPL information is required, an appropriate microphone with a compatible pistonphone or acoustic

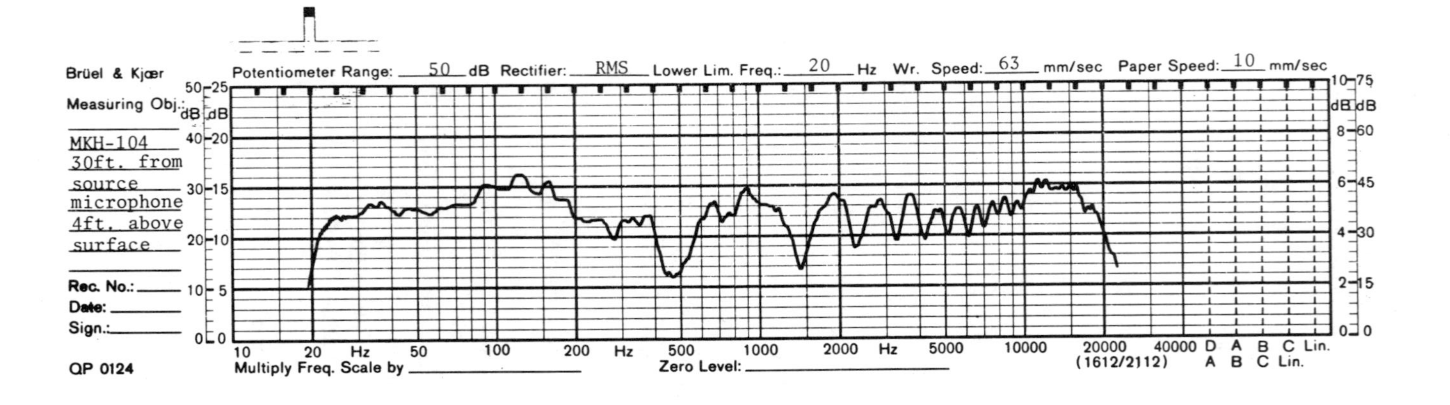

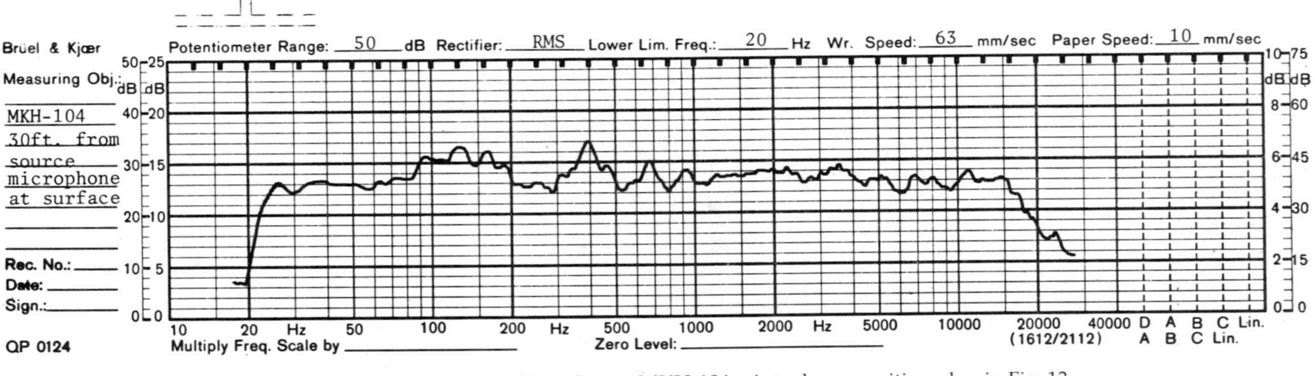

Fig. 14. Frequency response of Sennheiser MKH-104 microphone positioned as in Fig. 13.

calibrator should be carried (manufacturers: General Radio, Bruel and Kjaer).

For a rough check of microphone deterioration, the following techniques can be used. When the microphone is new and assumed to be operating correctly, make a recording of your own voice outdoors in a quiet nonreflective environment. Position yourself at a good working distance from the microphone. Note the settings of all the controls and your distance from the microphone. Save this recording and periodically repeat the test. By comparing the recordings, you can hear if something has changed. Of course, if a new microphone of the same model as the one in question is available, a simple A–B comparison can be performed. For a rough indication of polar response, you can walk a circle around the microphone and note your position on the tape. By listening to this recording, you can hear how angular incidence affects the recording. Any crucial parameter should be verified by further testing.

IV. THE TAPE RECORDER

A. Some Fundamentals and the Direct Recorder

There are four basic elements involved in recording and playing back magnetic tape.

1. The transport mechanism which moves and guides the tape.
2. The recording section which takes the signal from the microphone and records it on the tape.
3. The playback system.
4. The tape itself.

Magnetic recording is possible because certain materials can be magnetized by an external force and remain in that state after the force is removed. To make a recording, the magnetic material is moved past a transducer or "head" at a linear rate. The tape and the record head form a magnetic circuit. As the tape is moved past the head, the magnetic particles are magnetized in proportion to the current flowing in the record head at that instant.

One of the bothersome characteristics of the magnetic medium is that it does not acquire magnetism in direct proportion to the applied magnetic force. To overcome this problem, an ultrasonic signal, or bias, is added to the record signal to linearize the recording characteristic. The frequency chosen for the bias must be high enough so that it will not be recorded and not beat with harmonics in the signal. In addition, each magnetic material has its own magnetization curve. This means in practice that the tape and the recorder must be matched. If one is not sure about a particular machine/tape match, a qualified technician should be consulted.

To play back the signal, the tape is moved past another transducer similar to the record head (combined in some machines). As the tape is moved across the gap of the play head, the magnetic force causes an electrical signal to be produced. The level of the output produced is dependent on the rate of change across the gap. In other words, unless the tape moves, no signal is produced. As the output is produced by the rate of change of the signal, it can be seen that it must vary with frequency. If the bandwidth of interest is recorded on the tape at an equal level of magnetization and then played back, the output will rise with frequency at a rate of 6 dB per octave.

In any tape recorder, compromises exist between tape speed, gap width, high-frequency limit, and system noise. If the head gap is decreased, the high-frequency performance improves, but the low-frequency output decreases. As the tape speed is increased, the recorded wavelength of a specific frequency increases, but more tape is used. When the recorded wavelength reaches the same dimension as the gap width of the play head, there can be no change, hence no output.

If the rudimentary recorder just described were actually constructed, it would be called a "direct" recorder. The advantages of direct recording are simplicity as well as appropriateness to a wide variety of applications. Depending on tape speed, it is possible to obtain frequency response of at least 5 Hz to 500 kHz. A dynamic range of 50 dB is not unusual.

The major disadvantages of direct recording are limited low-frequency response, amplitude variation, and time variation. Limited low-frequency response is not a problem for most avian recording as most machines have dependable response to 30 Hz. Time variation, both short and long term, can be a problem if the work requires accurate frequency determination or a low residual FM.

Amplitude variation, or modulation deserves a little more discussion. Most often called dropout, the amplitude modulation (AM) in analog recording is caused by oxide variation and/or by the tape moving away from the head. The tape can be lifted from the head in a number of ways. The two most common are surface imperfections in the tape itself and surface contamination of either the tape or the heads. The effect can be quantified by the equation: Playback Drop in dB $= 54\ D/L$, where D equals the separation from the head, and L equals the wavelength. For example, if the tape is lifted from the head 0.0005 inch (human hair diameter, approximately 0.003 inch), the drop in level of a 10-kHz signal recorded at 7.5 inches/sec is 36 dB.

B. The Audio Recorder: A Specialized Direct Recorder

In order to minimize noise and use the maximum capacity of the tape for audio recording, circuits to add preemphasis and deemphasis to the signal are added to the basic direct recorder. Early on it was observed that in speech and orchestral music, the predominant energy is in the midband, decreasing at the extremes. Using this spectral distribution as justification, the signal is equalized to better distribute the energy. The level is increased with frequency during recording and attenuated during playback. The record equalization (preemphasis) is not the perfect inverse of the playback in that various record/play losses are addressed at the same time. These excess losses are greater at slower tape speeds.

Because of the preemphasis of the record signal and tape oxide characteristics, the saturation point for an audio recorder is not linear with frequency. In other words, the higher the frequency of a signal at a given input level, the closer the recorder is to severe distortion. At tape speeds of 15 inches/sec and above, which require less pre-emphasis, this effect is almost nonexistent. At lower speeds the amount of record equalization is substantial. This is particularly important to the avian recordist, because virtually no devices used to monitor levels are spectrally weighted and the spectral distribution of avian sound bears little resemblance to the speech and music curves used to justify the choice of the standard record/playback equalization curves.

The result is that, given a bird whose song has its predominant energy in the 4- to 6-kHz band, and a tape recorder running at 3.75 inches/sec, an unweighted meter reading could easily be in error by 10 dB. In this example, the actual recording level, due to the action of the record equalization, will be 10 dB higher than the level indicated by the meter. If the level is set to the meter's maximum, the actual record level will be well into the distortion range of most recorders. If the tape speed is slower, or the sound higher in frequency, the error will be greater. Figure 15 is a group of amplitude-versus-frequency charts made on one channel of a Nagra IV-S operating at different speeds, the input level being decreased in 10-dB increments. Figure 16 gives the results of the same tests using a Marantz PMD-220 cassette recorder. The loss of high-frequency response at the higher levels is due to saturation.

C. Other Types of Recorders

1. Instrumentation Recorders

The commercial realization of the simple direct recorder is available as one version of an instrumentation recorder, or a "scientific recorder." If one is recording a signal with a spectral distribution totally incompatible with an audio recorder, an instrumentation type of recorder should be considered.

Instrumentation recorders offer many options. Care should be taken during selection to make sure the correct options are purchased for the type of recording intended. Tapes do not necessarily interchange among different brands of instrumentation recorders, a capability we assume with audio recorders.

Along with direct recording capability, instrumentation recorders can offer

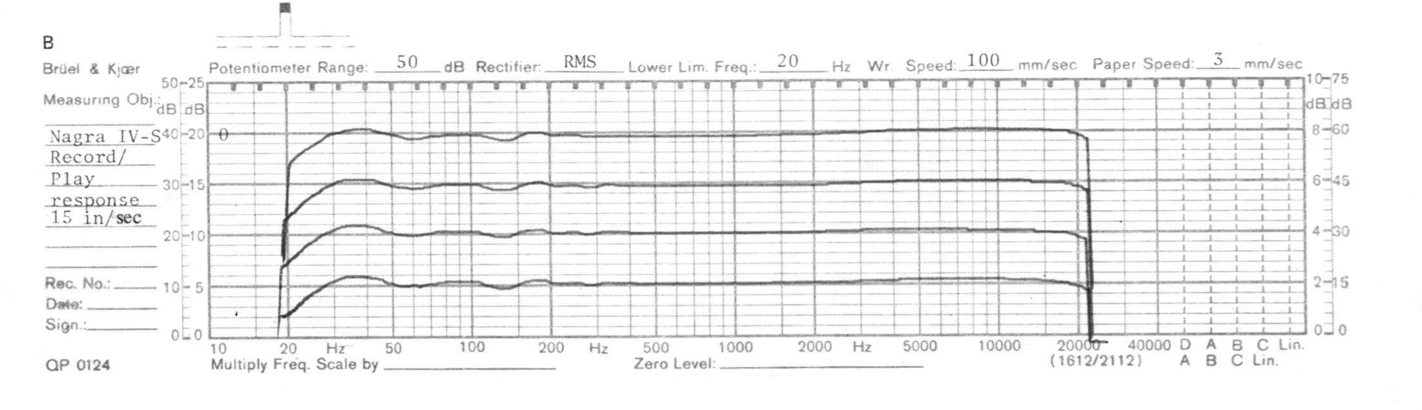

B
Brüel & Kjær
Measuring Obj.:
Nagra IV-S
Record/
Play
response
15 in/sec
Rec. No.:
Date:
Sign.:
QP 0124
Potentiometer Range: 50 dB Rectifier: RMS Lower Lim. Freq.: 20 Hz Wr. Speed: 100 mm/sec Paper Speed: 3 mm/sec
Multiply Freq. Scale by
Zero Level:
(1612/2112)

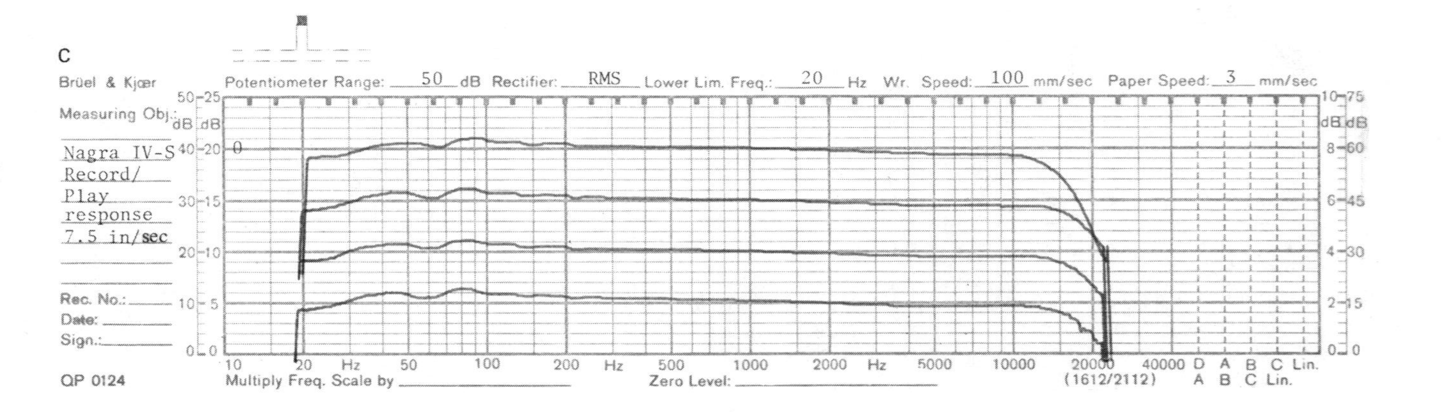

C
Brüel & Kjær
Measuring Obj.:
Nagra IV-S
Record/
Play
response
7.5 in/sec
Rec. No.:
Date:
Sign.:
QP 0124
Potentiometer Range: 50 dB Rectifier: RMS Lower Lim. Freq.: 20 Hz Wr. Speed: 100 mm/sec Paper Speed: 3 mm/sec
Multiply Freq. Scale by
Zero Level:
(1612/2112)

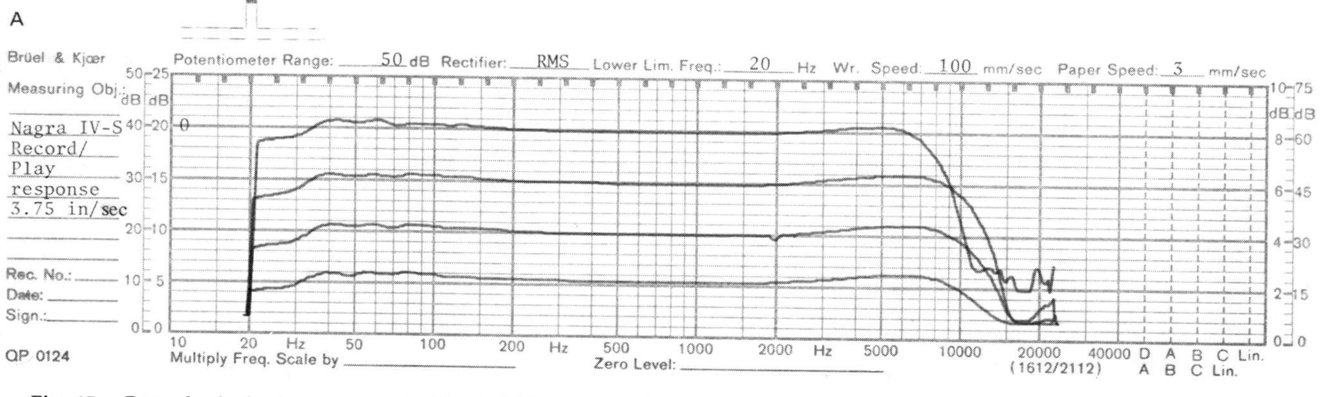

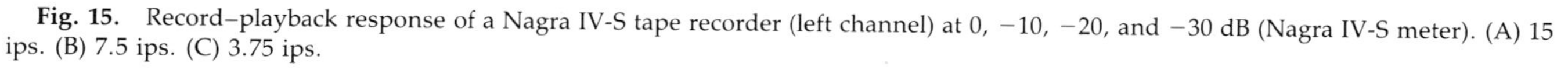

Fig. 15. Record–playback response of a Nagra IV-S tape recorder (left channel) at 0, −10, −20, and −30 dB (Nagra IV-S meter). (A) 15 ips. (B) 7.5 ips. (C) 3.75 ips.

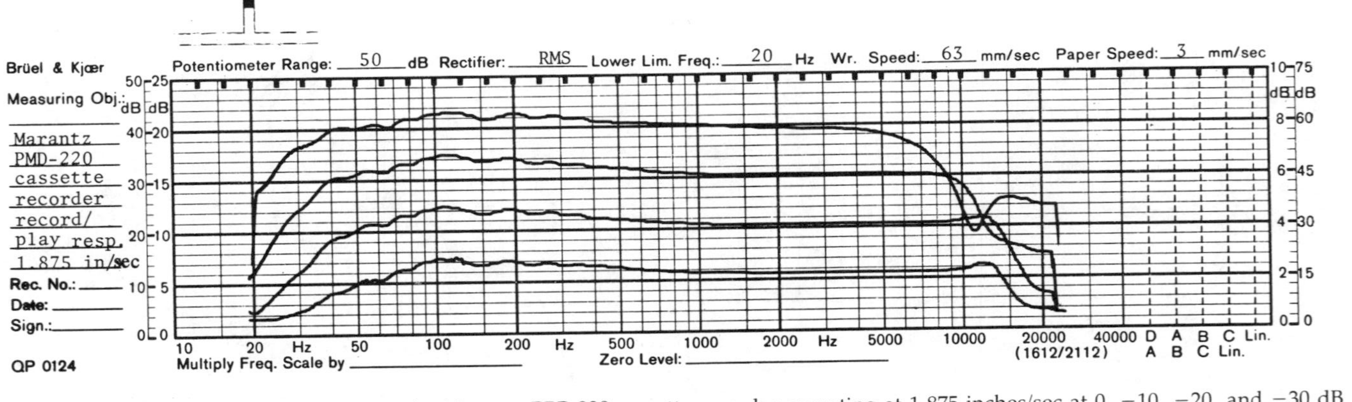

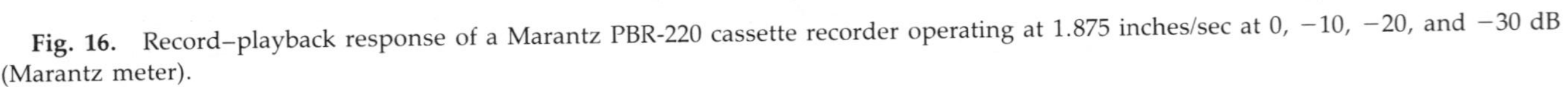

Fig. 16. Record–playback response of a Marantz PBR-220 cassette recorder operating at 1.875 inches/sec at 0, −10, −20, and −30 dB (Marantz meter).

FM and digital recording. FM recording differs from direct in that a carrier is modulated by the input signal. It is this modulated carrier that is recorded. This in turn is demodulated upon playback. FM recording evolved in answer to the two basic limitations of the direct process, the inability to record low frequencies (to dc) and amplitude instability.

One of the limiting factors of the FM system is the ability of the tape transport to maintain absolute tape speed since any variation in speed will manifest itself as spurious modulation and increase the noise. The FM system preserves the phase versus frequency characteristic of the input signal. The system is more complicated electronically and is far less efficient in its utilization of tape. Although it gives extremely low frequency response, its high-frequency capability is generally limited to about one-fifth that of direct recording at the same tape speed.

The FM instrumentation recorder is not a general purpose recorder, but if one needs to record extremely low frequencies, have faithful phase-versus-frequency response and good immunity to amplitude variations, it may have to be considered.

2. *Digital Recorders*

Digital recording uses the magnetic medium to store logic states. This type of recording is used extensively in the data processing industry, to a degree in instrumentation recorders, and is beginning to find application in audio recording. Since the signal consists of binary information the recording process itself is relatively simple.

To record sound, the signal must be converted to digital information. This is called an analog-to-digital (A-to-D) conversion. Once digitized, the digital bit stream is recorded. When played back it must be reconverted to an analog signal. This process is called digital-to-analog (D-to-A) conversion.

The A-to-D converter takes discrete samples of the analog waveform (typically 50,000 per second for commercial audio recorders) and quantizes them. The more often we sample, the wider the bandwidth of the system. In addition, the more bits used to quantify the sample, the more accurately the signal is represented, and the lower the noise. At the moment, systems in common use have a sampling rate of approximately 50 kHz and use 12- to 16-bit quantization.

To protect the recording from lost information, various error-correcting schemes have evolved. Discussing the relative merits of these is beyond the scope of this chapter. Suffice it to say, the better the error-correcting scheme, the more storage space it requires.

Digital audio recording requires compromises between bandwidth (or sampling rate), accuracy of each sample, signal-to-noise ratio, effectiveness of the error-correcting system, and amount of tape required. At the moment, there are no standards for these variables, with resulting problems of tape interchangeability. Given more time for the technology to evolve, digital audio recording will become a method in general use with capabilities for accurate recording beyond those available with most analog recording systems.

D. Metering

1. *Types of Meters*

No matter which method of recording is chosen, the machine will have a specific limit to the range of signals it can handle. To get the best signal-to-noise ratio one records at as high a level as possible without reaching the point at which the signal becomes grossly distorted. To do this some means of indicating the level of the signal, usually some form of meter, is required. The most familiar is the Volume Unit (VU) Meter. A 1-dB level change of a sine wave equals a change of 1 VU.

This meter was developed to answer the young audio industry's need for a standard operating level. Previous to this, each organization had its own standards and program interchange was difficult. In 1939, the characteristics of the VU meter were standardized along with a reference level of 1 mW into 600 ohms. The section of the standard that is particularly important to avian recording is the one specifying meter ballistics. The standard requires the meter to give a 100% reading with a pointer overshoot of not less than 1% and not more than 1.5% (0.15 dB). Furthermore, the pointer must reach 99% of the 0-VU mark in 300 msec (Tremaine, 1969) in response to the sudden application of a sine wave at a level equivalent to 0 VU.

It is a testimony to the design that after 40 years, the VU meter is still an accepted standard. The meter, however, was designed to indicate levels of human speech and orchestral music. Peaks in sound can occur in much less than 300 msec, making it possible for a VU meter to give an inaccurate reading of the true level. In answer to this potential deficiency, various peak-indicating meters have evolved. One standard for peak metering is the Peak Program Meter (PPM). Its standard calls for an integration time of 12 msec (Gordon and Wood, 1979). If the fallback time were the same as the integration time as in the case of the VU meter whose pointer returns to minimum scale in 300 msec, the pointer of the PPM would be a blur for most program material and the human eye would find it impossible to register the reading. To overcome this problem, a longer fallback time is specified. For the PPM it is 1 sec.

There are numerous metering schemes of varying efficacy. It is unfortunate that their integration characteristics are not always specified. In less expensive equipment, it is common to find a meter that, although labeled "VU," does not meet the standard.

A key word in all meter specifications is "program." The intent of the designers is not to provide an absolute measure of the minutest peak but to provide a means to easily set levels during a program. The predictable and periodic nature of most musical and spoken program material is what allows the VU meter to be the useful tool it is. In the case of bird sound where it is possible to find peaks in the 5-msec range occurring over relatively long intervals, it is possible to have substantial metering errors.

2. *Metering Errors*

There are two interlocked variables that cause metering errors. One is the duration of the pulse itself. The other is the time between succeeding pulses. In the case of the VU meter, the analogy of pushing a swing is appropriate. The swing will move a given distance either in response to one mighty shove or a succession of small shoves. The longer between shoves the less effective they are.

I carried out an experiment on three different meters using a continuous tone set to yield a reading of zero. The mode of the oscillator was then changed to generate a pulse of sine waves of varying length. Interpulse distance (the time between pulses) was also varied. Tables I and II show the metering error for various settings of the oscillator. If the pulse is short enough and/or the interpulse time is long enough substantial metering errors can result.

With any metering system, it is necessary to determine what a reading means relative to a given amount of distortion of the recorded signal. In the case of a recorder equipped with a VU meter, the zero calibration point cannot correspond to the saturation point since any signal occurring at a higher level for less than 300 msec will distort. The designer must decide how much headroom to leave between the maximum reading on the meter, and the point at which objectionable distortion occurs.

Meter specifications of some current recorders are: Nagra IV-S, meter integra-

TABLE I

Response of Three Meters to a Constant Amplitude 10 kHz Tone Burst of Varying Duration[a]

Burst length (msec)	50-msec Repetition rate			500-msec Repetition rate		
	Nagra	Uher	VU	Nagra	Uher	VU
100.0				0 dB	0 dB	−5 dB
50.0	0[b]	0[b]	0[b]	0	0	−10
10.0	0	0	−11	−1	−4	−19
5.0	0	0	−15	−4	−8	OS[c]
1.0	−7.5	−10	OS[c]	−16	OS[c]	OS
0.5	−12	−20	OS	−20	OS	OS

[a] Tone burst repeated every 50 and 500 msec.
[b] Continuous time.
[c] Off scale.

TABLE II

Response of Three Meters to a 0.1-msec Signal Repeated at Different Intervals[a]

Interpulse interval (msec)	Nagra	Uher	VU
0.05	0 dB	0 dB	−3 dB
0.10	0	0	−5
0.50	0	0	−12.5
1.00	0	0	−16
5.00	−6	−5	OS[b]
10.00	−9	−11	OS
50.00	−19	OS	OS
100.00	OS	OS	OS

[a] Amplitude set to 0 dB for a continuous 10-kHz signal.
[b] Off scale.

tion time 10 msec plus or minus 20%, 1% third harmonic distortion at 15 ips at +4 dB; Ampex ATR-101 studio recorder, 300-msec integration time (VU meter), less than 0.3% third harmonic distortion of a 1-kHz tone at 0 VU, less than 3% at +9 VU; Uher 4000 Report Monitor, meter integration time 30 msec, decay time 400 msec, 1% distortion at 0 on meter. The Uher's meter is equalized to give an indication of the effect of preemphasis. This is an excellent idea and it is unfortunate that it is not offered on more recorders.

A case can be made for the superiority of each of the above metering systems. What is more important, though, is to be aware of which metering system a recorder uses, and how close its 0 mark is to the point of objectionable distortion. Armed with this information, and a good monitor system, one can begin to evolve an operating procedure appropriate to a recorder.

In view of the variables involved in metering sounds, it should be apparent that the widespread custom among avian sound recordists of arbitrarily adjusting recording input levels to −10 or less on whatever meter their recorder is equipped with will not produce optimum recordings.

E. Recording Formats and Tape Recorder Performance

Figure 17 illustrates common track widths in current use. The terms full, half, and quarter track are used, but the actual track widths are not fully indicated by these terms. For this domestic manufacturer, the half track and quarter track use only 32% (not 50%) and 18% (not 25%), respectively, of the width of a full track on ¼-inch tape. The casette uses narrower tape (or 0.150 inch) and one track of a stereo casette is only 10% of the width of a full track recording on ¼-inch tape.

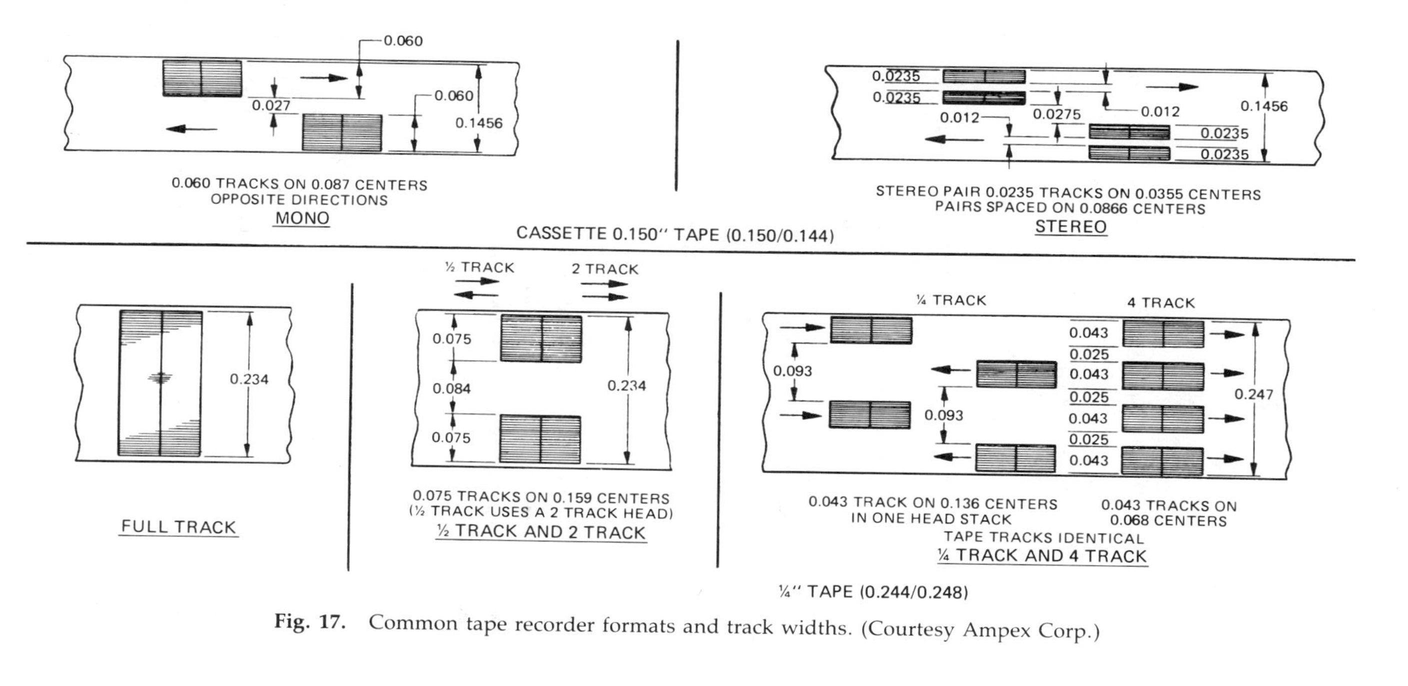

Fig. 17. Common tape recorder formats and track widths. (Courtesy Ampex Corp.)

These percentage differences result from the necessity of having a guard band between the tracks to prevent one track from bleeding into the other. This bleed always occurs to some degree. Most manufacturers specify the amount of bleed, called crosstalk.

A way to compare various formats is to consider the recording area per unit of time. One second of information at 7.5 ips full track uses a recorded area of 0.234 × 7.5 inches or 1.755 sq. inches. A cassette recorder, using one stereo track with a width of 0.0235 inch and running at a speed of 1.875 ips, will use a recording area of 0.044 sq. inch. The casette track uses about 2.5% of the area used by a full track recorder at 7.5 ips. Tape contamination that would obliterate the entire casette track would leave 97.5% of the information intact on a full track recording.

The choice of format affects the signal-to-noise ratio. The smaller the track width, the more the signal-to-noise ratio will deteriorate.

A stereo machine has two tracks available at a time for recording. A common field practice is to feed the signal from the single microphone to the two tracks thereby doubling the recording area. What is gained is not double the usable playback area but redundancy. It is virtually impossible to sum the two outputs of a stereo recorder without distorting the signal. This results from what are virtually normal head alignment problems in most recorders. For example, the wavelength of a 15-kHz signal at 1.875 ips is 0.000125 inch. If one track is offset from the other half of that amount or 0.0000625 inch, there will be a phase difference of 180° at 15 kHz. When the outputs are summed, there will be complete cancellation at that frequency. In practice, the phase difference between the tracks always varies slightly because of machine instability and/or tape variation. The result of summing the two tracks is to create a complex filter, the action of which is greater at high frequencies and the characteristics of which vary as the tape is moved across the head.

F. Some Tape Recorder Problems

Speed instability while the recorder is being moved can be a serious problem for the field recordist. Many designers assume that a portable tape recorder will not be in motion while it is being operated. The easiest way to check a machine for speed instability is to play a tape of a continuous tone. Move the machine about and listen for pitch variations.

Some recorders run off speed or run at different speeds under different conditions of temperature and battery capacity. The practice of recording a tone from a pitch pipe or tuning fork, common in the early days of field recording, is still a recommended procedure. The tone is recorded at the beginning of each recording and checked upon playback. Speed variation will manifest itself as frequency modulation of the signal. Be careful about attributing FM to the subject without first checking the recorder for flutter occurring at the same rate.

Audio recorders produce a high-frequency amplitude variance (Budelman, 1978). The program material itself will cause the high-frequency output to vary. It can amount to an amplitude change of 2–4 dB, and it increases as the tape speed gets slower. Functionally it is program-dependent AM.

G. Tape Recorder Maintenance

Maintenance is crucial for proper recorder performance. Routine maintenance should be performed at least every 10 hr of running time. Routine maintenance must include, as a minimum, a thorough cleaning and demagnetization of the tape path. If the proper test and calibration equipment are not available for verification of the recorder's performance, a qualified technician should be found and a maintenance schedule set. For machines in daily use, this verification should be performed at least once a week. If the machine is consistently meeting specifications at the 1-week interval, it may be possible to extend the service interval. Equipment seldom fails catastrophically, and gradual deterioration may go unnoticed.

If you are going to an area where it is impossible to obtain normal service, carry manuals and wiring diagrams for the equipment. If service is required, and a technician can be found, having this information is an invaluable aid. Where possible carry spares for failure-prone parts.

Be aware of the maintenance requirements of equipment and be realistic when purchasing. One tape recorder requires a series of screws to be removed to allow head cleaning and another uses heads that are out in the open. The easier a machine is to maintain, the greater the likelihood that it will be maintained.

V. THE MAGNETIC TAPE

The quality of the final recording is determined not only by the recorder but by the tape itself. There are three basic components to recording tape: the base film, the binder, and the magnetic material. The most common magnetic material is a type of iron oxide. The base film is usually a polyester material that has been prestretched to increase tensile strength. Until recently, cellulose acetate was in common use as a backing material. Paper has also been used as a base and many other materials have been tried.

A. The Tape and the Tape Recorder

1. *Matching the Tape to the Recorder*

The interface between the magnetic medium (tape) and the tape recorder is in every way a very sensitive one. Recording tape is the result of the skillful blending of many variables into a product with optimum operating parameters for a given application. Since needs vary, there are many different types available. The magnetic properties that are of most interest are bias requirement, output capability, and high-frequency sensitivity. To realize maximum performance from a tape, the recorder's bias level, record level, and record equalization must be adjusted to each tape stock's specific properties.

While there is some interchangeability among different tape stocks, one should be very careful. The recorder's performance should be verified by measurement after a change in tape. Good frequency response alone is not an indicator of correct adjustment. Because both bias level and record equalization will cause the high-frequency output to vary, it is possible to obtain a relatively flat frequency response through compensating errors in the two settings. Since bias level also affects tape output and distortion, flat frequency response may have been achieved at the expense of other parameters. Inaccurate bias settings cause increased noise and distortion.

2. *Contact of Tape with Recorder Head*

Intimate contact between the head and tape is achieved in two basic ways. One uses a pressure pad to press the tape against the head. Although used in some reel-to-reel machines this is most commonly found in cassettes where the pressure pad is an integral part of the cassette itself. Alternatively, the tape transport is designed to put the tape under tension and to pull the tape against the head(s). If the pressure (or tension) is too high, excessive tape and head wear occurs. If it is too low there is excessive amplitude variation.

For maximum high-frequency output and accurate interchangeability, the tape must be held so that the gap in the head is perfectly perpendicular to the edge of the tape. In reel-to-reel recorders the guides that accomplish this are held to close tolerances. In the cassette recorder the cassette itself supplies some of the tape guiding. As a result the mechanical quality of the cassette has a strong effect on overall recorder performance.

B. Tape Storage and Handling Problems

As magnetic recording is a relatively new process, we do not have substantial data on the effects of long-term storage. However, sufficient time has elapsed to indicate definite trends. The magnetic material itself is not likely to deteriorate. However, its magnetic state can be changed by an external field, destroying or modifying the recorded material.

Many problems of tape storage and handling can be traced to a failure or deformation of the base material or a failure or degradation of the binder. Most short-term tape deterioration is related to deformation of the tape. The most common causes are: too rapid winding, uneven winding, improper tape tension, bent reels, mixing tape stocks on a single reel, and poorly slit tape (recording tape is cut from wide rolls and sometimes the finished width varies). While dust does not attack tape directly, it can cause loss of tape to head contact and can scratch the tape or head. For this reason, areas where tape is handled should be clean. In the field, recorder covers should be kept closed whenever possible.

Print-through occurs on all recordings. It is caused by the signal on the tape being recorded onto the layers of tape directly adjacent to it and is heard as a preceding and following echo when the tape is played back. If there is an external magnetic field present, or the temperature is high, print-through becomes worse. The distance between layers is fixed by the thickness of the tape. This is one of the arguments against thinner tape. There are tapes on the market that are specifically formulated to reduce print-through and can be effective if used properly. Magnetic fields can cause effects ranging from increased noise levels to complete erasure. Temperature extremes can deteriorate the base.

It is common practice to store tape on the take-up reel as it comes from the machine after being played (tail out). This practice has many benefits. Most machines do their best job of tape spooling (evenness of wrap) at playback speed, and the tape tension is more consistent than when spooled under fast wind conditions. An additional benefit of tailout storage is that fast rewinding before playback can reduce print-through as much as 10 dB. Figure 18 shows a level/time representation of print-through. The second graph is of the same tape played just after rewinding. Notice that both pre- and postprint have been reduced.

For short-term storage, avoid extremes of environment and rapid changes of temperature and humidity. For long-term archival storage, professionals should be consulted. For polyester base type, the general rule is the cooler and drier the better. For acetate base type, the humidity should be around 50% since as the backing dries it loses its plasticisers, and the tape becomes brittle, cups, and

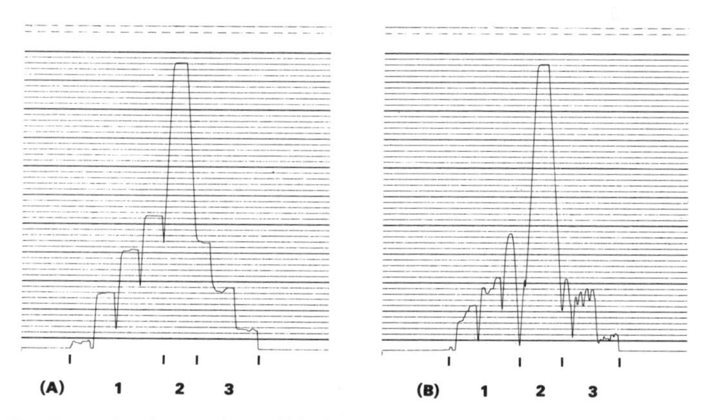

Fig. 18. Example of print-through before (A) and after (B) rewinding of the tape. (1) Preprint. (2) Signal. (3) Postprint.

cracks. On the other hand, acetate tapes stored in this fashion (high humidity) can be subject to fungus attacks. Polyester should be dry as the binder is least active at low humidity levels.

Long-term storage of magnetic tape requires a substantial commitment to climate control, beyond the range of the average recordist. If possible, the original or a safety copy of valuable field and research recordings should be archived.

Many archivists recommend periodic rewinding and respooling of the tape at 6-month to 1-year intervals. This reduces print-through and balances uneven tension which may have built up within the tape pack from temperature or humidity changes. After use, the tape is played off the reel, the end affixed, the recording acclimatized to the storage environment, placed in a plastic bag prior to being placed in its container, and then stored on edge in the archive.

One of the advantages of reel-to-reel tape is that it can be edited by cutting and splicing desired portions. Until recently, most splicing tape had an adhesive which flowed with temperature and eventually dried out. If at all possible, an unspliced copy should be archived as well.

Cassettes do not usually contain tape stock optimized for long-term storage. In addition, they cannot be mechanically edited. A cassette may be archived, but a high-quality reel-to-reel copy should be made and stored as well.

VI. SIGNAL MONITORING AND MODIFICATION

A. Monitoring the Recording

1. The Sense of Hearing and the Skill of Listening

The most critical elements in any monitor system are our ears. Good hearing by itself is not sufficient, however, and needs to be used in concert with other techniques.

A monaural recording does not supply the information necessary for our ears to locate a sound. Many recordings are not as good as they might have been because the recordist ignored extraneous sounds in the environment and fixated on the sound of interest. This is an ability that the recording system does not have. The effect is even more pronounced when there are corresponding visual cues to distract us. The best recordings are made by people who remain aware of the total sound environment whenever they are recording in the field.

Another problem is that our hearing fatigues and acuity degrades during long listening sessions. Our ears, while remarkably resistant to abuse, are not immune. If subjected to loud sounds, the ears' sensitivity changes. If this is temporary, it is called temporary threshold shift. If the sound is loud enough, the threshold shift and other hearing damage can be permanent. Protect your hearing! If you are around loud sounds—e.g., power tools, firearms—use hearing protectors.

Be aware that there is a time/intensity relationship in inducing hearing loss; in other words, a softer sound for a long time may accumulate as much damage as a louder sound for a short time. Seek the advice of a professional to obtain devices to protect your hearing as some on the market have little or no effect (cotton in your ears offers almost no protection). If you rely on your hearing in your work it is especially desirable to have an annual audiometric exam.

Listening is a skill that must be developed and must not be confused with hearing. Hearing is a sense. Listening is an active endeavor and requires the use of one's intelligence. In 1960, Dr. P. P. Kellogg described a simple experiment to illustrate how little of an unfamiliar sound we perceive.

> A simple but convincing experiment is to record your own voice, speaking your own name or some simple phrase or sentence. Reproduce these sounds in reverse. Then diligently try to reproduce with your voice what you hear. Record this jumble of sounds, and in turn, reproduce them in reverse. Normally what you hear will be astonishing, and an indication of how superficially you comprehend the intracacies of unfamiliar sounds. Usually great improvements are achieved with practice.
>
> It is much the same with vocalizations of birds and other animals. With experience we become sensitive to smaller and smaller variations.

Kellogg assumed that the recorder being used was full track, reel-to-reel.

2. *Headphones and Speakers*

For critical evaluation of a recording, it is necessary to present our ears with the best possible reproduction of the sound. A monitor system can use either headphones or a loudspeaker as the output transducer. Whichever is used, the quality is important. The loudspeaker built into the tape recorder is completely inadequate for judging the quality of the recording. For field recording, a good pair of headphones, properly matched to the recorder, is a must. For evaluating and editing, a good loudspeaker system is preferred. While not usually requiring maintenance as frequently as the rest of the recording system, headphones and loudspeakers are not immune to damage or drift and should be tested periodically.

B. Signal Modification: Filters, Equalizers, Limiters, and Noise Reducers

If one intends to manipulate a signal with various processing devices, it is necessary to understand how they operate. The following descriptions are intended as an introduction. Before using these devices, it is recommended that not only the manufacturer's literature and instruction manual be consulted, but that a broader perspective of their operating principles be acquired by reading appropriate literature (e.g., Tremaine, 1969; Woram, 1976).

A wide range of signal-modifying devices is available. Filters and equalizers can be used to modify the shape of the spectrum and minimize extraneous sounds. Band-pass filters allow a desired portion of the spectrum to pass. Band-reject filters reject a portion of the spectrum. High-pass filters allow the portion of the spectrum above their setting to pass through. Low-pass filters do the opposite.

Graphic equalizers are arranged so that the settings of their controls give an approximate graphic indication of how the amplitude response is being changed. A graphic equalizer is usually made up of filters of either full octave or one-third octave width. These filters can be arranged in such a way that when used together a combined response curve results. This is not always the case and if combining action is desired it should be verified by testing.

The parametric equalizers are so named because they have controls that allow every parameter of the equalizer to be varied. The center frequency is continuously adjustable as well as the bandwidth. The degree of gain or attenuation is also adjustable. If one intends to use equalization in the dubbing or analyzing of a recording, it is worthwhile to examine the available literature. It is very easy to cause distortion of the signal with an equalizer and the system should be tested carefully.

Limiters and compressors are special purpose audio devices used to control automatically the dynamic range of the signal. They are sometimes found on tape recorders under the guise of automatic level controls. As a rule their use is not appropriate in field recording. If there is a special application that demands automatic control of level, the literature on these devices should be consulted.

Noise-reduction systems can be divided into two groups: encode–decode systems, and single-pass systems. The two common encode–decode systems are Dolby_{tm} and DBX_{tm}. These two share the common goal of reducing tape noise and increasing the dynamic range. To accomplish this, they require that the signal be encoded during the recording process. Reciprocal processing is required for playback. Discussing the relative merits of these systems is beyond the scope of this paper, but it should be mentioned that, although both systems succeed in reducing noise, they do so at some expense to overall quality. If the application is at all critical and the use of noise reduction is anticipated, the characteristics of the system should be examined closely and its performance verified. It is this author's experience that some bird sounds fool all the systems some of the time.

Single-pass noise-reduction systems do not require encoding of the signal; rather, they act on the signal as it is played back. With the exception of computerized noise reduction, the single-pass systems use the signal itself to control one or more filters so arranged as to attenuate the portion of the spectrum unoccupied by the desired program. Effectiveness is governed by the filters themselves and the machine's ability to differentiate the desired signal from noise. From an archival standpoint, all noise reduction systems represent a modification of the recording, and their use should be noted. Single-pass systems, while sometimes helpful for a specific project, are not appropriate for general use.

By striving for the best possible recording in the first place, the necessity of postprocessing of the recording can be avoided.

VII. THE RECORDING SYSTEM

When a recording of a bird is made it is done with a recording system, not a group of modules. It is possible to assemble a system of excellent components that will produce abysmal results. Each component can have acceptable performance, but the combination may accumulate errors in a way that makes the final result unacceptable. The recording system should be tested and calibrated to make sure its errors are not significant relative to the sensitivity and type of the intended analysis.

A. Interconnection of the Components

Space does not permit a thorough discussion of interconnection techniques. Impedance, balancing, wire and connector types are all important. If electrical impedance and balanced and unbalanced lines are unfamiliar terms, a good reference such as Davis and Davis (1975) should be consulted. Wire and connectors present mechanical as well as electrical difficulties. Wire needs to be electrically correct for its application as well as durable and flexible at working temperatures.

My preference is for a system comprised of as few connections as possible. Those that are necessary should be of the same type. This minimizes the need for repair and makes it easier to carry spares.

B. Batteries

For field recording, power is generally supplied by batteries. Make sure the types used by your equipment are available in the area where you are recording. If the availability of replacement batteries is in doubt it is a good idea to carry a supply sufficient for the duration of the project.

Batteries are generally classified by the elements they use to produce electricity. Currently available types include: carbon–zinc, alkaline, nickel–cadmium, lead–acid, silver oxide, and mercury oxide.

The carbon–zinc battery is a low cost battery. Its output voltage falls gradually through its useful life. It is best used at room temperatures since its life decreases at high temperatures and its output falls at low temperatures, virtually stopping at 0°F (−18°C).

An alkaline battery costs more than a carbon–zinc but it has more available energy. The output voltage falls during its life, but at a slower rate. It works well at low temperatures, and is not unduly sensitive to high temperatures. This is the best battery for most field recording.

Nickel–cadmium and lead–acid batteries are types designed to be recharged. The "ni–cad" is relatively insensitive to temperature although it should not be be charged at temperatures below freezing and does not store well at high temperatures. One charge will not last as long as the life of a comparable nonrechargeable type.

The lead–acid battery is relatively economical if a rechargeable battery that can deliver large amounts of current is needed and size and weight are not a problem. High temperatures cause little problem, but output drops substantially at low temperatures. A car battery is a special type of lead–acid battery. Its life is shortened considerably when discharged to the limit of its capacity. The type of lead–acid battery that is built to withstand full discharges without damage is called a deep cycle battery.

Silver oxide batteries are used primarily in applications where the voltage needs to be constant throughout the battery life. They are generally available only in miniature sizes offering small current capability. They are not rechargeable.

Mercury batteries offer good efficiency in proportion to size, and have almost constant output voltage over their life. Their high temperature capability is good but their output is nil at freezing. They are quite expensive and are used where small size is important. The "button" batteries used in some microphones are mercury batteries. They cannot be recharged.

C. Selecting a System: A Sample Problem

As an example of problems encountered in choosing components for a field recording system, one potential system is examined. As a recorder, the system will use a Nagra IV-S, a stereo machine. For the purposes of discussion only one channel will be considered. The recorder has been equipped with a case to protect it. A copy of pertinent sections of the manual have been placed in the cover. The heads and pinch roller are out in the open permitting easy access for cleaning and demagnetizing. The stock recorder cover accommodates only 5-inch reels, and the cover should be closed whenever possible to reduce the chance of tape or machine contamination. The correct tape has been ordered on 5-inch reels. The recorder has been aligned, biased, and equalized for the intended tape stock.

The Nagra manual gives the meter integration time of 10 msec, which should catch most of the peaks in the sounds to be recorded.

The next problem is to select a microphone. If a humid working environment is anticipated, a dynamic shotgun offers the advantages of ruggedness and immunity to moisture. A possible candidate is the Electro-Voice DL-42. Its frequency response (50–12,000 Hz) is not exceptional and it will have some trouble tracking fast rise time sounds. However, a condenser would be trouble prone in the anticipated environment. The DL-42 comes with a windscreen and a handle, although the catalog does not show how the handle fits.

The specifications of the DL-42 give its output as −50 dB, but −50 dB relative to what? In the front of the E.V. catalog, it is stated that the reference level is 0 dB = 1 mW, delivered to a load impedance equal to the microphone's impedance, with a sound pressure level of 94 dB. The test frequency is 250 Hz. In other words, the microphone was presented with a 250-Hz test tone at the level of 94 dB SPL at the microphone. The output of the microphone was measured and referenced to 1 mW.

The frequency range of our intended subjects is between 1 and 6 kHz. The microphone's response is adequate at those frequencies; however, the second harmonic of 6 kHz will be attenuated. This limitation is not important for the intended analysis. The microphone's specified impedance is 150 ohms. Therefore, it will work with the recorder's 200-ohm input position. The 200-ohm input has a specified sensitivity of 0.28 mV for a 0-dB reading on the meter with the gain control at maximum.

To determine if the microphone has sufficient sensitivity when used with this recorder, it is necessary to reconcile the ratings. By current convention the microphone's impedance is not actually matched by the low impedance inputs of the recorder. Even though the input is labeled 200 ohms, the actual input impedance is on the order of ten times the stated value. Such an input is called a bridging input. A power reading is converted to an open circuit voltage reading by the formula

$$E = \sqrt{(0.001\ Z)\ 10^{[(\mathrm{dBm}\ +\ 6)/10]}}$$

where E is voltage, Z is impedance, and dBm is decibels referenced to 1 mW. For the DL-42, this formula gives the open circuit voltage as 2.44 mV. The difference in dB between 2.44 and 0.28 mV is 20 log (2.44/0.28) = 18.8 dB. Subtracting the 18.80 from the 94 dB required for the microphone to produce the 2.44 mV, it can be seen that it would require a SPL of a little over 75 dB at the microphone to produce a 0-dB reading on the meter with the gain control at maximum. This is slightly less sensitivity than is desired. Further examination of the Nagra manual indicates that the machine has a high sensitivity position that gives another 6 dB of gain. This extra gain reduces the SPL requirement at the microphone to 69 dB.

Since more sensitivity would be desirable for a worst-case situation, a Sennheiser MKH-816 condenser microphone is also examined. Its sensitivity is specified as 40 mV/Pascal. A Pascal is equivalent to 94 dB SPL. Conveniently both microphones were specified with the same SPL applied. The DL-42's output is 2.44 mV unterminated in comparison to the Sennheiser 40 mV, better than 24 dB difference. The Sennheiser is also available with the necessary windscreens and handle.

Because it is a condenser, the MKH-816 requires power to operate; this can be provided by the Nagra. The Nagra's input with T-standard powering has a sensitivity of 4.2 mV, which is 23.5 dB less sensitive than the 200-ohm input. Using the recorder in this configuration, there is no significant difference in usable sensitivity between the two microphones. The Nagra's 48-V phantom input has a sensitivity of 1.4 mV, and the microphone is offered in that version. That gives an additional 9.5 dB of gain.

If it were possible to power the microphone separately from the Nagra, the 200-ohm input could be used for maximum sensitivity. This can be done using the in-line battery supply available for the MKH-816. With the battery supply in the line between the microphone and the 200-ohm input, the system has the maximum possible sensitivity with this microphone. Unfortunately, the microphone has a certain amount of electronic noise that will be amplified as well. In this configuration it will take approximately 50 dB SPL at the microphone to produce a meter reading of 0 dB.

Unfortunately, the 1% distortion point for the recorder's 200-ohm input is specified as 54 mV. Since the microphone produces 40 mV at 94 dB SPL, it can be seen that any loud sound (above 9 dB SPL) will drive the recorder's input into distortion.

This exercise illustrates the need to be aware of the specific requirements for recording the intended subject. The above examples are not intended to recommend equipment, but to illustrate a thought process. This is the first step in determining how a given combination of components will work as a system.

VIII. SUMMARY

The serious avian sound recordist, someone who is recording bird sound for scientific or professional use, must possess adequate knowledge, not only of the bird itself, but of the behavior of sound and the capabilities and limitations of the recording system components and the system as a whole.

The concepts and the reasons for record preemphasis and the design philosophy of metering systems are particularly relevant to the avian recordist.

A recording that is not documented and is recorded on an uncalibrated system has limited usefulness. One that is properly made becomes a valuable research resource and should be archived.

The choice of equipment appropriate to the recording of avian sounds is especially difficult in that virtually all available equipment has been deliberately designed for other purposes. While it is important to choose components carefully for their specific merits, it is essential that their combination result in an optimally functioning system.

There is no shortcut to this knowledge. It must be gleaned from a variety of sources. The "standard" references, while helpful, are quickly outdated and sometimes in error. The "cookbook" approach, while of use to an amateur, can only be a basic introduction for the beginning professional.

Recordists must be aware of the requirements of their intended analysis and view the limitations of their recording system in that light. Lacking thorough verification and calibration of the recording system, the analysis is likely to display not only avian sound but recording system artifact.

ACKNOWLEDGMENTS

I would like to thank James L. Gulledge for encouragement and expert assistance in preparation of this text. Saul Mineroff kindly made available some of the equipment discussed.

REFERENCES

Beranek, L. L., ed. (1971). "Noise and Vibration Control." McGraw-Hill, New York.

Bruel and Kjaer (1977). "Condenser Microphones and Microphone Preamplifiers, Theory and Application Handbook." Bruel and Kjaer, Naerum, Denmark.

Budelman, G. A. (1978). High frequency variance: a program dependent signal mode in analog magnetic tape recording. Prepr. No. 1377(E-2). Audio Eng. Soc., New York.

Burroughs, L. (1974). "Microphones: Design and Application." Sagamore, Plainview, New York.

Davis, D., and Davis, C. (1975). "Sound System Engineering." Sams & Bobbs-Merrill, New York.

"Eveready" Battery Application and Engineering Data (1971). Union Carbide Corp., New York.

Fisher, J. B. (1977). "Wildlife Sound Recording." Pelham Books, London.

Gordon, J. K., and Wood, J. B. (1979). Bridging the gap between the vu meter and ppm. Prepr. No. 1518(D-7). Audio Eng. Soc., New York.

Greenewalt, C. H. (1968). "Bird Song: Acoustics and Physiology." Smithsonian Inst. Press, Washington, D.C.

Gulledge, J. L. (1977). Recording bird sounds. *Living Bird* **15,** 183–204.

Hassall, J. R., and Zaveri, K. (1979). "Acoustic Noise Measurement." Bruel and Kjaer, Naerum, Denmark.

Hetrich, W. L. (1976). The accu-peak level indicator. Prepr. No. 1125(B-4). Audio Eng. Soc., New York.

Kellogg, P. P. (1960). Considerations and techniques in recording sound for bio-acoustics studies. *In* "Animal Sounds and Communication" (W. E. Lanyon and W. N. Tavolga, eds.), Publ. No. 7, pp. 1–25. Am. Inst. Biol. Sci., Washington, D.C.

Knight, G. A. (1977). Factors relating to long term storage of magnetic tape. *Recorded Sound* **66/67,** 681–692.

McClurg, D. R. (1976). "Professional Recorders." Otari Corp., San Carlos, California.

Olson, H. F. (1957). "Acoustical Engineering." Van Nostrand, New York.

Olson, H. F. (1972). "Modern Sound Reproduction." Van Nostrand-Reinhold, New York.
Peterson, A. P. G., and Gross, E. E., Jr. (1972). "Handbook of Noise Measurement," 7th ed. General Radio, Concord, Massachusetts.
Peus, S. (1977). Microphones and transients. *Sound Eng. Mag.* **11,** 35–38.
Roederer, J. G. (1975). "Introduction to the Physics and Psychophysics of Music." Springer-Verlag, Berlin and New York.
Tall, J. T. (1958). "Techniques of Magnetic Recording." Macmillan, New York.
Temmer, S. F. (1979). Vu meter and peak program meter, peaceful coexistence. Prepr. No. 1474(G-5). Audio Eng. Soc., New York.
Toombs, D. (1981). "Sound recording." David & Charles, London and North Pomfret, Vermont.
Tremaine, H. M. (1969). "Audio Cyclopedia," 2nd ed. Sams & Bobbs-Merrill, New York.
Watkins, W. A. (1967). The harmonic interval; fact or artifact in spectral analysis of pulse trains. *In* "Marine Bioacoustics" (W. N. Tavolga, ed.), Vol. 2, pp. 15–43. Pergamon, New York.
Weber, P. J. (1967). "The Tape Recorder as an Instrumentation Device." Ampex Corp., Redwood City, California.
Woram, J. M. (1976). "The Recording Studio Handbook." Sagamore, Plainview, New York.

2

The Structural Basis of Voice Production and Its Relationship to Sound Characteristics

JOHN H. BRACKENBURY

I. INTRODUCTION

Birds share with other terrestrial vertebrates the ability to utilize airflow within the respiratory system as a source of energy for sound production. The evolutionary factors that led to the development of vocal organs within the respiratory tract can only be guessed, but one of the important results of this process is that it enables the energies of the large and powerful respiratory muscles to focus on a relatively small structure capable of generating sound. The means for this particular kind of amplification is not available to the other major group of sound-producing animals, the Arthropoda, because of fundamental differences in the structure of the respiratory system (though it can be claimed, of course, that use

ACOUSTIC COMMUNICATION IN BIRDS
VOLUME 1

ISBN 0-12-426801-3

of the leg muscles for mechanical stridulation in orthopterans, for instance, represents an equally effective amplification system).

The vocal organ in birds is the syrinx; it lies at the base of the trachea and is surrounded by the interclavicular air sac. This contrasts with the larynx of mammals, which lies at the top of the trachea, and it is logical to ask whether this fundamental difference in location of the vocal organs is related to the unique structure of the lung–air sac system in birds. Again, any answer would be speculative, but it can at least be said that the functioning of the syrinx is entirely dependent on its being situated within an air space, as will be shown below, for it could not work if it were surrounded by tissues.

This chapter is concerned with the structure and mechanism of the syrinx and the relationship between syringeal function and sound characteristics. Particular attention is directed to the intimate relationship between vocal function and the respiratory function on which it is founded. It is also shown that, at the control level, the syrinx may act quite independently of the respiratory mechanism. Although the innervation and peripheral control of the syrinx will be discussed, the rapidly growing subject of central nervous control of song is dealt with separately by Arnold (Chapter 3, this volume).

II. THE SYRINX

The syrinx consists of the specialized junction between the trachea and the two primary bronchi of the lungs (Figs. 1 and 3A). It lies within the interclavicular air sac, the only unpaired sac of the avian lung–air sac system. The structure of the passerine syrinx has received most attention and the following description is based largely on the Common Crow (*Corvus brachyrhynchos:* George and Berger, 1966; Chamberlain *et al.*, 1968; Ames, 1971).

The most cranial element of the syrinx is the tympanum, a cylinder of cartilage formed by the fusion of the last few tracheal rings (Fig. 1). The pessulus is a looped sagittal outgrowth from the caudal end of the tympanum that divides the

Fig. 1. Generalized passerine syrinx. (A) Ventrolateral view. (B) Dorsal view. (C) Ventrolateral view of tympanum, pessulus, and first bronchial semiring, BR 1. (D) Semidiagrammatic frontal section of syrinx in normal respiratory position (left) and sound-producing position (right). Only the left half of the syrinx is shown in D. During vocalization, contraction of the intrinsic syringeal muscles, indicated by the broken arrow in D, rotates the third bronchial semiring, BR 3, as indicated by the solid arrow. This brings the external labium closer to the internal tympaniform membrane, causing the air to accelerate over the surface of the membrane and produce suction forces which trigger it into vibration. BR 2 and BR 4 represent the second and fourth bronchial semirings, respectively. [A and C modified from George and Berger (1966); D modified from Chamberlain *et al.* (1968).]

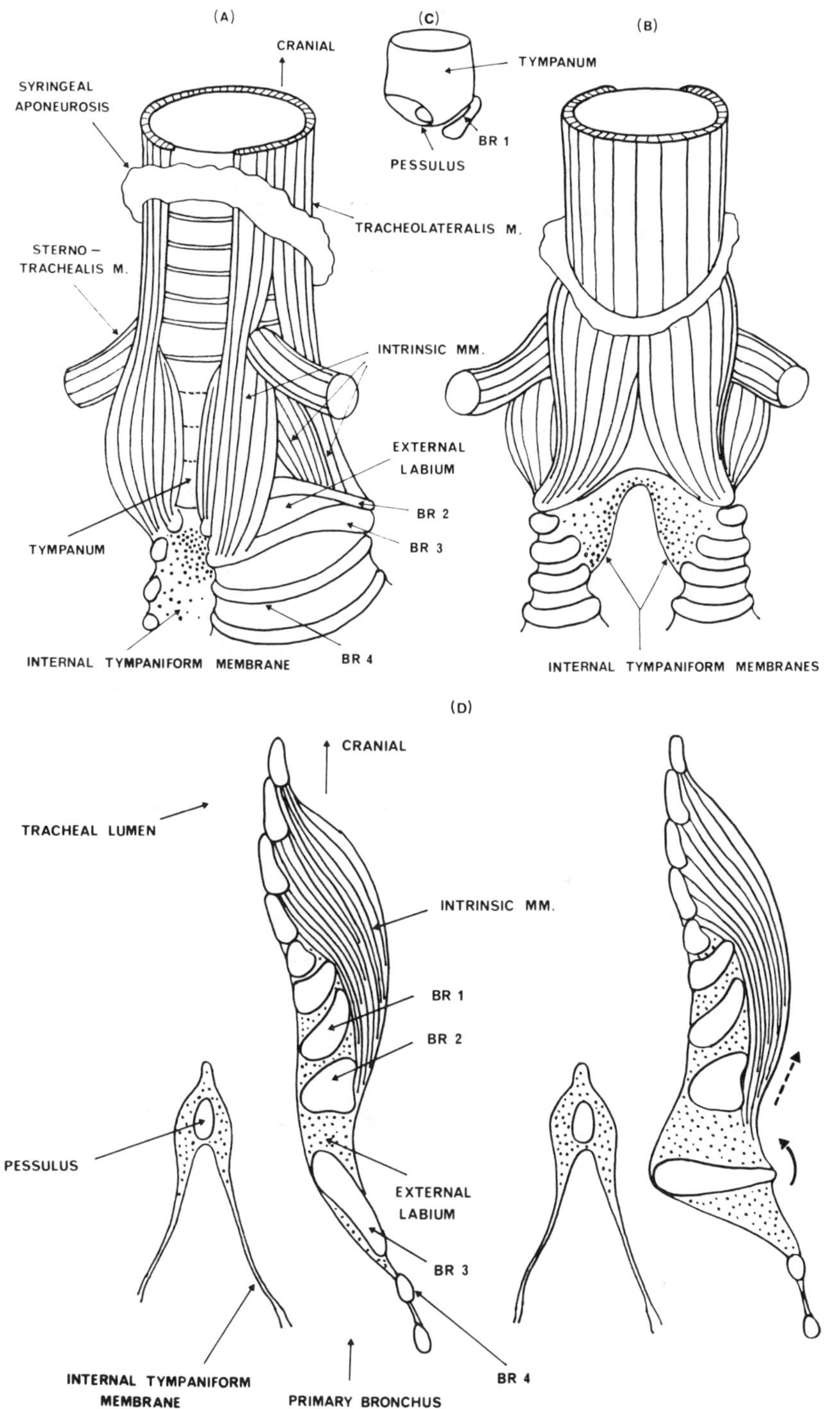
(A)
(C)
(B)
CRANIAL
TYMPANUM
SYRINGEAL
APONEUROSIS
BR 1
PESSULUS
TRACHEOLATERALIS M.
STERNO–
TRACHEALIS M.
INTRINSIC MM.
EXTERNAL
LABIUM
BR 2
TYMPANUM
BR 3
INTERNAL TYMPANIFORM MEMBRANE
BR 4
INTERNAL TYMPANIFORM MEMBRANES
(D)
CRANIAL
TRACHEAL LUMEN
INTRINSIC MM.
BR 1
BR 2
PESSULUS
EXTERNAL
LABIUM
BR 3
INTERNAL TYMPANIFORM
MEMBRANE
PRIMARY BRONCHUS
BR 4

tracheal lumen into the openings of the two primary bronchi. The vibratile elements of the syrinx are the paired internal tympaniform membranes consisting of the thin, medial walls of the primary bronchi just caudal to the tracheobronchial junction. The internal tympaniform membranes are attached to the pessulus at their cranial borders while their caudal boundaries merge into the cranial parts of the primary bronchi. Any gross longitudinal movements of the trachea and tympanum are thus relayed to the internal tympaniform membranes, which tense or relax in consequence (see below). The lateral walls of the syrinx contain three bars of cartilage, consisting of the first three bronchial semirings, between the second and third of which the wall is thickened by loose connective tissue to form the so-called external labium.

There is some dispute about the number of muscles associated with the syrinx, Ames (1971) recognizing six, Warner (1972) seven. Two of these, the sternotrachealis and the tracheolateralis, are extrinsic, arising on the costal process of the sternum and the cartilages of the larynx, respectively, and inserting on the trachea just cranial to the tympanum. The remaining four or five pairs of muscles are intrinsic, originating on or just cranial to the tympanum and inserting variously on the bronchial semirings of the syrinx.

The structure of the syrinx in non-oscine species differs from the above account mainly in the number or attachments of intrinsic muscles. According to Ames, the forerunner of the passerine syrinx resembles the syrinx of the present-day woodpeckers and allies (Order Piciformes) which, like the syringes of most non-passerines, lacks intrinsic syringeal muscles. "Pico-passerine" type syringes are found in several passerine families including Pittidae, Tyrannidae, Cotingidae, and Pipridae, but it is difficult to find direct correlations between "musical" ability and the presence or absence of intrinsic syringeal muscles in birds in general.

The precise roles of the syringeal muscles during vocalization have not been clearly identified except in a few cases discussed below in which the absence of intrinsic muscles simplifies the issue. Miskimen (1951) ascribed a dominant role to the extrinsic muscles in passerine sound production: contraction of the sternotrachealis pulls the tympanum caudally thereby relaxing the internal tympaniform membranes, and suction forces created by the expiratory airstream then draw the membranes into the bronchial lumen until they are triggered into vibration. Contraction of the tracheolateralis reverses this series of movements and abolishes sound. Chamberlain *et al.* (1968) proposed an entirely different scheme for the activation of the internal tympaniform membranes: contraction of the intrinsic muscles produces a rotation of the third bronchial semiring about its longitudinal axis, bringing the external labium closer to the internal tympaniform membrane (Fig. 1D). The resultant channeling of airflow between the labial and membranous surfaces creates a suction force that triggers the membranes. Both schemes therefore agree on the final phase of membrane excitation by fluid mechanical forces but they disagree on the events which produce the initial yielding of the membranes.

Gaunt *et al.* (1973) expressed the major differences between the two schemes as the "Passive" and "Active" models of syringeal closure. The passive model was used to explain sound production in two species that possess extrinsic syringeal muscles but lack intrinsic syringeal muscles, the Domestic Fowl (*Gallus domesticus:* Youngren *et al.,* 1974; Gaunt *et al.,* 1976; Gaunt and Gaunt, 1977; Brackenbury, 1980) and Mallard (*Anas platyrhynchos:* Lockner and Youngren, 1976). However, the form of interaction between the extrinsic muscles is not simply a case of mutual antagonism as Miskimen suggested, since electromyographic recordings have shown that both muscles may be active at the same time. The adjustment of tension within the syringeal membranes which leads to the production of sound depends on the finely graded interaction between the muscles. The Domestic Fowl syrinx presents a number of special features. Sound is produced not by the internal but by the external tympaniform membranes; moreover, the pessulus is separate from the tympanum and is attached by the syringeal ligament to the connective tissue sheath enclosing the sternotrachealis muscles (Fig. 2). This means that contraction of the sternotrachealis not only pulls the tympanum caudally but also draws the pessulus cranially, the combined movements leading to a more effective buckling of the external tympaniform membranes. Many passerine species, including some species of pittas, tyrant flycatchers, and larks, lack a pessulus (Ames, 1971); a rigid pessulus is also absent in pigeons (family Columbidae: Warner, 1971a). Warner suggests that this may reduce the ability to vary the tension of the internal tympaniform membranes but the songs of many of these species, including the Ringed Turtle-Dove (*Streptopelia risoria:* Mairy, 1976) and Skylark (*Alauda arvensis:* Brackenbury, 1978c; Csicsáky, 1978), are nevertheless rich in sound modulations.

III. INNERVATION AND CONTROL OF THE SYRINGEAL MUSCLES

The syringeal muscles are innervated by the descending cervical branch of the hypoglossal nerve. No information is available on the types of fiber present in this nerve although the hypoglossal is normally considered to contain only somatic motor fibers. It may be significant that the vocal muscles in birds do not form part of the branchial musculature from which the vocal muscles of the mammalian larynx are derived; the contractile properties of branchial muscle, innervated by visceral motor fibers of the vagus, may be unsuitable for the production of rapid modulations which are common in bird sounds. This is speculation but the importance of the syringeal muscles in vocalization warrants research into the topic.

The syrinx in all passerine species examined to date receives bilateral and ipsilateral innervation. In the Domestic Fowl the left hypoglossal provides part of the innervation of the right muscles (Youngren *et al.*, 1974) and a cross-connection between the left and right hypoglossals occurs in the Mallard (Lockner and Youngren, 1976). The left hypoglossal is usually dominant to the right, as shown by the differential effects of nerve section on song in the Chaffinch (*Fringilla coelebs:* Nottebohm, 1971), White-throated Sparrow (*Zonotrichia albicollis:* Lemon, 1973), Domestic Fowl (Youngren *et al.*, 1974), White-crowned Sparrow (*Zonotrichia leucophrys:* Nottebohm and Nottebohm, 1976), and Java Sparrow (*Padda oryzivora:* Seller, 1979). Bilateral hypoglossectomy abolishes vocal muscle activity but not necessarily the ability to produce sound. Seller (1979), for instance, found that some calls survived bilateral nerve section in the Java Sparrow and attributed these to respiratory pulsations.

Bilateral denervation reportedly produced loss of sound-producing ability in the Chaffinch (Nottebohm, 1971) and Red-winged Blackbird (*Agelaius phoeniceus:* Peek, 1972; Smith, 1977), but the primary effect in these cases may have been on respiratory function. The balance of evidence suggests that the syringeal muscles are not essential for sound production per se but that their primary role is molding the syllabic structure of the song. Severance of the sternotrachealis muscles in a number of passerine and non-passerine species did not lead to the abolition of sound (Gross, 1964; Youngren *et al.*, 1974; Smith, 1977; Brackenbury, 1978a). Perhaps the strongest evidence in favor of the independence of sound production from muscle activity is the observation that a unilaterally denervated bird can still produce two simultaneous sounds, one of them modulated (by the innervated side of the syrinx), the other unmodulated (Nottebohm, 1971). The latter sound must be produced by fluid mechanical forces in the denervated side of the syrinx.

Youngren *et al.* (1974) doubted whether the tracheolateralis in the Domestic Fowl was a true vocal muscle since it was also phasically active during respira-

---

Fig. 2. Syrinx of the Domestic Fowl *Gallus domesticus.* (A) Ventral view. (B) Left lateral view. (C) Dorsolateral view showing syringeal ligament. Intrinsic muscles are absent and sound is produced by the vibration of the external tympaniform membranes. A series of thin bronchial semirings lies within the external tympaniform membranes but these are not shown. According to the Passive Closure model (see text), contraction of the sternotrachealis muscles pulls the tympanum toward the pessulus, thereby slackening the membranes that bulge into the syringeal lumen; these membranes are then activated by the airstream and produce sound. The sternotrachealis also exerts an effect on the syringeal ligament, which draws the pessulus toward the tympanum. Contraction of the tracheolateralis muscles restores the syrinx to its resting position.

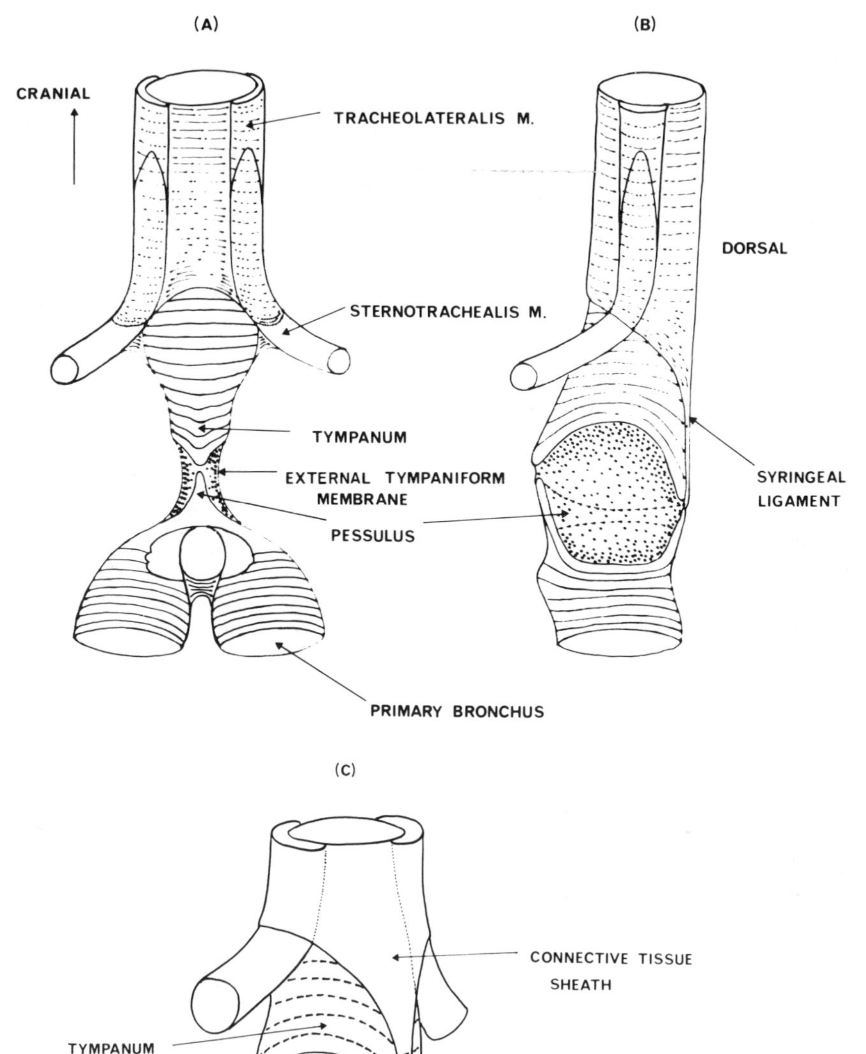
(A)
(B)
CRANIAL
TRACHEOLATERALIS M.
DORSAL
STERNOTRACHEALIS M.
TYMPANUM
EXTERNAL TYMPANIFORM MEMBRANE
PESSULUS
SYRINGEAL LIGAMENT
PRIMARY BRONCHUS
(C)
CONNECTIVE TISSUE SHEATH
TYMPANUM
SYRINGEAL LIGAMENT
PESSULUS
EXTERNAL TYMPANIFORM MEMB

tion. They suggested that its chief function was to prevent collapse of the external tympaniform membranes into the bronchial lumen during inspiration. Nottebohm (1971) also found evidence for a respiratory role of the syringeal muscles in Chaffinches, noting that bilateral denervation sometimes led to respiratory distress, particularly in birds that were excited. He reasoned that the increased Bernouilli force in the syrinx due to the hyperventilation of these birds was sufficient to collapse the internal tympaniform membranes. The sternotrachealis muscle, in contrast to the tracheolateralis, is active only during vocalization, and Phillips and Peek (1975) considered that it had achieved a high degree of "emancipation" from the respiratory mechanism.

Since unilateral nerve section does not impair the vocal ability of the contralateral side of the syrinx, it appears that both sides are independently controlled; this has been adduced by Nottebohm (1971) as experimental evidence of the "two-voice" theory of Greenewalt (1968) and Stein (1968). The theory of two independent acoustic sources within the syrinx is based on sonagraphic evidence showing double tones which are nonharmonically related. Greenewalt cited examples in 12 oscine and 7 non-oscine species. Miller (1977) found not only double notes in the Wood Duck (*Aix sponsa*) but also triple and quadruple notes, all non-harmonically related. The structural basis for these multiple notes is unknown and would repay detailed anatomical investigation.

IV. RESPIRATORY MECHANICS AND SOUND PRODUCTION

Sound is produced during expiration, the air sacs being compressed by the intercostal muscles of the thorax (enclosing the interclavicular, cranial thoracic, and caudal thoracic sacs) and by the abdominal muscles surrounding the large abdominal sacs (Fig. 3). Increased electromyographic activity in the abdominal muscles synchronous with vocalization has been recorded in the Domestic Fowl (Gaunt and Gaunt, 1977) and Mallard (Lockner and Youngren, 1976). The resultant increases in air sac pressure and/or airflow have been directly measured in the Evening Grosbeak (*Hesperiphona vespertina:* Berger and Hart, 1968), Starling (*Sturnus vulgaris:* Gaunt *et al.*, 1973), Domestic Fowl (Brackenbury, 1978a; Gaunt *et al.*, 1976; Gaunt and Gaunt, 1977), Gray Lag Goose (*Anser anser:* Brackenbury, 1978a), and Mallard (Lockner and Murrish, 1975). In the Domestic Fowl and Grey Lag Goose, peak airflow rates during vocalization may exceed 0.5 liter sec^{-1} (Fig. 4).

All air sacs have, in addition to their connections to the narrow gas-exchanging parabronchi of the lung, wider direct connections to the primary bronchus which enable them to deliver a rapid stream of air along a path of least resistance to the primary bronchus and syrinx during vocalization (Fig. 3B). Increased

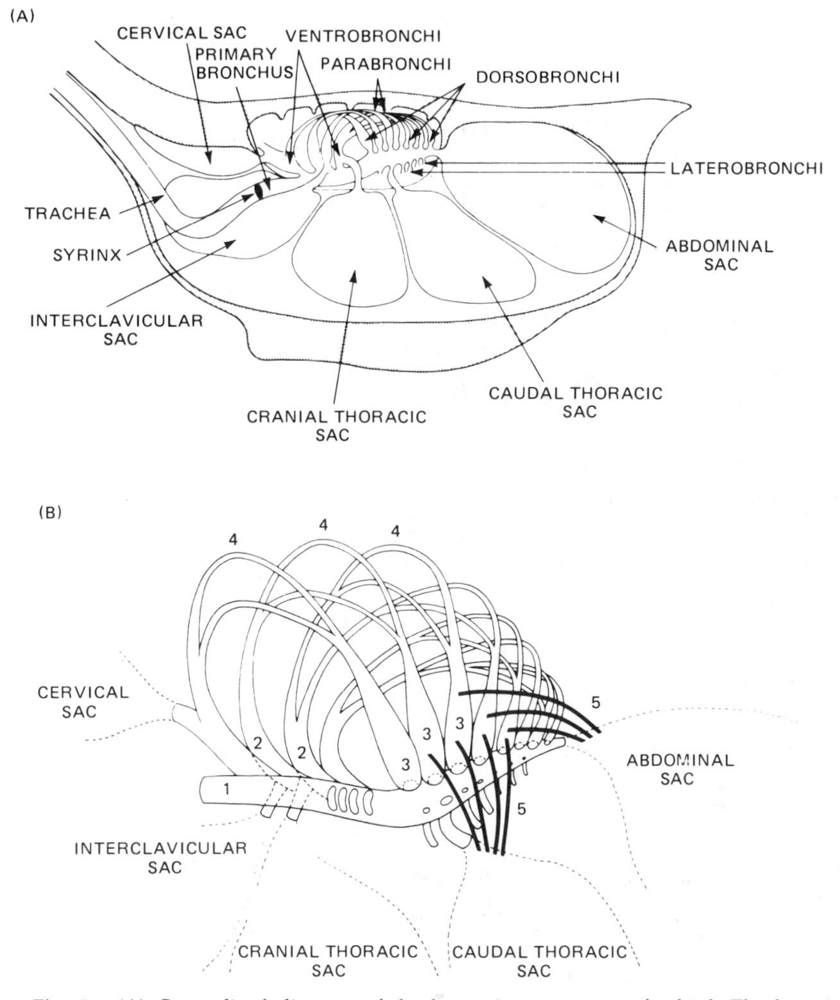

Fig. 3. (A) Generalized diagram of the lung–air sac system of a bird. The lung is inexpansible and ventilation is brought about by the bellows-like action of the air sacs. Gas exchange takes place in the narrow parabronchi that connect with the primary bronchus via the ventrobronchi and dorsobronchi. (B) Detailed diagram of lung air sac connections. (1) Primary bronchus; (2) ventrobronchi; (3) dorsobronchi; (4, 5) parabronchi. The interclavicular and cranial thoracic sacs have direct connections to the primary bronchus via the wide ventrobronchi; the caudal thoracic and abdominal sacs also have direct connections to the primary bronchus in addition to their parabronchial connections to the dorsobronchi.

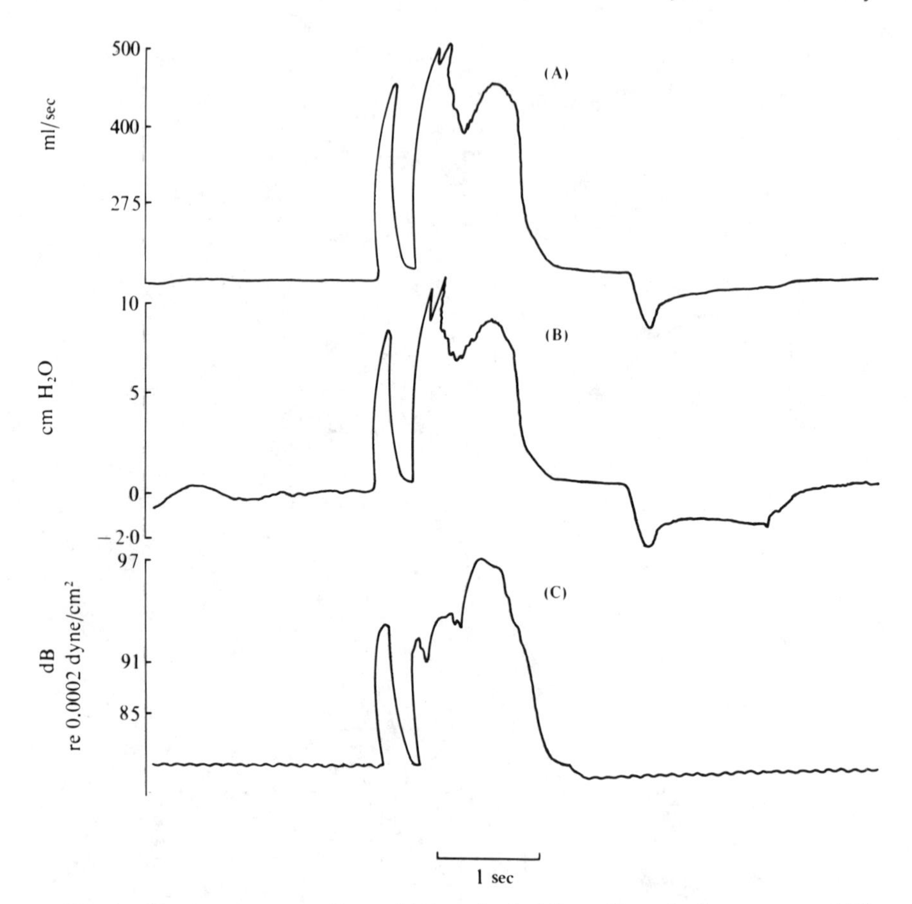

Fig. 4. Simultaneous recordings of (A) tracheal airflow, (B) tracheal pressure, and (C) sound pressure level during crowing in Domestic Fowl. Note the very high peak flow rates and the similarity of waveforms showing that sound pressure level is directly related to flow rate.

airflow past the tympaniform membranes produces local suction forces which draw the membranes into the bronchial lumen. This process is assisted by syringeal muscle activity as discussed in Section II. The interclavicular air sac plays an essential mechanical role by ensuring that the hydrostatic pressure gradient across the membranes from their outer to their inner surfaces is effectively zero. This allows the membranes to respond to the relatively small suction forces created by the airflow on their inner surfaces, without hindrance from the extremely large air sac pressures which are necessary to generate high airflow rates during sound production (Fig. 4). In order to fulfill the role of pressure equilibration, the interclavicular air sac must be able to relay internal pressure changes instantly to the syrinx; this short circuit is provided by the direct connection of the interclavicular air sac to one of the ventrobronchi (Fig. 3B).

V. ENERGY COUPLING BETWEEN AIRSTREAM AND SYRINGEAL MEMBRANES

The triggering of the membranes depends on the relationship between the flow-induced forces tending to draw the membranes into the bronchial lumen and the opposing membrane tension which tends to restore the membranes to their resting position. For a given airflow velocity, both these forces increase continuously as the membrane becomes displaced into the lumen; but while the suction force increases linearly with membrane displacement, the membrane tension increases curvilinearly (Fig. 5B) (Brackenbury, 1979a). As a result, below a certain threshold airflow velocity [marked (2) in Figs. 5B and 6] no stable membrane displacement can occur since the membrane tension is always greater than the suction force regardless of membrane position. At velocities above this threshold, for instance, that indicated by position (3) in Figs. 5B and 6, the situation reverses and any slight displacement of the membrane will cause it to be drawn further into the airstream until a point is reached at which the membrane tension again exceeds the suction force; at this point the membrane decelerates and eventually reverses in motion and returns toward its resting position. Once more, however, the suction begins to exceed the tension and in this way a self-generating oscillation becomes established. The amplitude of this oscillation, and thus of the sound produced, increases with airflow rate, and this is confirmed by direct recording in vocalizing birds (Fig. 4).

The membranes behave like a pair of flutter valves, each cycle of motion producing a local expansion and contraction of the syringeal lumen, thereby introducing periodic surges in the airflow. The resultant periodic variation in air particle velocity is effectively the sound which issues from the end of the trachea. Thus, only the alternating component of the airflow performs acoustic work, the "direct" component being acoustically idle (Fig. 5A). The rate at which sound energy is emitted from the beak is proportional to the amplitude of the alternating airflow component squared, multiplied by the radiation damping of the air (Brackenbury, 1979a). The total rate at which fluid mechanical energy is injected into the respiratory tract by the air sacs is proportional to the amplitude of the "direct" airflow component squared, multiplied by the respiratory airway resistance. Consequently, a comparison of sound energy and total fluid energy gives an indication of the efficiency of the syrinx as a mechanoacoustic energy converter. It is possible that in large, powerful species like the Domestic Fowl, which are capable of generating very high expiratory flow rates during vocalization, further enhancement of sound power output is achieved by "convective amplification,'' a process whereby the sound vibrations within the trachea are magnified by the velocity of the airstream in which they are being carried (Brackenbury, 1979a).

A. S. Gaunt and Gaunt (1980) and S. L. L. Gaunt and Gaunt (1980) have proposed a fundamentally different kind of sound-production mechanism in the Ringed Turtle-Dove. The primary source of sound is not the tympaniform membranes per se but aerodynamic vortices set up by the flow of air through the slit formed between the membranes when the syrinx is in the vocal configuration. It is not discounted that the membranes may vibrate in sympathy with the sound,

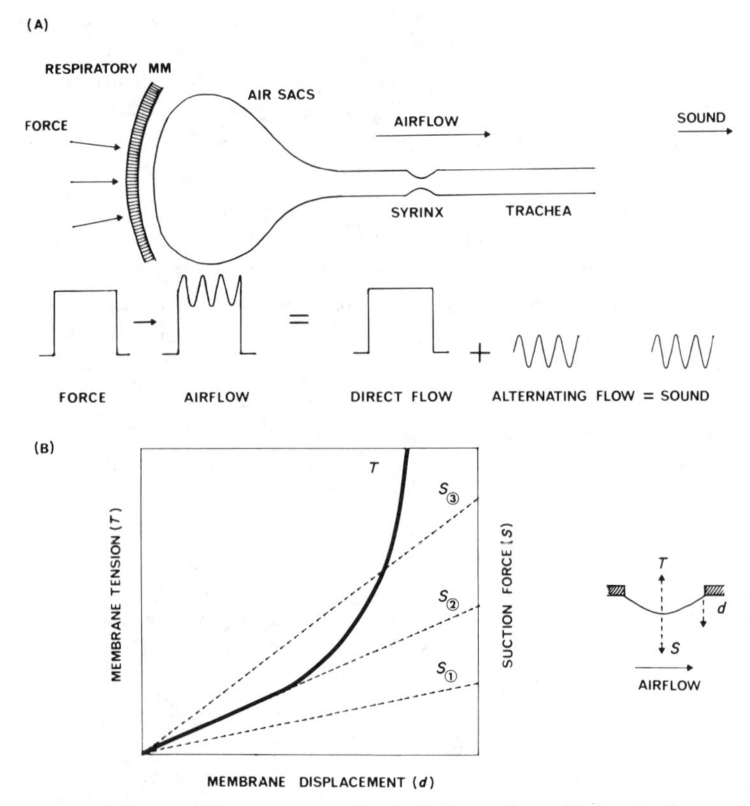

Fig. 5. (A) Diagram illustrating mechanics of sound production in the respiratory system. On the left a force is applied by the respiratory muscles. This produces a pulse of airflow from the air sacs which triggers the syringeal membranes. Each vibration of the membranes causes an oscillation of the airflow so that the pulse consists of a "direct flow" component on which is superimposed an "alternating flow" component. The latter is in effect the sound which emanates from the trachea. (B) Diagram illustrating the relationships between the mechanical forces within the syrinx that produce the vibrations of the tympaniform membranes. See text for explanation.

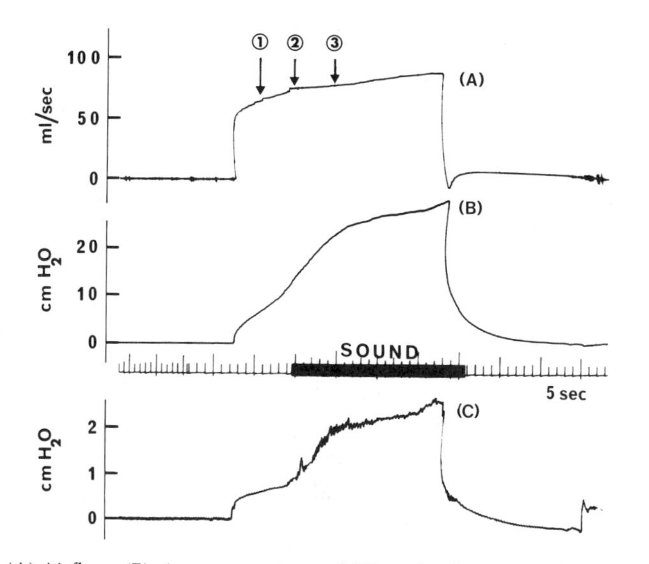

Fig. 6. (A) Airflow, (B) air sac pressure, and (C) tracheal pressure during experimental sound production in syrinx of Domestic Fowl. Air was blown into the interclavicular air sac and out of the trachea at a continuously increasing rate. At a threshold airflow rate, shown by arrow (2), sound production began. At subthreshold flow rates, for instance that shown by arrow (1), the suction forces in the syrinx were insufficient to trigger the external tympaniform membranes. Sound production continued at all airflow rates above threshold, for instance that shown by arrow (3).

but in this scheme of "whistled" sounds they are clearly not the chief determinant of sound quality. These authors believe that a whistle mechanism can more easily explain the production of pure tones and harmonic sounds than a conventional membrane vibration mechanism. The whistle mechanism may therefore be of widespread occurrence among birds. However, the distinction between the whistle and membrane vibration mechanisms poses considerable technical and theoretical problems, particularly concerning the "primary" or "secondary" nature of the membrane vibrations. A very detailed experimental examination of the mechanoacoustic events in the syrinx will be necessary in order to make a correct interpretation.

VI. ENERGY CONTENT OF SONG

Only a small amount (ca. 2–3%) of the total fluid energy generated by the compression of the air sacs during crowing in the Domestic Fowl is converted to sound, most being lost by friction in the syrinx and trachea (Brackenbury, 1977). Judged by this criterion, therefore, the Domestic Fowl syrinx appears to be a very inefficient machine for producing sound, particularly when compared to the sound organs of some insects which may operate at efficiencies of 30% (Bennet-Clark, 1971). However, the apparent inefficiency of the Domestic Fowl syrinx may be related to the high threshold airflow velocity required to trigger the external tympaniform membranes. This is a necessary safety device to avoid accidental triggering of the membranes during normal respiratory function, such as exercise or thermal panting, both of which incur large increases in ventilation.

Limited evidence suggests that some songbirds produce more acoustic power in relation to body size than roosters (Brackenbury, 1979b). The maximum sound power output of roosters in the Domestic Fowl is approximately 60 mW kg^{-1}; in comparison, the Song Thrush (*Turdus philomelos*), Winter Wren (*Troglodytes troglodytes*), and Robin (*Erithacus rubecula*) produce 870, 600, and 300 mW kg^{-1}, respectively. Gaunt *et al.* (1973, 1976) and Brackenbury (1977, 1978a) have suggested that the intrinsic muscles of songbirds may be capable of regulating the airflow through the syrinx in a way most favorable to the effective energy coupling between the airstream and the internal tympaniform membranes. However, not all songbirds are equally effective at producing sound power. Some relatively weak (in power terms) singers like the Sedge Warbler (*Acrocephalus schoenobaenus*), Willow Warbler (*Phylloscopus trochilus*), and Whitethroat (*Sylvia communis*) produce less than 60 mW kg^{-1} of sound power. This may be related to the range of frequencies utilized during singing in these species; the radiation damping of the air, which governs sound power (see above), is proportional to frequency to the fourth power (Brackenbury, 1979a). It would be misleading to place too much stress on the selective value to a species of any single acoustic characteristic of song, such as sound power, at the expense of other characteristics of equal or even greater informational importance, such as sound modulations. Nevertheless, any factor which served to increase the broadcasting range of an individual, such as the ability to use the syrinx at relatively high (but biologically realistic) frequencies, is likely to prove on balance to be an advantage.

VII. SOUND MODULATIONS

Many birdsongs contain amplitude and frequency modulations with repetition rates up to several hundred Hertz (Greenewalt, 1968; Stein, 1968). Greenewalt defined true modulation as the superposition, on a phrase of constant amplitude and frequency, of a modifying process which changes amplitude or frequency or both. He listed criteria for distinguishing true modulations from "spurious" modulations that are due to purely acoustic phenomena within the vocal tract, such as wave interaction.

Most spurious modulations result from either the harmonic vibrations of the tympaniform membranes or wave resonance in the trachea. If the membranes vibrate in a series of harmonics, the resultant sound waveform, as observed, for instance, on an oscilloscope, will appear to be modulated at the frequency of the fundamental membrane vibration. According to Greenewalt, examples of such modulations resulting from source-generated harmonic spectra are common in calls, though not so common in songs.

Tracheal resonance may also give rise to sound waveforms resembling those produced by harmonic membrane vibrations. These occur when each pulse of air produced by the oscillation of the membranes (see previous section) excites the trachea into damped resonant vibration. The tracheal resonances are governed by the tracheal length and occur in series of frequencies related as 1:2:3 etc., or 1:3:5 etc., depending on whether the trachea behaves as a tube open at both ends (the syrinx and the glottis) or only one end (the glottis). In general, the modulation frequency, which is the repetition frequency of the air pulses (equal to the fundamental frequency of membrane vibrations), is not related to the tracheal resonances, thus providing one means of distinguishing these two types of modulation.

True modulations, like spurious modulations, may also produce equally spaced lines of frequency when displayed sonographically. The frequencies lie above and below the fundamental or carrier frequency of the membrane, and the separation frequency is the same as the modulating frequency. Modulations are imposed on the membrane vibration by rhythmical muscular activity (Section IX); consequently, there is in general no multiple relationship between the modulation frequency and the carrier frequency. This latter feature distinguishes true modulations from both source-generated harmonic spectra and tracheal resonances.

VIII. TRACHEAL RESONANCE

Greenewalt cited a number of apparent instances of tracheal resonance but then dismissed them, primarily on the basis of nonconformity between predicted and measured tracheal lengths. However, it is probably incorrect to assume that the length of the trachea in morbid specimens is the same as its actual length during vocalization. White (1968) showed that the trachea of male Domestic Fowl was retracted several centimeters by muscular activity during crowing. This movement is produced by the tracheolateralis and tracheohyoideus muscles (Gaunt and Gaunt, 1977; Brackenbury, 1980). Harris *et al.* (1968) found that the harmonics of artificially produced sounds in the Domestic Fowl varied according to tracheal length and concluded that the trachea and primary bronchi combine to form a single resonant tube. Hinsch (1972) observed that the frequency spectra of some of the sounds in the Willow Warbler resembled the resonance curve of a damped vibration in a tube having the same length as the trachea and suggested that alterations to the curve in different notes might be related to active changes in tracheal length. Sound pulses in the Grasshopper Warbler (*Locustella naevia*) show a very sharp rise and an exponential decay (time constant = 0.6 msec), features which would be expected of a pulsed tracheal resonator (Brackenbury, 1978b). Warner (1971b) was of the opinion that one of the functions of the extrinsic syringeal muscles in ducks of the genera *Anas* and *Aythya* was to vary the length and resonance of the trachea. Males of these genera also possess at the base of the trachea resonant chambers or bullae which serve to amplify the relatively restricted vibrations of the unusually thickened internal tympaniform membranes. These limited examples therefore suggest that the trachea may be a more important determinant of sound characteristics than Greenewalt thought.

IX. MODULATIONS PRODUCED BY THE SYRINX AND RESPIRATORY MUSCLES

Greenewalt (1968) attributed true modulations to the action of the syringeal muscles on the tympaniform membranes and the external labium. Interclavicular air sac pressure forces the tympaniform membranes into the bronchial lumen and this movement is opposed by the syringeal muscles which, by increasing the tension of the membranes, draw them out of the lumen and permit an acceleration of the airflow which triggers the vibration of the membranes. Increased contractile force by the muscles produces greater withdrawal of the membranes and greater vibrational amplitude. Rhythmical contraction of the syringeal muscles thus provides a mechanism for directly coupled amplitude and frequency modulations. This mechanism of membrane excitation depends on a simultaneous increase in membrane tensional forces and flow-induced forces and is compatible with that described in Section V. Greenewalt's second model was designed to explain amplitude modulations which are unaccompanied by frequency modulations: Rhythmical intrusion of the external labium into the bronchial lumen produces changes in the area of the passage, and thus in the amplitude of vibration of the membranes, without altering the frequency.

Given the possibility that these two mechanisms exist, great uncertainty surrounds the phasic relationships between syringeal muscle activity and the modulated movements of the membranes and external labium. High-frequency modulations cannot result directly from the neurogenic contractions of the muscles because of the limitations on the speed of contraction of vertebrate fast muscle (Stein, 1968; Brackenbury, 1978b). Stein proposed that modulations with repetition frequencies greater than ca. 70 Hz are produced by the resonant vibration of the external labium. This vibration could be triggered by the airflow alone or energy could also be injected at intervals by the intrinsic muscles. Brackenbury (1978b) suggested that the myogenic contractions of insect muscle provide a possible parallel for the mode of action of the intrinsic syringeal muscles. For example, some species of cicada produce trains of pulsed sounds at rates of several hundred hertz by the resonant vibrations of a tymbal organ, consisting of a membrane driven by a muscle (Pringle, 1954). The oscillation frequency of the tymbal organ is governed entirely by its mass and elasticity, and nerve impulses are sent to it simply to maintain the contractile state of the muscle. In other species there is a 1:1 relationship between nerve impulse and muscle contraction, but these species produce sound pulses at lower rates (120 Hz; Hagiwara, 1959). However, none of these ideas concerning muscle contractile mechanisms has yet been tested experimentally and the electromyographic recordings of Youngren *et al.* (1974) and Gaunt and Gaunt (1977) in the Domestic Fowl and those of Lockner and Youngren (1976) in the Mallard remain the only direct demonstration of rhythmical activity of the (extrinsic) syringeal muscles during pulsed sound production.

Calder (1970) introduced a second category of sound modulations in addition to those produced by the syrinx. These result from the pulsatile activity of the respiratory system, and are based on his ideas of the observation of a synchronous relationship between trilled sound production and rapid respiratory movements in Canaries (*Serinus canaria*). These occurred at rates up to 27 Hz, which is presumably close to the maximum possible vibration frequency of the respiratory apparatus. Calder assumed that these respiratory movements represented complete miniature breaths, although it is difficult to distinguish them from pulsed expirations such as those which occur during vocalization in the Evening Grosbeak (Berger and Hart, 1968), Starling (Gaunt *et al.*, 1973), Mallard (Lockner and Murrish, 1975), Domestic Fowl (Gaunt *et al.*, 1976; Gaunt and Gaunt, 1977; Brackenbury, 1978a), and Gray Lag Goose (Brackenbury, 1978a).

I put forward a "minibreath" mechanism to explain the origin of pulsed sounds in Grasshopper and Sedge warblers and suggested that it would enable the former species to satisfy gaseous exchange while sustaining long sequences of uninterrupted song (Brackenbury, 1978b). Pulsed expirations can be discounted in the case of Grasshopper Warbler song since there are no detectable pauses in the sequence which would correspond to regular inspirations. Such pauses can be identified, however, between individual sequences of song in the Sedge Warbler and Skylark (Brackenbury, 1978b,c; Csicsáky, 1978). The repetition rates of these sequences vary from 0.3 to 3.0 Hz in the Sedge Warbler, which is close to the estimated resting respiratory rate of 1.5 Hz. The repetition rates of individual phrases in Skylark song vary from 4 to 10 Hz compared with an estimated resting respiratory rate of ca. 1 Hz, but an elevated respiratory rate would be expected during flight. Csicsáky found that the wing beat frequency in Skylarks was 16.3 Hz and was thus able to rule out flight movements as a possible cause of song fragmentation.

Further experimental work is clearly needed on the relationship between respiratory movements and sound production, but additional light can probably be shed on the subject by considering the different types of ventilatory patterns which have been observed in panting birds. To some extent the physiological problems experienced by panting birds and sustained singers like the Grasshopper Warbler are very similar; both are confronted by the need to hyperventilate the respiratory system, for the purpose of either increased respiratory evaporation or vocalization, respectively, without at the same time incurring severe carbon dioxide washout from the lungs. Panting birds have evolved at least two distinct ventilatory mechanisms to obviate this hazard. Some, like the Domestic Fowl and the Greater Flamingo (*Phoenicopterus ruber*), practice extended periods of very rapid shallow breathing, confining the hyperventilation to the tracheal dead space. One of the penalties incurred by this type of breathing is that carbon dioxide, although maintaining its normal concentration in the lungs, steadily accumulates within the caudal air sacs and must be removed by regular periods of deeper respiration which interrupt the thermoregulatory response (J. H. Brackenbury, P. Avery, and M. Gleeson, unpublished observations). It is possible that a similar kind of respiratory mechanism could explain the production of Grasshopper Warbler song, the shallow respiratory movements being associated with pulsed sounds, the deep respiratory movements with the brief silent intervals (so-called "breathing pauses") which interrupt the train of song on average every 27 sec (Hulten, 1959; Kliebe, 1966).

The second pattern of ventilation in hyperthermic birds is displayed by the Rock Dove (*Columba livia*), in which a continuous series of shallow, oscillatory breaths is superimposed upon a normal respiratory wave. The underlying deeper wave serves to maintain the normal gas exchange requirements of the lung while simultaneously the rapid panting movements result in a large increase in tracheal ventilation and evaporative heat loss from the upper respiratory tract (Ramirez and Bernstein, 1976). The temporal patterning of Sedge Warbler song could be based on an analogous form of breathing sequences of pulsed sounds being generated by puffs of expired air, and silences between sequences representing inspirations.

X. UNSOLVED PROBLEMS

Further experimental investigation on almost any aspect of sound production in birds would add significantly to present understanding. Most present knowledge concerning the function of the syrinx is based on a few species, some of which, including the Domestic Fowl and Mallard, are probably atypical of birds in general. Many of the ideas about the function of the passerine syrinx are based on informed guesswork. Virtually nothing is known about the physiological properties of the intrinsic syringeal muscles. The vibration of the syringeal membranes, the central assumption of syringeal physiology, has never been directly measured. The involvement of the trachea in vocalization is still in doubt; could the membranes produce sound if the trachea were not present, or is an acoustic coupling between the syrinx and trachea a necessity? On the problem of active sound modulations: what, if any, is the role of extrasyringeal structures such as the trachea, pharynx, and hyoid apparatus? What is the structural basis of the "three-" and "four-voice" phenomena? All of these areas are in need of experimental investigation. The situation at the moment with regard to the physiology of voice production in birds is that there is a wealth of phonetic and acoustic analysis and relatively little structural and functional knowledge on which to base it. The difficult problems await solution.

ACKNOWLEDGMENT

The work contained in this chapter was carried out with the support of the Science Research Council.

REFERENCES

Ames, P. L. (1971). The morphology of the syrinx in passerine birds. *Bull. Peabody Mus. Nat. Hist.* **37,** 1–194.

Bennet-Clark, H. C. (1971). Acoustics of insect song. *Nature (London)* **234,** 255–259.

Berger, M., and Hart, J. S. (1968). Ein Beitrag zum Zusammenhang zwischen Stimme und Atmung bei Vogeln. *J. Ornithol.* **109,** 421–424.

Brackenbury, J. H. (1977). Physiological energetics of cock crow. *Nature (London)* **270,** 433–435.

Brackenbury, J. H. (1978a). Respiratory mechanics of sound production in chickens and geese. *J. Exp. Biol.* **72,** 229–250.

Brackenbury, J. H. (1978b). A comparison of the origin and temporal arrangement of pulsed sounds in the songs of the Grasshopper and Sedge warblers, *Locustella naevia* and *Acrocephalus schoenobaenus*. *J. Zool.* **184,** 187–206.

Brackenbury, J. H. (1978c). A possible relationship between respiratory movements, syringeal movements and the production of song by Skylarks *Alauda arvensis*. *Ibis* **120,** 526–528.

Brackenbury, J. H. (1979a). Aeroacoustics of the vocal organ in birds. *J. Theor. Biol.* **81,** 341–349.

Brackenbury, J. H. (1979b). Power capabilities of the avian sound producing system. *J. Exp. Biol.* **78,** 163–166.

Brackenbury, J. H. (1980). Control of sound production in the syrinx of the fowl *Gallus gallus*. *J. Exp. Biol.* **85,** 239–251.

Calder, W. A. (1970). Respiration during song in the Canary (*Serinus canarius*). *Comp. Biochem. Physiol.* **32,** 251–258.

Chamberlain, D. R., Gross, W. B., Cornwell, G. W., and Mosby, H. S. (1968). Syringeal anatomy in the Common Crow. *Auk* **89,** 244–252.

Csicsáky, Von M. (1978). Über den Gesang der Feldlerche (*Alauda arvensis*) und seine Beziehung zur Atmung. *J. Ornithol.* **119,** 249–264.

Gaunt, A. S., and Gaunt, S. L. L. (1977). Mechanics of the syrinx in *Gallus gallus*. II. Electromyographic studies of *ad libitum* vocalisations. *J. Morphol.* **153,** 1–20.

Gaunt, A. S., and Gaunt, S. L. L. (1980). Phonation of the Ring Dove: implications and variants. *Am. Zool.* **20,** 758.

Gaunt, A. S., Stein, R. C., and Gaunt, S. L. L. (1973). Pressure and air flow during distress calls of the Starling, *Sturnus vulgaris* (Aves: Passeriformes). *J. Exp. Zool.* **183,** 241–262.

Gaunt, A. S., Gaunt, S. L. L., and Hector, D. H. (1976). Mechanics of the syrinx in *Gallus gallus*. I. A comparison of pressure events in chickens to those in Oscines. *Condor* **78,** 208–223.

Gaunt, S. L. L., and Gaunt, A. S. (1980). Phonation of the Ring Dove: the basic mechanism. *Am. Zool.* **20,** 757.

George, J. C., and Berger, A. J. (1966). "Avian Myology." Academic Press, New York.

Greenewalt, C. H. (1968). "Bird Song: Acoustics and Physiology." Smithsonian Inst. Press, Washington, D.C.

Gross, W. B. (1964). Voice production by the chicken. *Poult. Sci.* **43,** 1005–1008.

Hagiwara, S. (1959). Myogenic rhythm of sound producing muscles in cicadas. *Bull. Tokyo Med. Dent. Univ.* **1,** 113–124.

Harris, C. L., Gross, W. B., and Robeson, A. (1968). Vocal acoustics of the chicken. *Poult. Sci.* **47,** 107–112.

Hinsch, Von K. (1972). Akustiche Gesangsanalyse beim Fitis (*Phylloscopus trochilus*) zur Untersuchung der Rolle der Luftrohre bei der Singvogel. *J. Ornithol.* **113,** 315–322.

Hulten, M. (1959). Beitrag zur Kenntnis des Feldschwirls. *Regulus* **39,** 95–118.

Kliebe, K. (1966). Zur Gesangsdauer des Feldschwirls (*Locustella naevia*). *Ornithol. Mitt.* **18,** 109–110.

Lemon, R. E. (1973). Nervous control of the syrinx in White-throated Sparrow (*Zonotrichia albicollis*). *J. Zool.* **171,** 131–140.

Lockner, F. R., and Murrish, D. E. (1975). Interclavicular air sac pressures and vocalisation in Mallard ducks *Anas platyrhynchos*. *Comp. Biochem. Physiol.* **52A,** 183–187.

Lockner, F. R., and Youngren, O. M. (1976). Functional syringeal anatomy of the mallard. I. *In situ* electro-myograms during ESB elicited calling. *Auk* **93,** 324–342.

Mairy, F. (1976). Tentative models of sound-figure and coo production in the Ring Dove (*Streptopelia risoria*). *Biophon* **4,** 2–5.

Miller, D. B. (1977). Two-voice phenomenon in birds: further evidence. *Auk* **94,** 567–572.

Miskimen, M. (1951). Sound production in passerine birds. *Auk* **68,** 493–504.

Nottebohm, F. (1971). Neural lateralisation of vocal control in a passerine bird. I. Song. *J. Exp. Zool.* **177,** 229–262.

Nottebohm, F., and Nottebohm, M. E. (1976). Left hypoglossal dominance in the control of Canary and White-crowned Sparrow song. *J. Comp. Physiol.* **108,** 171–192.

Peek, F. W. (1972). An experimental study of the territorial function of vocal and visual display in the male Red-winged Blackbird (*Agelaius phoeniceus*). *Anim. Behav.* **20,** 112–118.

Phillips, R. E., and Peek, F. W. (1975). Brain organisation and neuromuscular control of vocalisation in birds. *In* "Neural and Endocrine Aspects of Behaviour in Birds" (P. Wright, P. G. Caryl, and D. M. Vowles, eds.), pp. 243–273. Elsevier, Amsterdam.

Pringle, J. W. S. (1954). A physiological analysis of cicada song. *J. Exp. Biol.* **31,** 525–560.

Ramirez, J. M., and Bernstein, M. H. (1976). Compound ventilation during thermal panting in pigeons: a possible mechanism for minimising hypocapnic alkalosis. *Fed. Proc., Fed. Am. Soc. Exp. Biol.* **35,** 2562–2565.

Seller, T. J. (1979). Unilateral nervous control of the syrinx in Java Sparrows (*Padda oryzivora*). *J. Comp. Physiol.* **129,** 281–288.

Smith, D. G. (1977). The role of the sternotrachealis muscles in birdsong production. *Auk* **94,** 152–155.

Stein, R. C. (1968). Modulation in bird sounds. *Auk* **85,** 229–243.

Warner, R. W. (1971a). The syrinx in the family Columbidae. *J. Zool.* **166,** 385–390.

Warner, R. W. (1971b). The structural basis of the organ of voice in the genera *Anas* and *Aythya* (Aves). *J. Zool.* **164,** 197–207.

Warner, R. W. (1972). The anatomy of the syrinx in passerine birds. *J. Zool.* **168,** 381–393.

White, S. S. (1968). Movements of the larynx during crowing in the domestic cock. *J. Anat.* **103,** 390–392.

Youngren, O. M., Peek, F. W., and Phillips, R. E. (1974). Repetitive vocalisations evoked by local electrical stimulation of avian brains. III. Evoked activity in the tracheal muscles of the chicken (*Gallus gallus*). *Brain Behav. Evol.* **9,** 393–421.

3

Neural Control of Passerine Song

ARTHUR P. ARNOLD

I. INTRODUCTION

The study of animal behavior can be divided into two major classes of investigation. On the one hand, we seek to know the adaptive significance of a behavior and explain it in terms of what selective pressures led to its evolution. Alternatively, we wish to determine the mechanisms underlying the behavior and explain the physiological mechanisms which are responsible for its occurrence. Of course, these two endeavors are related, since one long-range goal is to understand the evolution of the physiological mechanisms controlling behavior. For example, by comparing the physiological mechanisms underlying a class of behaviors (such as vocal communication) in related or unrelated species, we may eventually gain insight into what kinds of physiological mechanisms are selected for, and which types of neural organizations are most adaptive (e.g., because of economy of developmental mechanisms, economy of energy utilization, etc.).

ACOUSTIC COMMUNICATION IN BIRDS
VOLUME 1

ISBN 0-12-426801-3

But so far, there is very little information on the evolution of physiological mechanisms underlying behavior, largely because it takes a great deal of effort to begin to understand the physiological basis of a single behavior in any one species. Therefore comparison of species is a task which must proceed more slowly.

In the study of neural control of vocalizations in birds, understanding the evolution of neural mechanisms is a long-term but often implicit goal. The great variety in use of vocalizations among birds suggests that comparative studies of neural control of vocalization will be fruitful. However, there exists a number of shorter-term goals or theoretical issues, around which most investigations have centered. These issues serve as organizing principles for the review which follows. Specifically, we are interested in neural mechanisms responsible for song learning (Sections II,A and III,A), for lateralization of neural control of song (Sections II,B and III,B), for androgen influences on song (Section III,C), and for sexual differences in vocal behavior (Section III,D). The scope of this discussion is limited to studies performed on neural control of vocalization in passerine birds [see, e.g., Phillips and Peek (1975) for a review of neural mechanisms in other species, and Nottebohm (1980b) for another recent review on passerines].

II. RECENT ORIGINS OF THE STUDY OF NEURAL BASIS OF SONG

A. Auditory Feedback in Vocal Control

In many passerine birds, song is learned during an early critical period (Marler and Mundinger, 1971; Nottebohm, 1975). For example, in the White-crowned Sparrow (*Zonotrichia leucophrys*), a young male must hear conspecific song within the first 3 months of life in order to produce a good copy of it (Marler, 1970). Full adult song is produced first at about 8–10 months of life, but the male need not hear the model after 3 months of age to produce an accurate copy. Because males of this species typically begin to sing after 3 months of age, there must exist a neural representation (auditory memory or template) of the song model, which the bird retains during the early stages of song development (subsong) until full, crystallized song is produced. In some of the earliest physiological investigations of song development, Konishi (1965) and Nottebohm (1968) sought to determine whether auditory feedback is necessary in the song development process. White-crowned Sparrows deafened at 3 months of age (at a time when the auditory memory should be relatively complete) do sing as adults, but the song is grossly abnormal in structure, highly variable, and does not resemble the model song in any way. This result suggests that not only do White-crowned Sparrows learn to recognize the song model (auditory learning or template formation) but they also learn to match the auditory template during the period of subsong (Konishi and Nottebohm, 1969). In order to match the auditory memory of song, the bird must be able to hear himself, presumably to assess the accuracy of his vocal performance and compare motor output with auditory memory. These observations raise a number of questions about the neural basis of song. Where and how is the auditory memory encoded in the brain? What are the neural mechanisms which are responsible for the comparison of the auditory memory and vocal performance? What neural mechanisms are responsible for the selection and enhancement of certain motor patterns (i.e., those which more closely match the model), which gradually results in improvement in performance? These complicated questions are largely unanswered, but they have led to certain preliminary findings concerning the interface between auditory and motor systems in passerine brains, a subject which we discuss further in Section III,A.

A second interesting finding of Konishi (1965) is that if adult White-crowned Sparrows are deafened as adults after song crystallization, the operation has little effect on song. Similar results have been obtained in Chaffinches (*Fringilla coelebs;* Nottebohm, 1968), and Zebra Finches (*Poephila guttata;* Price, 1979). For periods of over a year after deafening, the structure of song is remarkably invariant. After crystallization, auditory feedback is no longer needed to assess the accuracy of vocal performance. Since song is a complex motor task, this is rather surprising, especially in light of the extreme dependence of vocal learning on auditory feedback. Two possibilities could account for the decline in the importance of auditory feedback during development: (1) other feedback mechanisms may come to play a role, for example, proprioceptive feedback from the syringeal muscles, respiratory muscles, or respiratory system may help assess the accuracy of vocal output; or (2) after song crystallization, there may cease to be a need for feedback. Feedback control of complex muscular movements is usually necessary because there is variability in the loads against which muscles contract. Thus, feedback is necessary to determine whether a certain degree of muscular effort or force has achieved the goal specified in higher motor centers. If there exists little variability in load, then specifying a certain muscular force is equivalent to specifying a particular result. The loads against which the syringeal muscles work are probably determined by elastic forces in the tissues of the syrinx, and by pressures generated in the trachea (on the inside of the syrinx) and in the interclavicular air sac (on the outside of the syrinx). If all of these loads are either invariant or predictable by sensory feedback from the respiratory system's proprioceptive neurons, then feedback from the muscle movements themselves may not be needed to predict motor performance.

B. The Discovery of Lateralization of Motor Control of Song

The above arguments are rather speculative, and data about the existence and importance of nonauditory feedback mechanisms in the control of song are needed. Little is known about the existence of sensory receptors from the syrinx. The syrinx has two sound sources, the right and left internal tympaniform membranes, and each source is controlled by a set of syringeal muscles (Greenewalt, 1968; Nottebohm, 1971). In Canaries (*Serinus canaria*), and Zebra Finches, each set of syringeal muscles receives innervation from the ipsilateral tracheosyringeal branch of the hypoglossal nerve, which carries the motor axons to the muscles (Nottebohm *et al.*, 1976; Arnold *et al.*, 1976). If sensory fibers run to the central nervous system from the syrinx, they may well also follow this nerve, although one cannot exclude the possibility that they would follow a small branch of the vagus which does reach the syrinx in some bird species (see Nottebohm, 1971). Partly to test for the presence of sensory fibers in these nerves, Nottebohm (1971, 1972) sectioned the tracheosyringeal branch unilaterally in Chaffinches and observed the effect on song. Surprisingly, sectioning the nerve on the left resulted in greater deficits than sectioning the right, i.e., more song elements were disrupted after cutting the left nerve than after cutting the right. The temporal patterning and acoustic form of the remaining song elements were, however, virtually identical to those of the preoperative song. Nottebohm interpreted these results by suggesting that each side of the syrinx sings certain song elements which are temporally interspersed with each other and closely coordinated in time. Most of the song elements are sung on the left side of the syrinx, controlled by the left nerve. However, Nottebohm was careful not to eliminate the possibility that some elements lost after left nerve section could have been sung on the right side, but they depended on feedback along the cut left nerve for their occurrence. This possibility seems relatively unlikely. If proprioceptive feedback from the syrinx flows back to the brain along the tracheosyringeal nerve, then eliminating this feedback from one side of the syrinx has little effect on the song elements which remain, sung by the contralateral side, since they are sung in the same acoustical form and in the same temporal order as before the operation.

These studies suggest a minimal role for syringeal proprioceptive feedback from one side of the syrinx on the motor performance of the other half, but they do not address the possibility of this type of sensory modulation of ipsilaterally generated song elements. More important, however, is the discovery that neural control of bird song in some passerines is similar to human speech in that neural control is lateralized (Nottebohm, 1970). The pattern of left hypoglossal dominance is also found in Canaries, White-crowned Sparrows, White-throated Sparrows (*Zonotrichia albicollis*), and Java Sparrows (*Padda oryzivora*) (Nottebohm and Nottebohm, 1976; Lemon, 1973; Seller, 1979). The lateralization in neural control of human speech is known to exist in the cerebral cortex, whereas lateralization of Chaffinch and Canary song was discovered at the level of the motor neurons. This contrast led Nottebohm to search for neural regions which are important in the control of song and other vocalizations, so that one could determine at what neural levels lateralization could be found.

C. Central Motor Control of Song

To search for brain regions important in vocal control, Nottebohm *et al.* (1976) made electrolytic lesions in the brains of Canaries and observed the effects on song. Severe deficits in singing occurred after lesions were placed in two nuclei in the telencephalon [the caudal nucleus of the hyperstriatum ventrale (HVc) and robust nucleus of the archistriatum (RA) (Fig. 1)]. The effects of bilateral HVc lesions are sometimes dramatic. The bird no longer sings, although he may adopt the singing posture in a sequence of behaviors named "silent song" by Nottebohm *et al.* (1976). Nottebohm and colleagues then used silver stains to locate axons which degenerated when neurons were destroyed by lesions in HVc or RA. They showed that HVc neurons project caudally to RA and RA projects to the nucleus intercollicularis (ICo) in the midbrain and to a group of hypoglossal motor neurons in the medulla. These final two projections are not surprising, assuming an important role for HVc and RA in motor control of vocalizations. ICo has long been known as one of the lowest threshold sites in the brain for eliciting vocalization with electrical stimulation (Brown, 1965, 1971, 1973). The hypoglossal motor neurons (nXIIts) which receive input from RA

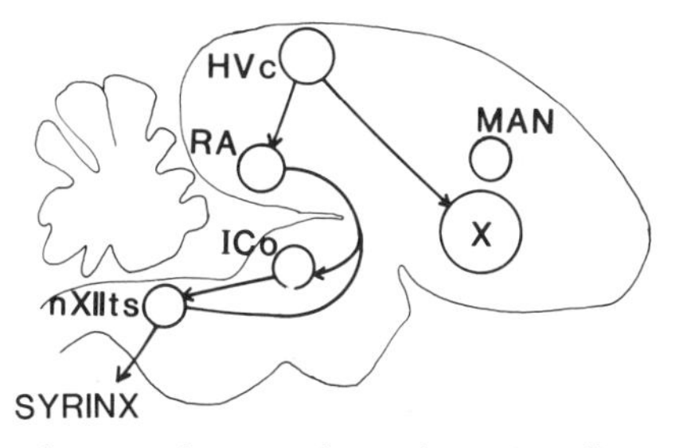

Fig. 1. This highly schematic drawing shows the major efferent pathways thought to be involved in control of song in passerine birds. Circles represent individual brain regions, and anatomical projections are indicated by arrows. The hyperstriatum ventrale, pars caudale (HVc), and robust nucleus of the archistriatum (RA) are intimately involved in song. The HVc projects to RA, which in turn projects to the nucleus intercollicularis of the midbrain (ICo) and to the hypoglossal motoneurons (nXIIts), which innervate the syrinx. The dorsomedial nucleus of ICo also projects to nXIIts. [Nottebohm *et al.* (1976) and Gurney and Konishi (1980).]

supply the axons which travel along the tracheosyringeal nerve to innervate the syringeal muscles. Thus the main descending pathways involved in motor control of song are thought to be HVc to RA to ICo and nXIIts.

In their original report Nottebohm *et al.* (1976) also established a massive projection rostrally in the brain from HVc to Area X in the lobus parolfactorius. In subsequent studies, Nottebohm and Kelley (1978) demonstrated projections from another rostral nucleus, the magnocellular nucleus of the neostriatum (MAN) to both HVc and RA (Fig. 2). Because of their heavy connections with HVc and RA, Area X and MAN are thought to be important parts of the network of neurons involved in vocal control. The large sexual differences seen in these two nuclei are compatible with this suggestion (see Section III,D). Paradoxically, lesions of Area X in Canaries (Nottebohm *et al.*, 1976) have relatively little effect on the form of song sung by adults. Finally, two other neural projections into HVc should be mentioned, from nucleus interface (NIF) near Field L, and nucleus uva in the midbrain (Nottebohm and Kelley, 1978) (Fig. 2). Anatomically, all of these brain regions are similar in Zebra Finches and Canaries (e.g., Arnold, 1980a) and the connections among the areas are similar also (Gurney and Konishi, 1979; Lewis *et al.*, 1981). In addition, Gurney (1980) reports a projection from the dorsomedial nucleus (DM) of ICo to the hypoglossal motor neurons in Zebra Finches (Fig. 1).

The discovery of this vocal control system, now only 5 or 6 years old, has been of considerable heuristic value. It has served as the foundation for new, rapidly accumulating information on several theoretical issues, each of which is discussed in the following section.

III. CURRENT STATUS OF THEORETICAL ISSUES ON NEURAL CONTROL OF SONG

A. The Interaction of Auditory and Motor Systems

The HVc–RA–nXIIts pathway is presently thought to be predominantly motor, mainly because of its close synaptic connections to the syringeal muscles, and because destruction of HVc and RA results in apparent motor deficits. Yet we know that, at least during development, auditory feedback is crucial to normal functioning of the vocal motor system. Where does this feedback take place? Although this question is far from answered, anatomical and physiological studies have made a significant start. In a number of bird species, auditory information is known to enter via the eighth nerve from the ear and pass from the cochlear nuclei to the dorsal nucleus of the lateral mesencephalon (MLd), to the nucleus ovoidalis, and finally to the telencephalic auditory projection area, Field L (e.g., Cohen and Karten, 1974). The complex response properties of Field L auditory neurons are compatible with the idea that it is a higher-order auditory

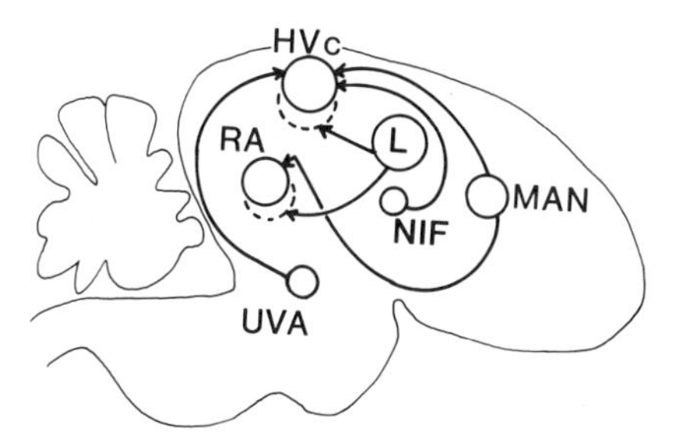

Fig. 2. This highly schematic drawing shows major afferents to HVc and RA. The magnocellular nucleus of the anterior neostriatum (MAN) projects to both HVc and RA. Field L is the neostriatal auditory projection area, which projects not to HVc or RA, but to regions immediately adjacent to these areas. Nucleus UVA and NIF (nucleus interface) project to HVc, but little is known of these two small nuclei. [Nottebohm and Kelley (1978), Kelley and Nottebohm (1979), and Nottebohm (1980b).]

processing region capable of detailed analysis of specific vocalizations (Leppelsack, 1974). Kelley and Nottebohm (1979) discovered that in Canaries, Field L sends fibers to a neostriatal shelf immediately medial and ventral to HVc and to an archistriatal region on the border of RA (Fig. 2). Projections into HVc and RA proper were not observed. The proximity of these areas makes it easy to imagine some flow of information into HVc and RA via short interneurons or perhaps directly from Field L to HVc or RA, if HVc or RA neurons possess long dendrites extending outside of the nuclei. Recent Golgi studies of RA neurons in Canaries (DeVoogd and Nottebohm, 1980) do not demonstrate any dendrites of RA neurons outside of this nucleus. However, Katz and Gurney (1981) have demonstrated that in Zebra Finches some HVc neurons which project to Area X also receive auditory input. Furthermore, auditory neurons just ventral to HVc, in the neostriatal shelf described by Kelley and Nottebohm (1979), send axon collaterals to both HVc and Field L. Thus HVc may be an important station for auditory feedback on motor control of song. Finally, Kirsch *et al.* (1980) have discovered auditory units in the anterior neostriatum near MAN and Area X in the Starling (*Sturnus vulgaris*), confirming reports in non-passerine birds (see Delius *et al.*, 1979). Thus far there is no evidence for auditory responses in MAN or Area X. These studies call for more intense anatomical and physiological investigations to characterize the auditory properties of these regions, in adulthood and during song learning, in the search for the neural basis of auditory model acquisition and its influence on song learning.

B. The Neural Basis of Lateralization

All anatomical projections described thus far in the song system are ipsilateral, except for some rather minor projections of ICo. This would imply that the left vocal regions control the function of the left syringeal muscles, and that the right brain regions control the right syringeal muscles. This possibility is only partially supported by evidence of the behavioral effects of unilateral brain lesions. Nottebohm *et al.* (1976) found that left HVc or left RA lesions in Canaries are more disruptive than those on the right, but the degree of difference between the two sides is less than that found at the level of the hypoglossus. For example, left hypoglossal lesions eliminate all but one or two song elements. Left HVc lesions disrupt the form of all song elements, but a significant number of song elements remain in altered form. In addition, left HVc lesions disrupt phrase structure. Right hypoglossal lesions selectively eliminate only one or two syllables (out of about 25–30) leaving the majority undisturbed in form and temporal patterning. A right HVc lesion in one bird disrupted or eliminated over half of the song syllables. This implies that right HVc plays some role in the control of the left half of the syrinx, but the pathways and mechanisms by which it does so are unknown.

The overwhelming predominance of ipsilateral projections is also paradoxical, since the left and right syringeal halves normally sing their separate song elements in an exquisitely interwoven, precisely timed sequence. The two brain halves must "talk" to each other to effect this timing. Electrophysiological evidence indicates that, at least in the Zebra Finch, HVc and RA are indirectly connected with the contralateral syringeal muscles, as expected (Arnold, 1980a). When microelectrodes were inserted into either RA or HVc, and short trains of microstimuli applied, syringeal contractions were detected at very low stimulus thresholds. The response in the ipsilateral syringeal half was always at a lower threshold and shorter latency compared to that on the contralateral side, but the presence of a small contralateral response indicates some crossing of information, as predicted by Nottebohm's behavioral observations.

Although Nottebohm and colleagues have argued convincingly for a larger role in song control for the left side of the brain, there are thus far no indications of anatomical asymmetries, except at the hypoglossal level. The left syringeal muscles and nXIIts are larger than the right in Canaries, but there are no right–left asymmetries in volume of higher centers (Nottebohm, 1977; Nottebohm and Arnold, 1976; Nottebohm *et al.*, 1981). Even using rather fine-grained Golgi analysis of one type of RA neuron in Canaries, DeVoogd and Nottebohm (1980) found no lateral asymmetries.

Nottebohm (1971, 1972) and Nottebohm *et al.* (1979) have investigated the ontogeny of left hypoglossal dominance in male Canaries and Chaffinches. Soon after hatching in Canaries, each hypoglossus has the potential to control song development and production, since cutting the left hypoglossus results either in right hypoglossal dominance or in a right–left sharing of song control, depending on the age when the nerve is sectioned. The ability of the right nerve to develop control of song decreases with age. If the left XII is sectioned at 1 year of age in Canaries, the initial severe loss of nearly all song syllables is followed by a period of recovery in which the male acquires the ability to sing new syllables, presumably with the right nerve. However, in such late lesions, the male never regains the number of song elements of intact birds. The story concerning reversal of hypoglossal dominance in Chaffinches is similar to that seen in Canaries, with one major exception. Once song is crystallized in Chaffinches, there is no major recovery from sectioning the left nerve. This contrast (some recovery in 1-year-old Canaries, none in 1-year-old Chaffinches) correlates well with the ability to learn new song elements. At this age Canaries normally do learn new song elements, and Chaffinches do not (Nottebohm and Nottebohm, 1978). Thus the neural plasticity needed to reverse hypoglossal dominance may be related to the plasticity needed to learn song.

When adult male Canaries receive unilateral HVc lesions at about 1 year of age, such birds have a new song repertoire when recorded 7 months later. This new repertoire has about as many syllable types as the song of intact birds, but a greater proportion of syllables in the new repertoire match preoperative syllables after right than after left lesions. When song is redeveloped after left HVc lesion, this may happen under a now dominant right HVc. There is evidence compatible with this, since such males lose all of their song elements if the right syringeal half is denervated (Nottebohm, 1977).

C. Androgen Control of Song

In most passerines, male song plays a courtship or territorial function and is closely related to reproduction. It has been known for many years that gonadal growth and androgen secretion lead to increased singing (e.g., Collard and Grevendal, 1946; Collias, 1950; Carpenter, 1932; Arnold, 1975). In some species, such as the Chaffinch, castration appears to abolish male song (Nottebohm, 1969), and in others, such as the Zebra Finch, song persists after castration but at a greatly reduced frequency of occurrence (Pröve, 1974; Arnold, 1975).

Because of our anatomical knowledge of brain regions involved in song, we can ask where in the brain do testicular androgens act to exert effects on song. This question was approached initially in the Chaffinch and Zebra Finch using steroid autoradiography (Zigmond *et al.*, 1973, 1980; Arnold *et al.*, 1976; Arnold, 1979, 1980d). Since steroid hormones are often accumulated by their target cells, it is possible to identify cells in the brain which might be sites of hormone action by determining which ones accumulate steroids. After injection of radioactive testosterone, cells in HVc, MAN, RA, ICo, and nXIIts accumulate hormone. Cells in other brain regions were also labeled, especially in the hypothalamus.

The accumulation of androgens is thought to be the prerequisite for at least certain types of androgen action on cells (McEwen, 1976), but demonstrating accumulation is not sufficient evidence to prove that such cells are androgen targets. So far, additional evidence for androgen action on these brain regions exists only in the case of the hypoglossal syringeal motor neurons (nXIIts). The neurotransmitter used by these neurons is acetylcholine (ACh), whose synthesis is catalyzed by the enzyme choline acetyltransferase (CAT), and whose degradation is catalyzed by acetylcholinesterase (AChE). Luine *et al.* (1980) found that castration reduced the levels of both of these enzymes in the tracheosyringeal nerve in Zebra Finches, and that testosterone administration reversed the drop in AChE and increased CAT levels to a level intermediate between castrates and intact males. Castration reduces syringeal weight in both Canaries and Zebra Finches, and reduces the levels of AChE and CAT measured in the syrinx as a whole. (AChE is present in both the hypoglossal nerve and the syringeal muscle fibers, whereas CAT is probably present predominantly in the neurons.) When testosterone is given, AChE and syringeal weight are increased to normal levels, but CAT is not. These studies indicate that androgen accumulation in the syringeal motor neurons is associated with androgen-dependent changes in AChE in these nerves. The presence of specific, high-affinity androgen receptors in the syrinx (Lieberburg and Nottebohm, 1979) is also associated with androgen dependence of AChE in the syrinx. These studies reinforce the idea that the demonstration of androgen accumulation strongly suggests androgen action on the target cells. However, it is also possible that androgens exert these effects on enzyme levels by acting at other sites. For example, if androgens act on higher brain regions (e.g., MAN) to increase the frequency of song, this increased use of the syringeal muscles could cause increased AChE levels, independently of direct androgen actions on the hypoglossal motor neurons or the syrinx. Because androgen effects on AChE in Canaries are seen both on the left side of the syrinx (which may bear the largest increase in usage since almost all song elements are sung on the left) and on the right side (which should experience relatively modest effects of increased usage since it sings few elements), Luine *et al.* (1980) argue that both increased use and direct androgen effects are occurring to regulate AChE.

The distribution of presumptive androgen target cells in the song system suggests two tentative conclusions about androgen control of this complex behavior (Arnold, 1981). First, it appears that androgens modify song via actions at many different levels in a neural heirarchy. Although we do not yet have very clear ideas about the function of each of the presumptive androgen target areas, it is likely that each plays a different role. This leads us to the conclusion that androgen actions may well modify a variety of functions (e.g., pattern generation of song, coordinating song with other behaviors, sensory feedback in the control of song, etc.). One of the challenges in the coming years is to determine which functions are altered. Because of our anatomical knowledge of the song system, we can be optimistic about the possibility of explaining the variety of roles played by androgens in modifying this vertebrate behavior pattern.

A second conclusion is that androgens act in the neural hierarchy at certain levels which at first blush seem paradoxical, i.e., at the final common path, in the motor neurons and syringeal muscles. The paradox arises if we consider that the main role played by androgens is to "switch on" singing behavior, i.e., to increase the probability of the behavior at the expense of other behaviors. It would seem that if this were the only function of androgens, one would expect this to occur at some high level in the hierarchy, at a place close to neurons which initiate the behavior, not at the motor neurons which are the last neurons to carry information before it is expressed as behavior. The resolution of this paradox is that androgens probably play both a switching role and other roles. For example, the "switching" function may be exerted by androgens in other target areas (MAN, HVc, hypothalamus?), but that when androgen levels are high and the bird is singing more, the motor neurons must be modified merely to handle the increase in activity, since they are driven more frequently by the higher centers. The increased enzyme levels in the hypoglossal nerve are perhaps a reflection of this type of androgen effect. However, other measures of androgen-induced changes are needed to confirm this hypothesis.

D. Sexual Differentiation of the Brain

In Zebra Finches and Canaries, males sing and females normally do not. Female Canaries treated with androgens do sing a song which is less complex than that of males, but even androgen treatment fails to induce song in female Zebra Finches. In both of these species, there are large sexual differences in brain regions controlling song, but not in other brain regions. HVc, Area X, RA, and nXIIts are all smaller in volume in females (Nottebohm and Arnold, 1976) (Fig. 3). Recently a portion of ICo, the dorsomedial nucleus (DM), has been added to this list in the Zebra Finch (Gurney and Konishi, 1980). The large sexual differences in size suggest that the reason for the inability of females to sing is the small size of their brain regions controlling song. This hypothesis is strengthened by the observation that female song areas are more developed in female Canaries, which sing, than in female Zebra Finches, which never sing.

The large sexual differences (male HVc is five to six times as large as female HVc in Zebra Finches) came as a surprise. Sexual differences in behavior, studied predominantly in mammals, had been thought to be associated with small, rather subtle differences in neuronal morphology and function (e.g., Raisman and Field, 1973). The idea that sexual differences in behavior can be related to large morphological differences in the nervous system has now been confirmed in mammals as well (Gorski *et al.*, 1978; Breedlove and Arnold, 1980). How do these sexual differences in morphology develop? There is extensive literature on mammals which suggests that sexually dimorphic behaviors and neuroendocrine functions are not determined strictly by genetic sex, but by the

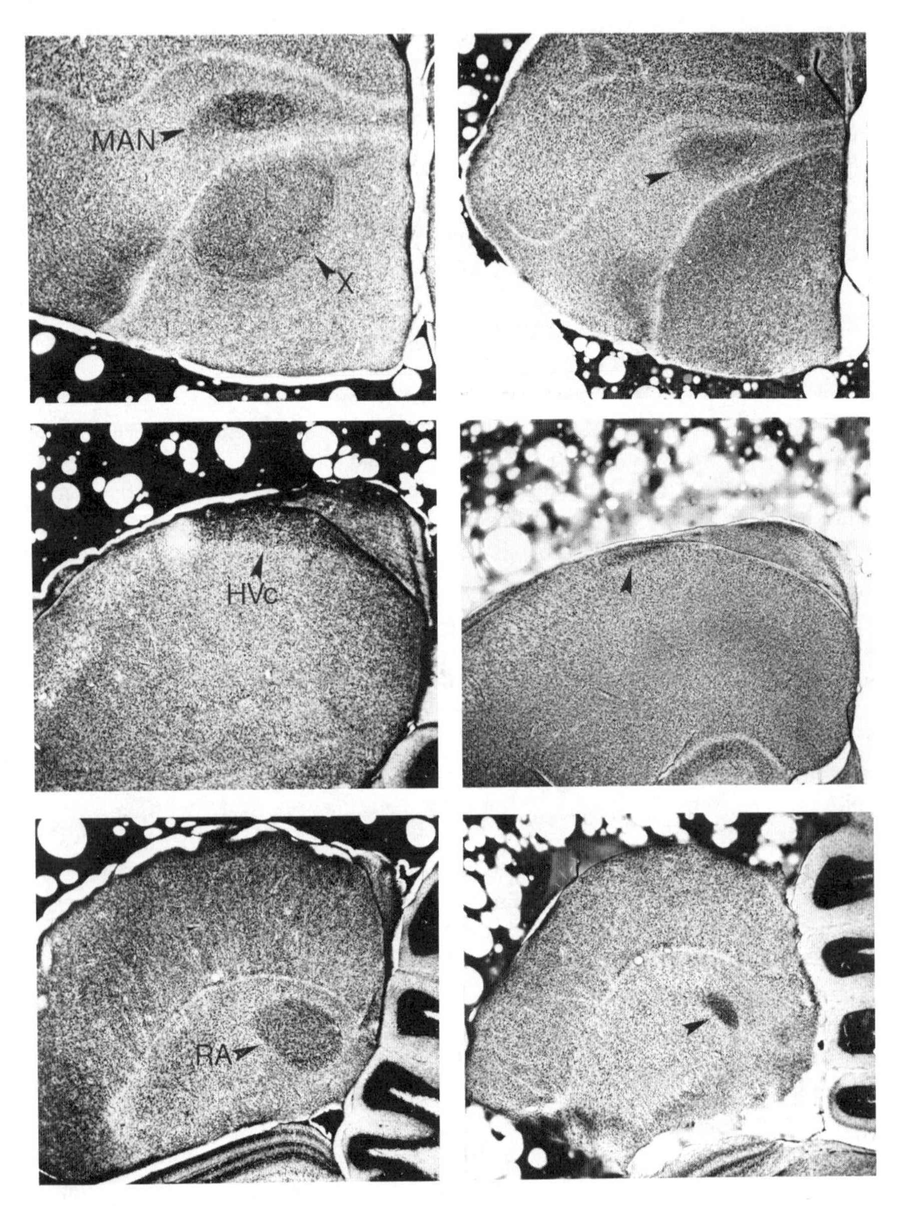

Fig. 3. Photomicrographs of Nissl-stained section containing vocal control regions in the brain of the Zebra Finch (*Poephila guttata*) in males (left) and females (right). MAN, HVc, and RA are seen in both sexes but are dimorphic, and area X is seen only in males. [Arnold (1980a).]

presence or absence of androgens during a perinatal critical period (Goy and McEwen, 1980). Female mammals given androgens or androgen metabolites during this time are permanently masculinized and defeminized, and males castrated at this time are permanently demasculinized and feminized. Because of this literature, and because of the observed androgen concentration by cells of vocal control regions in the adult songbird brain, we might ask whether gonadal steroids act on the latter areas in adulthood or during development to cause the avian sexual differences observed.

The first of these two possibilities has been investigated in Zebra Finches (Arnold, 1980c; Gurney and Konishi, 1980) and Canaries (Nottebohm, 1980a). Castration of adult male Zebra Finches has no specific effect on the volume of their vocal control regions, and androgen treatment of adult females appears to increase the volume only of nXIIts and DM. As noted above, androgen-treated female Zebra Finches do not sing. In Canaries, androgen treatment of adult females increases the volume of HVc and RA, although not enough to eliminate the sexual differences in brain volumes. Castration of males at ages 5–18 days reduces the volume of these two nuclei when measured in adulthood. These results point to interesting species differences. The androgen-induced morphological changes found in the brains of adult Canaries have not been observed in other androgen-sensitive neural systems in adult animals, and this suggests that there are steroid-induced changes in neurons in adult vertebrates larger than previously thought.

From the literature on sexual differentiation of behavior and neuroendocrine function, one would expect that embryonic or perinatal steroid levels might influence the differentiation of the brain and song behavior of songbirds. Gurney and Konishi (1979, 1980; Gurney, 1980) have analyzed early steroid effects on the song system. Female Zebra Finches were implanted on the first day of life with Silastic pellets containing either testosterone (T) or one of its two major metabolites: dihydrotestosterone (DHT), a 5α reduced androgen, or estradiol (E). After these birds reached adulthood, Gurney and Konishi used a variety of measures of sexual differences in the song system to see which were affected by these three steroids. In females given neonatal E, the adult volumes of HVc and RA were about five times as large as those of normal females, although still smaller than normal males. The nuclei DM and nXIIts were not affected by early E treatment. When DHT was implanted into 1-day-old females, the volumes of HVc and RA were increased only slightly, but DM and nXIIts were as large as those of males. This suggests that androgens and estrogens act on different neural regions to exert a permanent masculinizing influence.

Sexual differences have been analyzed most intensively in neurons of RA. In Canaries, DeVoogd and Nottebohm (1980) detected four classes of neurons stained by the Golgi method. In one of these classes, male neurons had much longer dendrites than female neurons. In Zebra Finches, Gurney (1980) found that RA neurons were larger, spaced further apart, and about 2.4 times more numerous in males. In adult female Zebra Finches which had been given T implants on the first day of life, RA neurons were much larger in somal diameter (equivalent in size to male RA neurons), spaced much further apart (nearly equivalent to male spacing), and occurred about twice as much as normal female RA neurons. DHT implanted into neonatal females mimicked the action of T by increasing neuronal number in RA, but did not increase somal diameter or spacing. Neonatal E implants did increase somal diameters and the spacing between neurons, but did not have a pronounced effect on number of neurons. Thus, Gurney (1980) suggests that testosterone secreted by the neonatal testis may be metabolized to androgens and estrogens, and that each of these classes of steroid has independent effects on the development of RA neurons.

Neonatal estradiol implants also appear to induce a second phase of androgen sensitivity in the song system later in life. In normal female Zebra Finches, androgens (testosterone propionate or DHT) in adulthood have little effect on volumes of HVc or RA, but nXIIts and DM are increased slightly in volume (Gurney and Konishi, 1980; Arnold, 1980c). This is compatible with the paucity of steroid-accumulating cells in HVc and MAN in female Zebra Finches compared with males (Arnold and Saltiel, 1979; Arnold, 1980b). Hypoglossal motor neurons accumulate steroid about equally in both sexes. The situation in estradiol-masculinized females may well be different. Females which are given implants of E on the day of hatching and treated with androgens (DHT seems more potent than T) in adulthood have HVc and RA nuclei which are even larger than in females given only neonatal E, and are as large as normal male HVc and RA. Gurney and Konishi (1980) suggest that the neonatal estradiol modifies the sensitivity of these neurons to androgens, hence allowing androgens to complete the masculine differentiation of HVc and RA by increasing their size.

The neural and behavioral effects of early hormone administration can be compared with interesting results. Since different steroids affect different neuronal parameters, we can ask which masculinizing effects are needed for female Zebra Finches to sing. Gurney and Konishi (1980) report that the only females to sing were those implanted with E as chicks and given T or DHT as adults. Thus, the partial masculinization produced by neonatal E or DHT implants alone was not sufficient for song in adulthood. It is not known whether neonatal DHT implants followed by adult androgen would produce song.

Finally, certain important neural characteristics seem not to be sexually dimorphic, and hence are not modified by neonatal steroids. Gurney and Konishi (1979; Gurney, 1980) found that if one divides neurons into classes according to their projections (e.g., RA neurons projecting to DM, HVc neurons projecting to Area X), then members of each projection class are found in both sexes. They suggest that sexual differentiation does not involve the elimination of certain neuron types in one sex, although sexual differences in numbers of neurons in one class may occur. Certain observations in our laboratory are compatible with this suggestion. Electrical stimulation of HVc in Zebra Finches of both sexes elicits syringeal contractions, thus the neurons connecting HVc indirectly with the syrinx are present in both sexes (Arnold, 1980a). Histochemical characterization of the vocal control system has thus far shown the presence of AChE-containing neurons in MAN, HVc, RA, DM, nXIIts, NIF, and Area X (Ryan and Arnold, 1979), dense catecholamine terminals in Area X (Lewis *et al.*, 1981) and immunoreactive enkephalin-containing afferents to HVc, MAN, NIF, RA, and ICo (Ryan *et al.*, 1980). Cells containing these substances are present in these regions in both sexes, so that we have yet to find cell types (defined by projection or histochemical characteristics) that are present only in one sex.

The neonatal T- or E-induced masculinization of female song behavior and related neural circuits is unexpected because it was not predicted by earlier work on steroid influences on sexual differentiation of behavior and morphology in birds. In the Common Quail (*Coturnix coturnix*), treatment of embryos with estrogen causes marked demasculinization in adult sexual behavior of males (Adkins, 1975, 1976; Whitsett *et al.*, 1977) and this corresponds to estrogen-induced demasculinization and feminization of morphology in various bird species (Burns, 1961; Witschi, 1961). There is evidence that demasculinization occurs even when adult female *Coturnix* are given estrogen, provided they were ovariectomized at hatching (Hutchison, 1978). The estradiol-induced masculinization of neonatal female Zebra Finches is precisely the opposite type of effect found in other birds. Is it possible that the mode of sexual differentiation of behavior in passerine birds differs from that in other bird species?

E. The Relation of Structure to Function in Vocal Control

In a number of studies cited above, there is a clear correlation between the size of brain regions in the song system and the capacity for vocal performance. This correlation is seen when comparing male and female Zebra Finches and Canaries (Nottebohm and Arnold, 1976), and when comparing vocal performance in female Zebra Finches who receive one of several different hormone treatments during ontogeny (Gurney and Konishi, 1980). In each of these studies, birds with the largest song areas sang best. Our anatomical knowledge of the song system should lead to even finer correlations between morphology and function in the near future. For example, when comparing different species, morphological differences are already apparent [Area X is much larger in Zebra Finches than in Canaries (Nottebohm and Arnold, 1976)], but we do not understand the significance of this yet. The incredible variety of vocal behaviors in passerine birds should provide a fertile ground for investigations of brain–behavior relationships. How does the song system differ among species which do and do not learn song, which have large and small repertoires? These same questions can also be addressed intraspecifically, since we now possess some powerful tools for selective modification of certain cellular characteristics of song regions. For example, neonatal DHT treatment of female Zebra Finches selectively increases neuronal number in RA, but leaves RA neurons with a relatively feminine morphology (Gurney, 1980). By perturbing selected morphological measures, it should be possible to refine our understanding of the functional importance of each.

Nottebohm *et al.* (1981) discovered a correlation between repertoire size (number of different song syllables sung) in male Canaries and the size of HVc and RA. They suggest that the size of these brain regions places a limit on the ability of a male to learn. Alternatively, greater song learning could induce growth of HVc and RA. This exceptional study raises many fascinating questions. What cellular properties account for the volume difference among males with small and large repertoires? Do they differ in the number of neurons involved in song, in the number of synapses between neurons, etc.? The importance of androgens also needs to be examined, because males with larger song nuclei tended to have larger testes at the time of sacrifice (although the correlation between testis weight and repertoire size was near zero). Do androgens influence the size of male song regions as they do in adult female Canaries (Nottebohm, 1980a), and is the ability to acquire large repertoires dependent on androgens? We can look forward to important studies on the relationships among androgen levels, size of repertoire, and size of neural regions involved with song.

In all of the above studies, there is a basic assumption that morphology reflects function: anatomy and physiology are one. If brain regions differ in function, this should be reflected at some level in the morphology of that region. One surprising aspect of these studies is that the morphological differences reflecting behavioral differences are often very large. This unexpected quality of the passerine song system has led to the discovery of important brain–behavior relationships not seen in other neural systems, and this has made the song system an important model for studying neural control of species-specific and learned behaviors in general.

ACKNOWLEDGMENTS

The author's research discussed in this article was supported by NSF Grants BNS 7705973 and BNS 8006798 to A.P.A., and USPHS Grant RR07009 to UCLA.

REFERENCES

Adkins, E. K. (1975). Hormonal basis of sexual differentiation in the Japanese Quail. *J. Comp. Physiol. Psychol.* **89,** 61–71.

Adkins, E. K. (1976). Embryonic exposure to an anti-estrogen masculinizes behavior of female quail. *Physiol. Behav.* **17,** 357–359.

Arnold, A. P. (1975). The effects of castration and androgen replacement on song, courtship, and aggression in Zebra Finches (*Poephila guttata*). *J. Exp. Zool.* **191,** 309–326.

Arnold, A. P. (1979). Hormone accumulation in the brain of the Zebra Finch after injection of various steroids and steroid competitors. *Soc. Neurosci. Abstr.* **5,** 434.

Arnold, A. P. (1980a). Sexual differences in the brain. *Am. Sci.* **68,** 165–173.

Arnold, A. P. (1980b). Quantitative analysis of sex differences in hormone accumulation in the Zebra Finch brain: Methodological and theoretical issues. *J. Comp. Neurol.* **189,** 421–436.

Arnold, A. P. (1980c). Effects of androgens on volumes of sexually dimorphic brain regions in the Zebra Finch. *Brain Res.* **185,** 441–444.

Arnold, A. P. (1980d). A technique for simultaneous steroid autoradiography and retrograde labelling of neurons. *Brain Res.* **192,** 210–212.

Arnold, A. P. (1981). Logical levels of steroid hormone action in the control of vertebrate behavior. *Am. Zool.* **21,** 233–242.

Arnold, A. P., and Saltiel, A. (1979). Sexual difference in pattern of hormone accumulation in the brain of a songbird. *Science* **205,** 702–705.

Arnold, A. P., Nottebohm, F., and Pfaff, D. W. (1976). Hormone accumulating cells in vocal control and other brain regions of the Zebra Finch (*Poephila guttata*). *J. Comp. Neurol.* **165,** 487–512.

Breedlove, S. M., and Arnold, A. P. (1980). Hormone accumulation in a sexually dimorphic motor nucleus of the rat spinal cord. *Science* **210,** 564–566.

Brown, J. L. (1965). Vocalizations evoked from the optic lobe of a songbird. *Science* **149,** 1002–1003.

Brown, J. L. (1971). An exploratory study of vocalization areas in the brain of the Red-winged Blackbird (*Agelaius phoeniceus*). *Behaviour* **39,** 91–127.

Brown, J. L. (1973). Behavior elicited by electrical stimulation of the brain of the Steller's Jay. *Condor* **75,** 1–16.

Burns, R. K. (1961). Role of hormones in the differentiation of sex. *In* "Sex and Internal Secretions" (W. C. Young, ed.), 3rd ed., Vol. 1, pp. 76–160. Williams & Wilkins, Baltimore, Maryland.

Carpenter, C. R. (1932). Relation of the male avian gonad to responses pertinent to reproductive phenomena. *Psychol. Bull.* **29,** 509–527.

Cohen, D. H., and Karten, H. J. (1974). The structural organization of avian brain: an overview. *In* "Birds: Brain and Behavior" (I. J. Goodman and M. W. Schein, eds.), pp. 29–76. Academic Press, New York.

Collard, J., and Grevendal, L. (1946). Etudes sur les caractères sexuels des Pinsons, *Fringilla coelebs* et *F. montifringilla. Gerfaut* **2,** 89–107.

Collias, N. E. (1950). Hormones and behavior with special reference to birds and the mechanisms of hormone action. *In* "Steroid Hormones" (E. S. Gordon, ed.), pp. 277–329. Univ. of Wisconsin Press, Madison.

Delius, J. D., Runge, T. E., and Oeckinghaus, H. (1979). Short-latency auditory projection to the frontal telencephalon of the Pigeon. *Exp. Neurol.* **63,** 594–609.

DeVoogd, T. J., and Nottebohm, F. (1980). Sex differences in dendritic morphology of a song control nucleus in the Canary: A quantitative Golgi study. *J. Comp. Neurol.* **196,** 309–316.

Gorski, R. A., Gordon, J. H., Shryne, J. E., and Southam, A. M. (1978). Evidence for a morphological sex difference within the medial preoptic area of the rat brain. *Brain Res.* **148,** 333–346.

Goy, R. W., and McEwen, B. S. (1980). "Sexual Differentiation of the Brain." MIT Press, Cambridge, Massachusetts.

Greenewalt, C. H. (1968). "Bird Song: Acoustics and Physiology." Smithsonian Inst. Press, Washington, D.C.

Gurney, M. E. (1980). Sexual differentiation of brain and behavior in the Zebra Finch (*Poephila guttata*): A cellular analysis. Ph.D. Thesis, California Inst. Technol., Pasadena.

Gurney, M. E., and Konishi, M. (1979). A cellular analysis of androgen- and estrogen-induced sexual differentiation in the Zebra Finch song system. *Soc. Neurosci. Abstr.* **5,** 446.

Gurney, M. E., and Konishi, M. (1980). Hormone-induced sexual differentiation of brain and behavior in Zebra Finches. *Science* **208,** 1380–1383.

Hutchison, R. E. (1978). Hormonal differentiation of sexual behavior in Japanese Quail. *Horm. Behav.* **11,** 363–387.

Katz, L. C., and Gurney, M. E. (1981). Auditory responses in the Zebra Finch's motor system for song. *Brain Res.* **211,** 192–197.

Kelley, D. B., and Nottebohm, F. (1979). Projections of a telencephalic auditory nucleus–Field L–in the Canary. *J. Comp. Neurol.* **183,** 455–470.

Kirsch, J., Coles, R. B., and Leppelsack, H.-J. (1980). Unit recordings from a new auditory area in the frontal neostriatum of the awake Starling (*Sturnis vulgaris*). *Exp. Brain Res.* **38,** 375–380.

Konishi, M. (1965). The role of auditory feedback in the control of vocalization in the White-crowned Sparrow. *Z. Tierpsychol.* **22,** 770–783.

Konishi, M., and Nottebohm, F. (1969). Experimental studies in the ontogeny of avian vocalizations. *In* "Bird Vocalizations" (R. Hinde, ed.), pp. 29–48. Cambridge Univ. Press, London and New York.

Lemon, R. E. (1973). Nervous control of the syrinx in White-throated Sparrows (*Zonotrichia albicollis*). *J. Zool.* **171,** 131–140.

Leppelsack, H.-J. (1974). Funktionelle Eigenschaften der Hörbahn im Field L des Neostriatum caudale des Staren (*Sturnus vulgaris* L., Aves). *J. Comp. Physiol.* **88,** 271–320.

Lewis, J. W., Ryan, S. M., Arnold, A. P., and Butcher, L. L. (1981). Evidence for a catecholaminergic projection to Area X in the Zebra Finch. *J. Comp. Neurol.* **196,** 347–354.

Lieberburg, I., and Nottebohm, F. (1979). High-affinity androgen binding proteins in syringeal tissues of song birds. *Gen. Comp. Endocrinol.* **37,** 286–293.

Luine, V., Nottebohm, F., Harding, C., and McEwen, B. S. (1980). Androgen affects cholinergic enzymes in syringeal motor neurons and muscle. *Brain Res.* **192,** 89–107.

McEwen, B. S. (1976). Gonadal steroid receptors in neuroendocrine tissues. *In* "Subcellular Mechanisms in Reproductive Neuroendocrinology" (F. Naftolin, K. J. Ryan, and I. J. Davies, eds.), pp. 277–304. Elsevier, Amsterdam.

Marler, P. (1970). A comparative approach to vocal learning: song development in White-crowned Sparrows. *J. Comp. Physiol. Psychol. Monograph* **71,** 1–25.

Marler, P., and Mundinger, P. (1971). Vocal learning in birds. *In* "Ontogeny of Vertebrate Behavior" (H. Moltz, ed.), pp. 389–450. Academic Press, New York.

Nottebohm, F. (1968). Auditory experience and song development in the Chaffinch, *Fringilla coelebs. Ibis* **110,** 549–568.

Nottebohm, F. (1969). The critical period for song learning. *Ibis* **111,** 386–387.

Nottebohm, F. (1970). Ontogeny of bird song. *Science* **167,** 950–956.

Nottebohm, F. (1971). Neural lateralization of vocal control in a passerine bird. I. Song. *J. Exp. Zool.* **177,** 229–261.

Nottebohm, F. (1972). Neural lateralization of vocal control in a passerine bird. II. Subsong, calls, and a theory of vocal learning. *J. Exp. Zool.* **179,** 35–49.

Nottebohm, F. (1975). Vocal behavior in birds. *In* "Avian Biology" (D. S. Farner and J. R. King, eds.), Vol. 5, pp. 289–332. Academic Press, New York.

Nottebohm, F. (1977). Asymmetries in neural control of vocalizations in the Canary. *In* "Lateralization in the Nervous System" (S. R. Harnad, R. W. Doty, L. Goldstein, J. Jaynes, and G. Krauthammer, eds.), pp. 23–44. Academic Press, New York.

Nottebohm, F. (1980a). Testosterone triggers growth of brain vocal control nuclei in adult female Canaries. *Brain Res.* **189,** 429–436.

Nottebohm, F. (1980b). Brain pathways for vocal learning in birds: A review of the first ten years. *Prog. Psychobiol. Physiol. Psychol.* **9,** 85–124.

Nottebohm, F., and Arnold, A. P. (1976). Sexual dimorphism in vocal control areas of the songbird brain. *Science* **194,** 211–213.

Nottebohm, F., and Kelley, D. B. (1978). Projections to efferent vocal control nuclei of the Canary telencephalon. *Soc. Neurosci. Abstr.* **4,** 101.

Nottebohm, F., and Nottebohm, M. E. (1976). Left hypoglossal dominance in the control of Canary and White-crowned Sparrow song. *J. Comp. Physiol.* **102,** 171–192.

Nottebohm, F., and Nottebohm, M. E. (1978). Relationship between song repertoire and age in the Canary, *Serinus canarius. Z. Tierpsychol.* **46,** 298–305.

Nottebohm, F., Stokes, T. M., and Leonard, C. M. (1976). Central control of song in the Canary, *Serinus canarius. J. Comp. Neurol.* **165,** 457–486.

Nottebohm, F., Manning, E., and Nottebohm, M. E. (1979). Reversal of hypoglossal dominance in Canaries following unilateral syringeal denervation. *J. Comp. Physiol.* **134,** 227–240.

Nottebohm, F., Kasparian, S., and Pandazis, C. (1981). Brain space for a learned task. *Brain Res.* **213,** 99–109.

Phillips, R. E., and Peek, F. W. (1975). Brain organization and neuromuscular control of vocalization in birds. *In* "Neural and Endocrine Aspects of Behavior in Birds" (P. Wright, P. G. Caryl, and D. M. Vowles, eds.), pp. 243–274. Elsevier, Amsterdam.

Price, P. H. (1979). Developmental determinants of structure in Zebra Finch song. *J. Comp. Physiol. Psychol.* **93,** 260–277.

Pröve, E. (1974). Der Einfluss von Kastration und Testosteronsubstitution auf das Sexualverhalten männlicher Zebrafinken (*Taeniopygia guttata castanotis* Gould). *J. Ornithol.* **115,** 338–347.

Raisman, G., and Field, P. (1973). Sexual dimorphism in the neuropil of the preoptic area of the rat and its dependence on neonatal androgen. *Brain Res.* **54,** 1–29.

Ryan, S. M., and Arnold, A. P. (1979). Evidence for cholinergic mechanisms in brain regions related to bird song. *Soc. Neurosci. Abstr.* **5,** 146.

Ryan, S. M., Arnold, A. P. and Elde, R. P. (1980). Immunohistochemical localization of enkephalins in song-related brain regions of a passerine bird. *Soc. Neurosci. Abstr.* **6,** 616.

Seller, T. J. (1979). Unilateral nervous control of the syrinx in Java Sparrows (*Padda oryzivora*). *J. Comp. Physiol.* **129,** 281–288.

Whitsett, J. M., Irvin, E. W., Edens, F. W., and Thaxton, J. P. (1977). Demasculinization of male Japanese Quail by prenatal estrogen treatment. *Horm. Behav.* **8,** 254–263.

Witchi, E. (1961). Sex and secondary sexual characters. *In* "Biology and Comparative Physiology of Birds" (A. J. Marshall, ed.), Vol. 2, pp. 115–168. Academic Press, New York.

Zigmond, R. E., Nottebohm, F., and Pfaff, D. W. (1973). Androgen-concentrating cells in the midbrain of a songbird. *Science* **179,** 1005–1007.

Zigmond, R. E., Detrick, R. A., and Pfaff, D. W. (1980). An autoradiographic study of the localization of androgen concentrating cells in the Chaffinch. *Brain Res.* **182,** 369–381.

NOTE ADDED IN PROOF

Since this review was written, the following relevant papers have appeared.

Arnold, A. P. (1981). Model systems for the study of sexual differentiation of the nervous system. *Trends in Pharmacological Sciences* **2,** 148–149.

Bottjer, S. W., and Arnold, A. P. (1982). Afferent neurons in the hypoglossal nerve of the Zebra Finch (*Poephila guttata*): Localization with horseradish peroxidase. *J. Comp. Neurol.* (in press).

DeVoogd, T., and Nottebohm, F. (1981). Gonadal hormones induce dendritic growth in the adult avian brain. *Science* **214,** 202–204.

Gurney, M. E. (1981). Hormonal control of cell form and number in the Zebra Finch song system. *J. Neurosci.* **1,** 658–673.

Gurney, M. E. (1982). Behavioral correlates of sexual differentiation in the zebra finch brain. *Brain Res.* **231,** 153–172.

Konishi, M., and Akutagawa, E. (1981). Androgen increases protein synthesis within the avian brain vocal control system. *Brain Res.* **222,** 442–446.

McCasland, J. S., and Konishi, M. (1981). Interaction between auditory and motor activities in an avian song control nucleus. *Proc. Nat. Acad. Sci. (U.S.A)* **78,** 7815–7819.

Nottebohm, F. (1981). A brain for all seasons: Cyclical anatomical changes in song control nuclei of the Canary brain. *Science* **214,** 1368–1370.

Nottebohm, F., Kelley, D. B., and Paton, J. A. (1982). Connections of vocal control nuclei in the canary telencephalon. *J. Comp. Neurol.* **207,** 344–357.

Ryan, S. M. and Arnold, A. P. (1981). Evidence for cholinergic participation in the control of bird song: Acetylcholinesterase distribution and muscarinic receptor autoradiography in the Zebra Finch brain. *J. Comp. Neurol.* **202,** 211–219.

Ryan, S. M., Elde, R. P., and Arnold, A. P. (1981). Enkephalin-like immunoreactivity in vocal control regions of the Zebra Finch brain. *Brain Res.* **229,** 236–240.

4

Auditory Perception in Birds

ROBERT J. DOOLING

I. INTRODUCTION

Birds have undergone an extensive and complex adaptive radiation into a wide range of habitats (Brodkorb, 1971). Acoustic information transmitted through these habitats plays a crucial role in a variety of avian behaviors including individual and species recognition, mate selection, territorial defense, and song learning (Brooks and Falls, 1975a,b; Falls, 1963; Kroodsma, 1976; Marler, 1970a,b; Thorpe, 1961). The premium on hearing well is high. The study of avian audition can be expected to provide a window on the evolution and adaptation of auditory systems to different environments as well as illuminating mecha-

ISBN 0-12-426801-3

nisms of sound detection and discrimination and the theoretical principles involved.

The sensory function of birds can be studied in several ways. Perhaps most familiar are anatomical and electrophysiological approaches seeking to uncover details of the mechanisms of sensory function (Knudsen and Konishi, 1978a,b,c, 1979; Sachs *et al.*, 1980; Schwartzkopff, 1973; Takasaka and Smith, 1971; Tanaka and Smith, 1978). Another approach involves the study of conditioned responses of awake, intact organisms to sensory stimulation. This "psychophysical" approach requires the participation of the whole conscious organism and, consequently, thresholds obtained in this way can be closely and directly related to an animal's functioning in its natural environment. In examining (1) species' differences in hearing abilities, (2) the relation between auditory perceptual capabilities and the information potentially available in bird sound signals, and (3) the relation between sensory capability and potential environmental constraints, the present review emphasizes psychophysical data. I will consider only those studies which meet the "criteria" of rigorous animal psychophysics (Stebbins, 1970). These criteria include good control of the acoustic stimulus, establishment of a reliable response measure, and adequate evidence of behavioral control (e.g., the use of sham trials to determine the rate of false-alarm responding). The reliability and validity of the psychophysical approach have been clearly established even though the training and testing procedures employed are often far removed from events in the animal's natural environment.

Consider some of the acoustic tasks facing a bird in its natural environment. It must be able to detect another bird's songs and calls against a background of environmental noise. A bird must be capable of discriminating between its own species' song and that of other species and between songs of individual members of its own species. A bird must be able to localize sound from predators, prey, companions, and rivals not only in azimuth but also in elevation. Songbirds, in which the development of normal song depends on learning, must be able to focus attention on the important, fine-grained acoustic details of their species' song while ignoring variation in other nonessential acoustic dimensions. The stereotypy of many songbird vocalizations is ample evidence for the success and precision of this last process.

Data from psychophysical studies of hearing in birds over the past decade bear directly on these and other issues. There are now at least partial answers to some of the more important questions. How well can birds hear in the quiet where there is no masking from background noise? Are there differences in auditory sensitivity between songbirds and non-songbirds? How well can birds discriminate changes in the frequency, intensity, or temporal aspects of an acoustic signal? How does background noise affect hearing? How well can birds localize a sound source? Is there a relation between auditory discriminatory abilities and the accuracy of vocal production? Is song learning guided by an innate perceptual bias toward conspecific song?

Comparison between songbirds and non-songbirds is useful since songbirds have long been suspected of possessing unusual auditory capabilities on the basis of their complex vocalizations (Greenewalt, 1968; Pumphrey, 1961). Finally, the comprehensive data available for humans are clearly the most rigorously obtained and precisely defined in all the animal kingdom. It is thus valuable to compare avian auditory thresholds with human auditory thresholds obtained in the same or a similar fashion.

II. BASIC AUDITORY PROCESSES IN BIRDS

A discussion of auditory perception must necessarily begin with the determination of the minimum audible sound pressure at frequencies throughout the range of hearing. This defines the audibility curve or threshold in the quiet and represents the starting point for comparative statements of auditory capability. There are now behavioral audibility curves available for 16 species of birds. These behavioral studies of absolute auditory sensitivity are about equally divided between the oscines and non-oscines (Fig. 1).

Note that human thresholds are lower at all frequencies than are the median thresholds for either oscines or non-oscines. In comparison with non-oscines, oscines tend to hear better at high frequencies and worse at low frequencies. Both

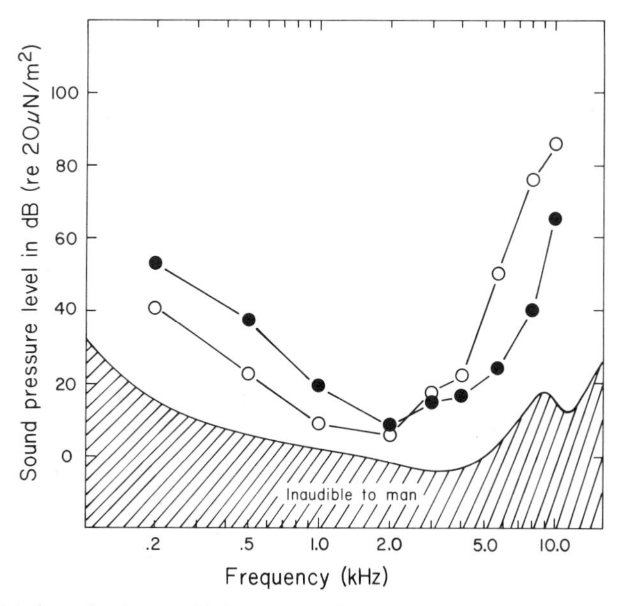

Fig. 1. Median absolute audibility thresholds of nine species of oscine birds (●) and seven non-oscine species (○). (These data are replotted from Dooling, 1980.)

groups of birds show lowest thresholds at about 2.0 kHz with a moderate decline in sensitivity below this frequency and a sharp decline above it.

While providing an excellent picture of avian versus human absolute sensitivity, this comparison conceals clear species variation among birds. There are, for instance, exceptions to the notion that humans have lower absolute thresholds than birds. Two nocturnal predators, the Barn Owl (*Tyto alba*) and the Great Horned Owl (*Bubo virginianus*) show very sensitive absolute thresholds (Konishi, 1973a; Trainer, 1946), clearly more sensitive than man. To facilitate a more quantitative comparison of audibility curves among the 16 avian species, I chose six arbitrary descriptive parameters similar to those used by Masterton *et al.* (1969) for comparing audibility curves of mammals (Fig. 2, Table I).

Values for each of these parameters for nine oscines and seven non-oscines have been previously published by Dooling (1980) and are summarized in Table I. The descriptive parameter of best frequency is not included since the range of

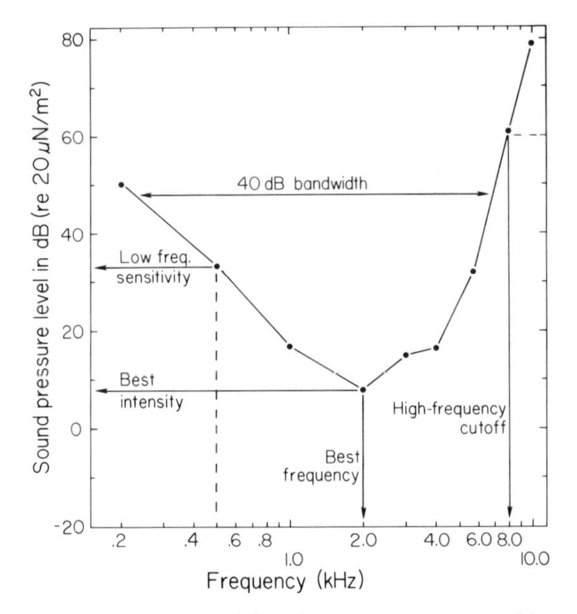

Fig. 2. Schematic representation of the arbitrary parameters used for comparing avian audibility curves. Bandwidth (in octaves) is the frequency range in octaves at an intensity level 40 dB above the most sensitive point in the audibility function; best intensity (in dB) is the lowest point on the audibility function; low-frequency sensitivity is the threshold (in dB) at 500 Hz; high-frequency cutoff (in kHz) is the highest frequency a bird can hear at a sound pressure level of 60 dB (SPL); low-frequency slope (in dB per octave) is the rate at which sensitivity declines for frequencies below the most sensitive region of the audibility function; and high-frequency slope (in dB per octave) is the rate at which sensitivity declines for frequencies above the most sensitive region of the audibility function.

TABLE I

Comparison of Oscines and Non-Oscines on Six Descriptive Parameters[a]

	40-dB bandwidth (octave)	Best intensity (dB)	Low-frequency sensitivity (dB)	High-frequency cutoff (kHz)	Low-frequency slope (dB/oct)	High-frequency slope (dB/oct)
Oscines (nine species)						
Mean	4.80	5.34	35.22	9.17	18.34	49.47
SD	0.75	9.58	10.31	1.69	5.60	31.30
Non-oscines (seven species)						
Mean	5.20	2.45	19.39	7.68	14.81	59.91
SD	0.42	14.48	13.53	2.23	2.53	20.40

[a] See Fig. 1. From Dooling (1980).

best frequencies of all the birds is fairly restricted, but not all birds were tested at the same frequencies within this range. Oscines and non-oscines do not differ tremendously on any of these measures but there are several trends in the data which are worth a closer look.

Figure 3A compares oscines and non-oscines on the basis of high-frequency cutoff. The Barn Owl shows the best high-frequency sensitivity of all species tested and the Rock Dove (*Columba livia*) the worst. The Canary (*Serinus canaria*) shows the poorest low-frequency sensitivity and the Great Horned Owl the best. For most songbirds [except the Canary and the Common Crow (*Corvus brachyrhynchos*)], the differences among species in low-frequency sensitivity are probably within the range of experimental error. Non-oscines show greater variability on this measure (Fig. 3B). The unusually good sensitivity shown by the two owl species at 500 Hz reflects the excellent absolute sensitivity shown throughout their hearing range (Konishi, 1973a; Trainer, 1946).

In terms of audibility curves or ''auditory space'' available for vocal communication, several points are clear. The great similarity among avian audibility curves suggests that one species does not filter out the vocalizations of another species by virtue of its absolute auditory sensitivity to particular frequencies. Several authors have also noted, however, that high-frequency sensitivity may be related to the highest frequencies contained in the species song (Dooling *et al.*, 1971; Konishi, 1970) and the data reviewed in Table I support this notion. It is also true that for most species the long-term average power spectrum of the species' song falls in the region of maximum auditory sensitivity. There is a good correspondence between hearing sensitivity and the spectral characteristics of vocal output for four bird species (Fig. 4). This simple comparison also indicates that each species can hear the vocalizations of all other species quite well. Thus, two species inhabiting the same territory must hear the vocalizations of the other

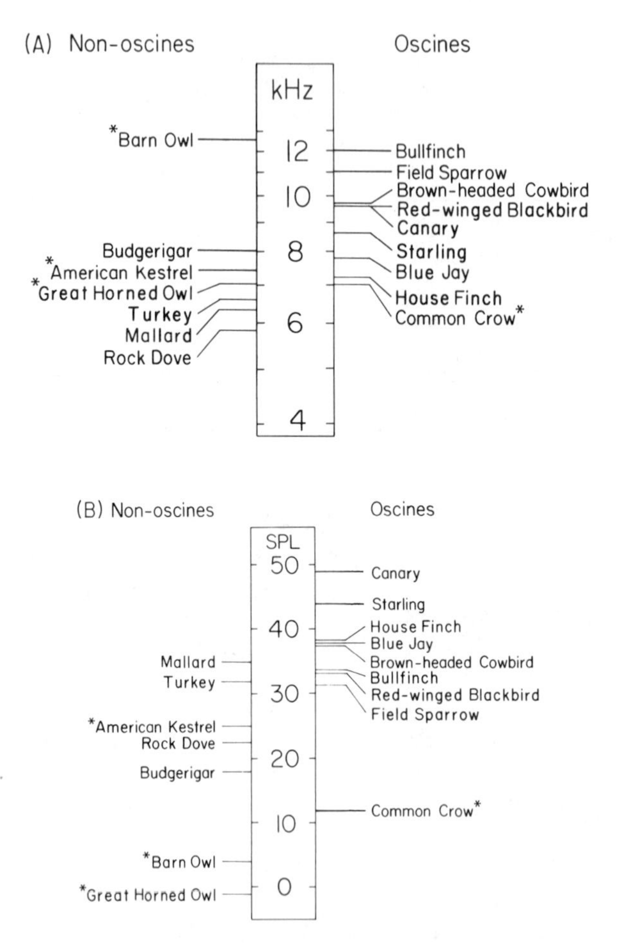

Fig. 3. (A) Comparison of oscines and non-oscines on the basis of high-frequency cutoff (in kHz) at 60 dB. Asterisks indicate data are derived from a single animal. (B) Comparison of oscines and non-oscines on the basis of low-frequency sensitivity (in dB) at a 500-Hz threshold (re 20 $\mu N/m^2$). Asterisks indicate data are derived from a single animal. (The Turkey, American Kestrel, Mallard, Starling, and Blue Jay are *Meleagris gallopavo, Falco sparverius, Anas platyrhynchos, Sturnus vulgaris,* and *Cyanocitta cristata,* respectively. All other scientific names are in the text.)

species almost as well as they can hear their own. This correspondence between hearing and vocal output does not hold as well for some non-songbirds such as the owls whose hearing may be specialized for prey-catching rather than vocal communication (Knudsen, 1980).

Several investigations of hearing in birds deserve special attention. Data for the Rock Dove reveal that this species is unusually sensitive to very low frequency sounds (Quine, 1978; Yodlowski, 1980). The Rock Dove is 50 dB more sensitive than man in the frequency region of 1–10 Hz (Kreithen and Quine,

1979). Since low-frequency sounds having long wavelengths propagate over tremendous distances, this information may be useful to the Rock Dove in homing. Whether low-frequency sensitivity is unique to Rock Doves, or common to other birds or even to other vertebrates is currently unknown. It does appear that low-frequency sensitivity is mediated by the auditory system in that columellar destruction raises infrasound thresholds about 40 dB and total cochlear destruction completely abolishes the sensitivity (Quine, 1978; Yodlowski, 1980).

Several species of birds, including Oilbirds (*Steatornis caripensis*), are known to echolocate (Griffin, 1954; Griffin and Suthers, 1970; Novick, 1959). Recent experiments suggest that the hearing sensitivity of Oilbirds is not different from that of other avian species. Evoked potential audiograms show a region of maximum sensitivity between 1 and 3 kHz with a fairly steep decline in sensitivity above and below this region (Konishi and Knudsen, 1979). Furthermore, the sound energy in the echolocation pulses of this species is predominantly in the frequency range of 1.5–2.5kHz, corresponding well with the region of maximum auditory sensitivity in the Oilbird as well as that of other birds.

In summary, birds appear relatively uniform in terms of absolute sensitivity. The most sensitive region of hearing for all species tested thus far is in the region of 1–5 kHz, with a dramatic decrease in sensitivity above this region. Nocturnal predators among birds have extremely good absolute sensitivity similar to that of the cat (*Felis domesticus*), also a nocturnal predator. Based on the species tested

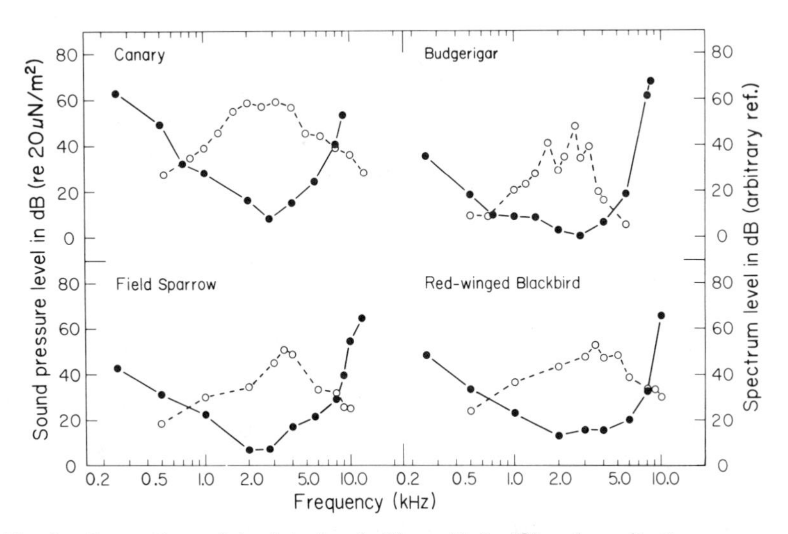

Fig. 4. Comparison of absolute threshold sensitivity (●) and vocalization power spectra for four avian species. [Absolute threshold curve data for Canaries are from Dooling *et al.* (1971), for Budgerigars from Dooling and Saunders (1975a), for Field Sparrows from Dooling *et al.* (1979), and for Red-winged Blackbirds from Hienz *et al.* (1977).]

to date, oscines hear better at high frequencies and poorer at low frequencies than non-oscines. Rock Doves may be particularly unusual in their sensitivity to infrasound, but no other avian species has been tested. The Oilbird, in spite of its echolocating ability, appears to have an audibility function which is similar to that of other birds. The message here is a simple one. In searching for mechanisms of species recognition among species which live within earshot of one another, one probably cannot appeal to differences in hearing threshold curves. Audibility curves between avian species—songbirds and non-songbirds alike—overlap to such an extent that each species is almost surely capable of hearing the vocalizations of other species as well as those of its own.

III. SENSITIVITY TO CHANGES IN AN ACOUSTIC SIGNAL

Although the audibility curve is a fundamental measure of auditory function, it is of limited value in relating auditory capability and natural behavior. Organisms are rarely called upon to detect the presence of a sound in an environment that is as quiet as that obtained in a testing booth. A more realistic measure of auditory capability would be the ability to detect whether or not any change has occurred in an otherwise constant background of sound. Such discrimination measures or thresholds are surely more representative of the kind of auditory problem confronting a bird or any organism faced with the task of discriminating important biological sounds from other less critical sounds or environmental noises. A number of acoustic discrimination tests have been conducted on the ability of birds to discriminate changes in either the frequency, intensity, temporal, or spatial aspects of a stimulus. In addition, there are several comprehensive investigations on Budgerigars (*Melopsittacus undulatus*) of a special case of differential sensitivity called masking. Results from masking studies have particular relevance for investigations of vocalization transmission distance in natural or noisy environments.

A. Frequency Discrimination

Playback studies in the field have implicated frequency as an important cue for song recognition in a number of species (Brémond, 1968, 1975; Falls, 1963; Fletcher and Smith, 1978). Psychophysical studies of frequency discrimination in five species support the notion that birds are quite sensitive to changes in the frequency of acoustic signals. The results from five species are compared with human sensitivity in Fig. 5. In terms of ΔF (frequency change), a Weber fraction of 0.010 to 0.020 means that birds can discriminate a 1% change in frequency. This corresponds to approximately 10–20 Hz at 1000 Hz, 20–40 Hz at 2000 Hz, and 40–80 Hz at 4000 Hz. Note that the five species perform best in the

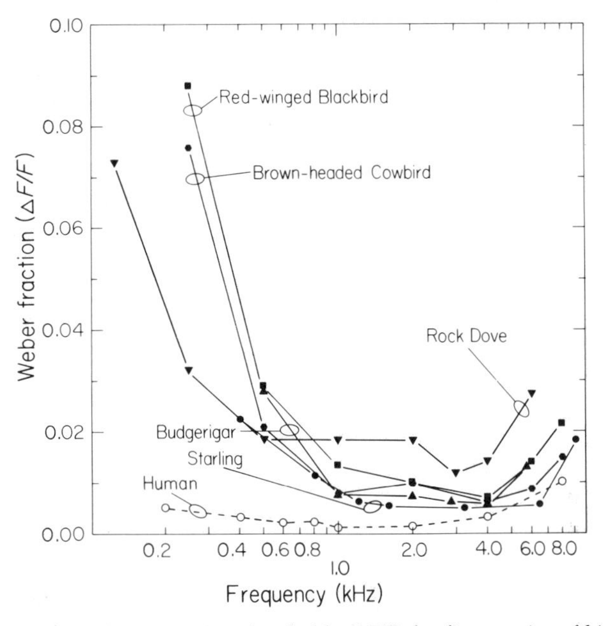

Fig. 5. Frequency discrimination thresholds ($\Delta F/F$) for five species of birds are compared to similar measures for man. [Data for Budgerigars are from Dooling and Saunders (1975a), for Rock Doves, cowbirds, and blackbirds from Sinnott *et al.* (1980), and for Starlings from Kuhn *et al.* (1980). Data for man are from Wier *et al.* (1977).]

frequency range of 1–4 kHz and much worse above and below this frequency range. Human thresholds for detecting changes in frequency are lower than those of the five birds at all test frequencies. Perhaps what is most surprising about these data is the fact that birds do so well. The basilar papilla of most birds is an order of magnitude shorter than the human basilar membrane (Schwartzkopff, 1973), yet over the narrow range of 1–4 kHz, birds are only about one-half to one-third as sensitive to changes in frequency as the human. These threshold values are considerably smaller than the frequency differences responded to in playback studies and are also near the limits of resolution of ordinary sonographic analysis. Given this degree of auditory resolving power, it is not surprising that songbirds can reproduce the same syllable type with equal precision in the frequency domain.

B. Intensity Discrimination

The greater number of hair cells per unit length of basilar papilla in the bird ear compared to the mammalian ear has led to speculation that the avian ear may be capable of a greater power of intensity discrimination compared to mammals

TABLE II

Intensity Difference Limens

Species	Intensity difference at 60 dB SL (in dB)	Reference
Red-winged Blackbird (*Agelaius phoeniceus*)	3.2	Sinnott *et al.* (1976)
Rock Dove (*Columba livia*)	2.5	Sinnott *et al.* (1976)
Canary (*Serinus canaria*)	1.5	R. J. Dooling and M. H. Searcy (unpublished data)
Brown-headed Cowbird (*Molothrus ater*)	1.3	Sinnott *et al.* (1976)
Budgerigar (*Melopsittacus undulatus*)	2.8	Dooling and Saunders (1975b)
Man	1.0	Jesteadt *et al.* (1977)

(Pumphrey, 1961). Psychophysical data from five species of birds suggest that this is not the case, however. Intensity difference limens or thresholds for a change in intensity (ΔI) are given for five avian species in Table II. These data indicate that humans are again more sensitive to changes in intensity than are birds. Birds show intensity discrimination thresholds which are similar to those measured in other nonhuman vertebrates, namely, cats (Raab and Ades, 1946), house mice (*Mus musculus*) (Ehret, 1975a), white rats (*Rattus norvegicus*) (Hack, 1971), and goldfish (*Carassius auratus*) (Jacobs and Tavolga, 1967). All of these vertebrates have relative intensity difference limens in the range of 1–4 dB. In other words, an intensity difference between two successive sounds must be greater than about 3 dB for a bird to detect the difference. It is perhaps not surprising that birds do not excel in intensity discrimination when one considers a bird's life style and habitat. For most birds, nonhomogeneous environments along with constant variation in distance, elevation, and azimuth between source and receiver must surely render sound intensity an unreliable acoustic dimension with which to encode species or individual identity. However, the kind of repetitive amplitude modulation so characteristic of bird song (e.g., a trill) is undoubtedly a useful strategy for encoding information by virtue of signal redundancy (Richards and Wiley, 1980).

C. Temporal Discrimination

Perhaps more than any other sensory system, the auditory system is time oriented. Thus, important questions for the auditory system concern the shortest time interval that can be perceived and how the auditory system might cope with environmentally induced temporal smearing of important biological sounds. The stereotyped, punctate nature of bird song magnifies these issues, and evidence obtained from song learning (Greenewalt, 1968), cochlear anatomy (Pumphrey, 1961; Schwartzkopff, 1968, 1973), and single-unit recordings (Konishi, 1969) supports the notion that birds may be capable of an unusual degree of temporal resolving power.

A fundamental issue here concerns the definition of temporal resolving power. There are many ways to vary the temporal characteristics of an acoustic signal and it is not immediately clear which measure might be most relevant to the kinds of acoustic problems facing birds in their natural environment. Furthermore, the experimental difficulty in varying only the temporal aspects of a signal while leaving amplitude and spectral characteristics unchanged is a formidable one (Green, 1971). Over the years, several psychophysical approaches have been developed that assess the sensitivity to temporal changes in an acoustic signal while minimizing or eliminating concomitant spectral or amplitude changes. Several of these tests have been performed on birds and the results are rather surprising. There is yet no evidence that birds are significantly better than other vertebrates at detecting changes in the temporal aspects of an acoustic signal. In fact, the similarity between avian and human sensitivity to temporal change is quite striking. Comparative data from birds are available on five different measures of temporal auditory processing.

1. Temporal Integration

Temporal integration usually refers to the relation between the threshold for detection of a sound and the duration of that sound. It is a measure of the auditory system's ability to sum acoustical energy over time. For a number of vertebrates, including man (Watson and Gengel, 1969), house mice (Ehret, 1976), chinchillas (Henderson, 1969), and bottle-nosed dolphins (*Tursiops* spp.) (Johnson, 1968), the ability to detect the occurrence of a sound improves as the duration of the sound is increased from a few milliseconds up to about 200 msec. Increasing the duration of sound beyond 200 msec does not make the sound easier to hear. In other words, all other things being equal, a 200-msec or longer sound is more easily detected than is a 50-msec sound.

Recent results from an experiment with Budgerigars and Field Sparrows (*Spizella pusilla*) show that thresholds for hearing a pure tone improve as the duration of the tone is increased from a few milliseconds to 200–300 msec (Dooling, 1979, 1980). Thus the "time constant" of the avian ear is similar to that found for higher vertebrates. The practical implication of these results is that in order to maximize audibility or sound transmission distance, a sound (vocalization) should be at least 200 msec in duration. All other aspects of the acoustic signal remaining constant, a vocalization only 50 msec in duration will result in a threshold 3–5 dB higher than that of a 200-msec vocalization.

2. *Gap Detection*

Gap detection threshold is a measure of the minimum detectable interval between two sounds. This is analogous in some ways to more familiar measures of sensory acuity in the visual system involving the minimum separation between two lines. Because turning a pure tone on and off quickly results in energy spreading and possible spectral cues which could aid in detection, broad-band clicks or noises are the stimuli of choice in such studies. Gap detection thresholds for man and chinchillas using broad-band noise are in the range of 2–3 msec (Giraudi *et al.*, 1980; Green, 1971; Plomp, 1964). For birds, two-click thresholds have been measured in Rock Doves, Bullfinches (*Pyrrhula pyrrhula*), and Greenfinches (*Carduelis chloris*) (Wilkinson and Howse, 1975) and gap detection thresholds have been measured in House Finches (*Carpodacus mexicanus*) (Dooling *et al.*, 1978). The avian data indicate a level of temporal resolution for birds which is very similar to that found for humans (Green, 1971). In terms of the perception of complex vocal signals, these results indicate that internote or intersyllable intervals of less than 2–3 msec may not be perceived by birds any more than they are perceived by humans.

3. *Duration Discrimination*

Another fundamental measure of the perception of auditory temporality is duration discrimination. The precision and stereotypy of avian vocalizations would suggest that birds can do well at such a task. Recent results for Budgerigars indicate that this species can detect a 10–20% change in the duration of a signal. Figure 6 compares the results for Budgerigars with results for humans using similar stimulus conditions. The thresholds for Budgerigars and man are similar over a range of stimulus durations from 20 to 200 msec. If Budgerigars are representative, birds can discriminate changes in duration as well as humans. In terms of the perception of complex vocal signals, these results indicate that a note or internote interval of 40 msec may not be discriminably different from a note or internote interval of 44 msec. Recall, however, that these results are only for Budgerigars. Songbirds have not yet been tested and may very well show different levels of sensitivity.

4. *Nonsimultaneous Masking*

Nonsimultaneous masking refers to masking effects which either precede or follow a masker in time. The mechanisms responsible for nonsimultaneous masking effects are complex and not fully understood. Presumably, when a tone follows a masker, improvement in thresholds is due to a decrease in auditory fatigue following the masker (Plomp, 1964). The masking of a tone which precedes the masker suggests more central processes, since the masker obscures some events initiated by the tone but not completed by the time of onset of the

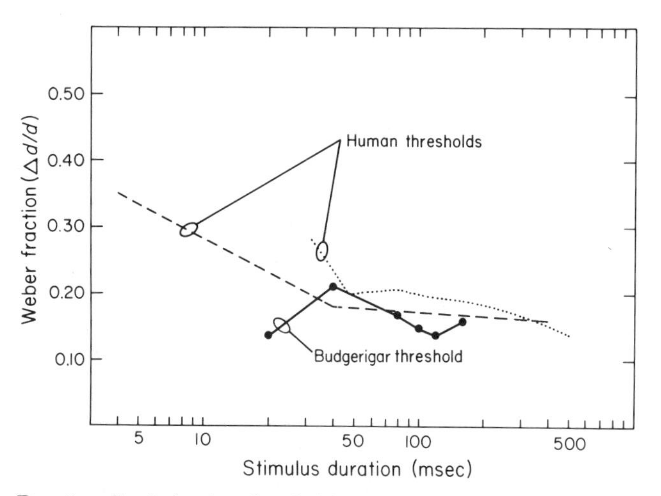

Fig. 6. Duration discrimination thresholds ($\Delta d/d$) for two Budgerigars are compared to results from similar experiments on humans (replotted from Dooling and Haskell, 1978). Budgerigars can discriminate a 10–20% change in the duration of pure tones 20–200 msec long.

masker (Lynn and Small, 1977). The repetitive, punctate nature of bird song suggests that the masking effect on a sound from a temporally adjacent sound could be important. Nonsimultaneous masking effects have been examined in Budgerigars, and these results are shown in Fig. 7. Note that for tones preceding a 70-dB masker (backward masking), the Budgerigar shows no masking at intervals of 10 msec or greater. For tones following the same masker (forward masking), however, thresholds do not return to unmasked levels until the interval between masker and tone is 100 msec or longer. In terms of the perception of complex vocal signals, these results suggest that masking effects on the initial portion of a note from a preceding note may be significantly greater than masking effects on the terminal portion of a note from a succeeding note.

5. Minimum Integration Time

Another measure of temporal resolving power is minimum integration time. Rather than asking how long a period of time the ear can sum energy, we determine the shortest integration time an organism can achieve. One procedure resembles that for measuring the familiar flicker fusion frequency threshold in visual psychophysics. Again broad-band noise is used to minimize spectral cues. This noise is amplitude-modulated at frequencies between four and several thousand Hertz with threshold defined as the amount of modulation required for detection. The essential task, then, is to detect the intensity change occurring between the peak and trough of the amplitude-modulated noise. As modulation

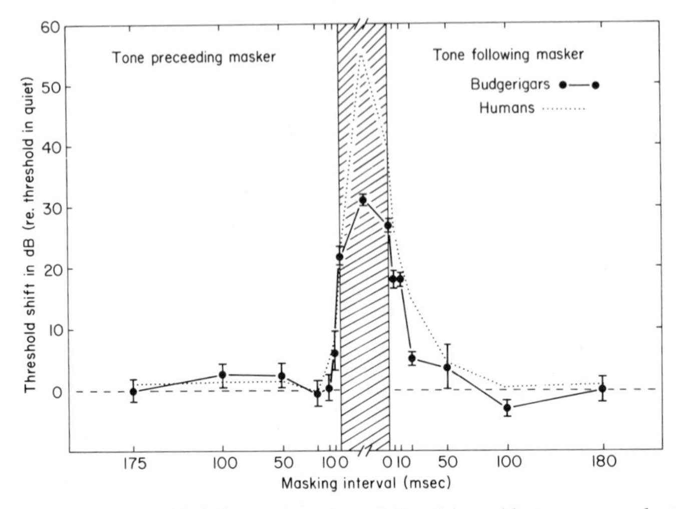

Fig. 7. Mean threshold shift as a function of time interval between a probe tone and a 40-msec noise burst masker (▨). Time interval is the time between the noise masker and the tone signal. (After Dooling and Searcy, 1980b.)

frequency increases, the auditory system has difficulty following the rapid intensity changes and thus requires a greater depth of modulation for detection.

At low modulation frequencies man is considerably more sensitive to changes in intensity than are Budgerigars (Fig. 8). However, as modulation frequency is increased, human sensitivity begins to decrease sooner than that of Budgerigars. Interpretation of these results is complicated. One method involves computing a time constant from modulation functions such as these, thereby treating the auditory system as a low-pass filter. From our data in Fig. 8, the low-pass bandwidth of the human auditory system is calculated to be about 30 Hz while that of the Budgerigar is about 130 Hz. Since there is a reciprocal relation between frequency (bandwidth) and time (Gabor, 1947), a time constant can be obtained from the formula:

$$\text{time constant} = 1/(2 \cdot \pi \cdot \text{bandwidth})$$

Viewed in this way, the time constant of the human auditory system is about 5.9 msec while that of the Budgerigar is about 1.2 msec. These results are another way of saying that the Budgerigar's auditory system begins to falter when trying to resolve events happening faster than once every 1 or 2 msec. Humans, by contrast, lose sensitivity to events happening faster than once every 5 or 6 msec. Thus, the Budgerigar is about 2–3 times better than humans tested under similar conditions. Since the Budgerigar is significantly less sensitive to intensity changes, it appears that the avian auditory system may have sacrificed good intensity-resolving power to obtain slightly better temporal resolving power when compared to humans. The net effect is that at modulation frequencies above about 50–60 Hz humans and Budgerigars show similar levels of sensitivity.

Many bird songs can be acoustically characterized as being rapidly amplitude modulated. An important question might be how well a bird can detect a change in the rate of amplitude modulation. Again using white noise to eliminate spectral cues, Dooling and Searcy (1981) found that Budgerigars can detect about a 10% change in modulation rate over the range of modulation frequencies from 40 to 320 Hz.

A number of playback studies in the field have implicated the temporal aspects of a song as providing important cues for song recognition (Emlen, 1972; Falls,

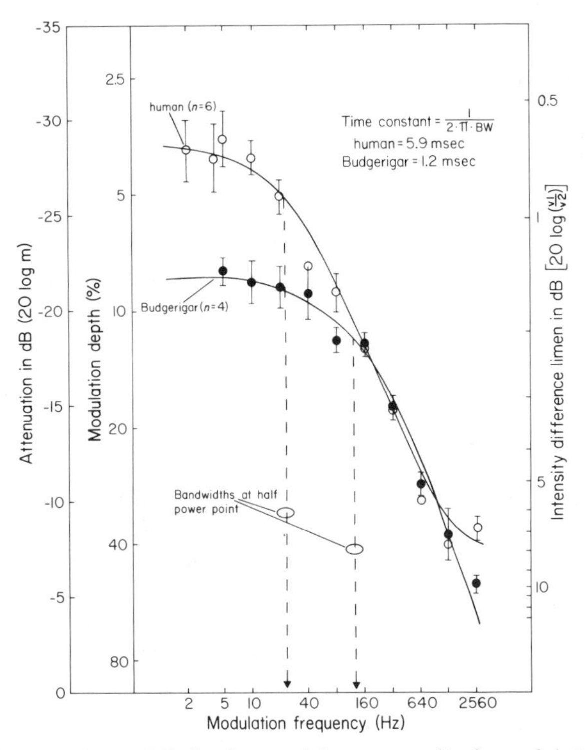

Fig. 8. A comparison of Budgerigar and human amplitude modulation functions. Vertical bars represent one standard error on each side of the mean. Differences between Budgerigars and humans in the bandwidths (point of inflection) of these functions are shown by the dotted lines extrapolated to the abscissa for both species. These bandwidths are used to compute time constants of 1.2 and 5.9 msec for Budgerigars and humans, respectively. Threshold is given as attenuation in dB of the modulating tone (20 log m), percent modulation depth, and the intensity difference between the peak and the trough of the modulated noise (ΔI). (After Dooling and Searcy, 1981.)

1963; Goldman, 1973; Peters *et al.*, 1980). The psychophysical studies summarized above indicate that birds do well on a number of perceptual tests of temporal discrimination. As with discrimination studies in the frequency domain, the threshold values obtained from psychophysical studies of temporal discrimination are smaller than the temporal differences usually responded to in playback studies. Since the data indicate that birds can discriminate much finer differences in the frequency and time domains than indicated by playback studies performed under natural conditions, an important question remains as to why acoustically discriminable differences are ignored in some situations and attended to in others.

D. Sound Localization

The ability to localize sound is crucial to the survival of songbirds and nonsongbirds alike. Again there is considerable evidence from field and playback studies that birds can and do localize the vocalizations of their own and other species (Emlen, 1972; Gottlieb, 1971; Krebs, 1976; Shalter, 1978). Furthermore, many species of owls have developed specializations for the purpose of prey catching (Knudsen *et al.*, 1979; Knudsen and Konishi, 1979; Payne, 1962, 1971). In one sense, it is perhaps unfortunate that the most detailed investigations of sound localization concern owls in view of their obvious and unique specializations for detection of prey. Recent behavioral and neurophysiological data from Barn Owls are the most complete yet available for any vertebrate (Knudsen, 1980). Whether or not these data for owls are relevant to the study of sound localization in other birds remains to be seen. Almost certainly there are parallels between the mechanisms of sound localization in owls and those operating in other birds with less specialized auditory systems.

Early work with human listeners established the relation between head size and the trade-off between time of arrival cues and sound shadow or interaural intensity cues (Mills, 1972; Stevens and Neuman, 1936). In these traditional terms, it is difficult to understand how a bird with such a small head and closely spaced ears can localize sound at all considering their narrow hearing range and poor sensitivity to high frequencies (Dooling, 1980). The heads of most birds are too small to allow any appreciable sound shadow for frequencies below about 8 kHz and closely spaced ears require a sensitivity to time of arrival cues better than about 10 μsec (Knudsen, 1980). Whether or not the interaural temporal resolution of birds is sufficiently developed to use this cue has not been tested but there is behavioral and physiological evidence that Barn Owls can use time-of-arrival cues in sound localization (Knudsen and Konishi, 1979; Moiseff and Konishi, 1981). On the other hand, standard measures of temporal resolving power in birds suggest a level of sensitivity similar to that shown for humans (Dooling, 1980). Man's threshold for interaural time resolution is about 10 μsec (Harris, 1960; Yost *et al.*, 1971).

The problem of avian sound localization has recently been reviewed in detail with special emphasis on the predatory behavior of owls (Knudsen, 1980; Lewis and Coles, 1980). Central to a theory of sound localization in birds is the recent rediscovery of the presence of interaural pathways in birds which acoustically couple the two middle ears (Coles *et al.*, 1980; Hill *et al.*, 1980; Rosowski and Saunders, 1980). Sound entering one ear travels through the trabeculated skull bones and is incident on the inner surface of the tympanic membrane of the opposite ear (Rosowski and Saunders, 1980). If the sound incident on the outer surface of the tympanic membrane is closely matched in intensity and phase to that acting on the inner surface of the membrane, there is cancellation. If the two sounds are out of phase, there is amplification.

Polar response curves can be generated by recording the cochlear microphonic potential from one ear while moving the sound source around the head in a horizontal plane. For both Common Quail (*Coturnix coturnix*) and Domestic Fowl (*Gallus domesticus*), nulls as deep as 20–25 dB occur in these polar response curves when forces on opposite sides of the tympanic membrane cancel (Lewis and Coles, 1980; Rosowski and Saunders, 1980). Obviously these effects are frequency dependent. What this means is that birds may in fact be able to use interaural intensity differences for both narrow and broad band signals but without the constraint imposed by the relation between head size and wavelength. Frequency is not a necessary constraint in this model, thus eliminating the relatively narrow range of hearing in birds as a handicap in sound localization.

Behavioral studies of hearing in Barn Owls clearly indicate that this species can localize acoustic signals with incredible accuracy, surpassing even man in this regard (Knudsen and Konishi, 1979; Konishi, 1973b; Mills, 1972; Payne, 1971). Noise signals are more easily localized by Barn Owls than are tonal signals, consistent with data from other vertebrates (Brown *et al.*, 1978a,b; Gourevitch, 1965; Knudsen and Konishi, 1979). There have been two behavioral studies of sound localization conducted on birds with auditory systems not obviously specialized for sound localization. These results for Common Quails (Gatehouse and Shelton, 1978) and Rock Doves (Jenkins and Masterton, 1979) are shown in Fig. 9. The dependent variable in these studies is percentage correct responding and the independent variable is the spectrum of the signal with speaker separation held constant at 120°. Both species are poorest at tonal frequencies around 1 kHz and better at frequencies above and below this frequency region. For both species, white noise signals are more easily localized than are pure tones. There are several curious aspects to these patterns of results. First, both species show the poorest sound localization in the frequency region where most birds show greatest auditory sensitivity and that contains a substantial amount of the energy in species-specific vocalizations (Dooling, 1980). Second, the results for both species are reminiscent of the results obtained from human studies, where poorest performance occurs at the transition between use of interaural phase cues and reliance on interaural intensity cues (Mills, 1972). Small

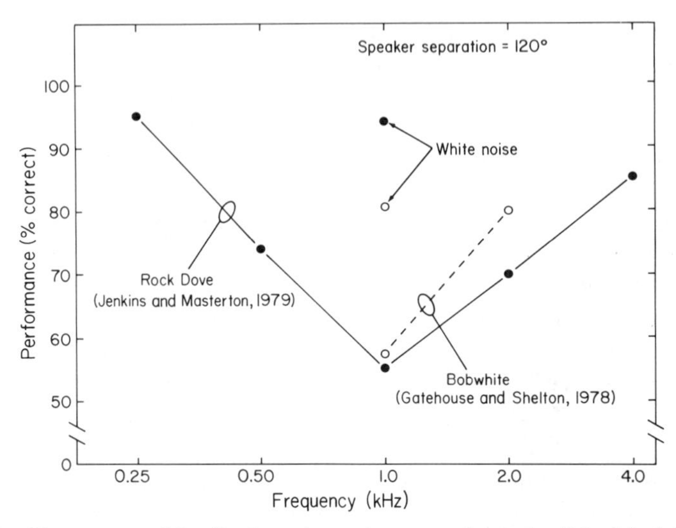

Fig. 9. Data on sound localization of pure tones are shown for Bobwhite (after Gatehouse and Shelton, 1978), and Rock Dove (after Jenkins and Masterton, 1979). Both species were also tested with white noise centered at 1.0 kHz. Chance performance is approximately 50%.

head size and the acoustic connection between middle ear cavities of birds preclude such a simple explanation.

Playback studies in the field provide numerous examples of birds flying toward and attacking a speaker in order to defend their territory against intruders (e.g., Peters *et al.*, 1980). A major question in the evolution of vocal signals concerns the issue of locatability. Marler (1955, 1959), for instance, has pointed out that some bird calls may have evolved to be ventriloquial to other birds while other calls seem to maximize locatability. Narrow-band, high-frequency alarm calls having gradual onset and offset minimize the availability of time-of-arrival cues for localization. By contrast broad-band or frequency-swept signals having sharp onset and offset and produced in a repetitive fashion (e.g., syllables in a song) provide a maximum number of acoustic cues for localization. This hypothesis has yet to be adequately tested but there is some evidence that avian aerial predators can localize narrow-band high-frequency alarm calls to some degree (Rooke and Knight, 1977; Shalter, 1978; Shalter and Schleidt, 1977). Unfortunately, the meager psychophysical data available on avian sound localization are not sufficient to specify the critical cues involved in simple acoustic signals, let alone complex natural vocalizations. Nevertheless, two general statements can be made with some confidence. First, available psychophysical data do suggest that broad-band acoustic signals are more easily localized than are narrow band or pure tone signals. This is consistent with the ethological observation that "mobbing calls" are more easily localized than "aerial predator" alarm calls (Marler, 1955). Second, it is inappropriate to appeal to mechanisms of human sound localization in attempting to understand or predict avian sound localization performance. The presence of an interaural pathway in birds precludes such extrapolation. Psychophysical studies of avian sound localization using natural vocalizations are desperately needed to resolve this critical issue of the relation between locatability and the evolution of vocalization structure.

IV. HEARING IN NOISE

A. Comparative Aspects of Masked Auditory Thresholds

How well the auditory system can extract a signal from noise is biologically important for two different reasons. First, measures of signal-to-noise ratio at different frequencies throughout the range of hearing provide information about the mechanisms of tuning or frequency selectivity in auditory systems (Dooling, 1980). Second, the relation between signal level and background noise level has obvious implications for the perception of vocal signals under natural conditions. Typically in a masking experiment, the threshold for a pure tone is determined in the presence of a known level of background noise. The ratio between the power in the pure tone and the power per hertz (spectrum level) of the background noise is called the critical ratio. Critical ratio data for Budgerigars and nine other vertebrates are summarized in Fig. 10. It is well established that the orderly increase of about 3 dB in critical ratio with each doubling of frequency is tied to the mechanics of peripheral auditory system function; presumably this reflects the logarithmic organization of traveling wave maximum displacement along the basilar membrane (Greenwood, 1961a,b). The departure from this pattern for horseshoe bats represents a remarkable specialization in the peripheral auditory system. It is related to a thickening of the basilar membrane at a point 4.5 mm from the base (Bruns, 1976a,b). Budgerigars also show a departure from the typical pattern of results shown for other vertebrates, but the reason for this departure remains unknown. Perhaps, as with the horseshoe bat, the unusual shape of the critical ratio function in the Budgerigar is also due to a peripheral, mechanical process (Dooling, 1980; Dooling and Saunders, 1975a; Dooling and Searcy, 1979).

B. Relevance to the Design of Vocal Signals

The auditory system is constantly faced with the task of detecting important biological signals against a background of environmental noise. There has been a recent surge of interest in the possibility of environmental influences on the

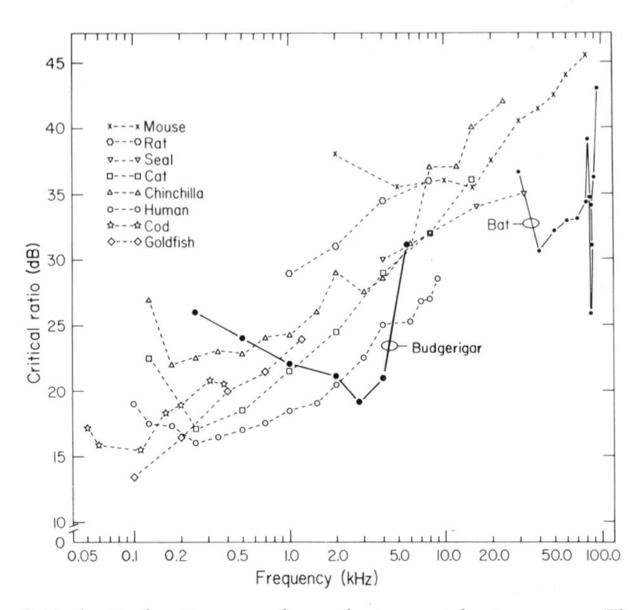

Fig. 10. Critical ratio functions are shown for ten vertebrates: mouse (Ehret, 1975b), white rat (Gourevitch, 1965), ringed seal (*Phoca hispida*) (Terhune and Ronald, 1975), cat (Watson, 1963), chinchilla (*Chinchilla laniger*) (Miller, 1964), man (Hawkins and Stevens, 1950), cod (*Gadus morhua*) (Hawkins and Chapman, 1975), goldfish (Fay, 1974), horseshoe bat (*Rhinolophus ferrumequinum*) (Long, 1977), and Budgerigar (Dooling and Saunders, 1975a). Operating with logarithms, the critical ratio in dB refers to the sound power in a tone minus the spectrum level of the background noise. (From Dooling, 1980.)

design of animal vocalizations (Morton, 1970, 1975; Waser and Waser, 1977; Wiley and Richards, 1978). To be sure, there are many factors influencing the maximum distance over which a biologically meaningful sound may be heard and whether or not certain characteristics of the sound are more prone to environmental influences than others. Consideration must be given to: (1) the intensity with which the signaler vocalizes, (2) the sound-attenuating and sound-modifying characteristics of the environment, and (3) the sensitivity of the receiver.

Assessing the relative importance of these three factors is a difficult problem, but clearly the issue of environmental influence on vocal signals cannot be adequately examined unless data are available for all three factors. Those field studies purporting to show effects of environmental variables in the evolution of vocal characteristics are correlational in nature and have not yet adequately dealt with either the intensity with which the signaler vocalizes nor the sensitivity of the receiver (e.g., Morton, 1975; Roberts *et al.*, 1980). Both of these issues are perhaps best tackled in the laboratory where the level and spectral characteristics of the masker and signal can be precisely controlled. A considerable amount is known about masking in Budgerigars from such laboratory studies.

The critical ratio data of Fig. 10 and previous experiments on critical bands (Saunders *et al.*, 1978) clearly indicate that noise in the spectral region of the signal is the most effective in masking a signal for Budgerigars and other vertebrates including man (Dooling, 1980; Hawkins and Stevens, 1950). The critical ratio for Budgerigars at 2.86 kHz (best frequency in Fig. 10) means that the power in a signal must be 18–20 dB greater than the spectrum level of background noise in the same frequency region in order to be detected. This is true whether the signal is a pure tone or a natural Budgerigar vocalization (R. J. Dooling and M. H. Searcy, unpublished data). Knowing the sensitivity of the receiver in noise, the next question concerns the intensity of the signaler's vocalizations.

The vocal performance of a number of songbirds and non-songbirds has been measured and maximum peak sound pressure levels of 90–95 dB are not uncommon (Brackenbury, 1979). Similar results have been obtained for Canaries, Swamp Sparrows (*Zonotrichia georgiana*), and Song Sparrows (*Zonotrichia melodia;* R. J. Dooling and M. H. Searcy, unpublished data). Knowing the maximum intensity at which the signaler can vocalize and the sensitivity of the receiver in a noisy environment, it becomes possible to estimate maximum transmission distances for a wide variety of natural environmental conditions.

The study of the physical constraints on acoustic communication under natural conditions is incredibly complex. Attention must be given to a wide range of effects, some linear and some nonlinear with distance. A review of these important variables is given by Wiley and Richards (Chapter 5, this volume).

In general there are two broad categories of effects causing attenuation of sound with distance. First, there is spherical spreading where, in the ideal case, there is a quartering of power or halving of pressure with each doubling of distance. Known as the inverse square law, the effect is a decrease of approximately 6 dB with each doubling of distance. Second, in a homogeneous environment, there are also constant attentuation effects which represent a deviation from the attenuation expected from the inverse square law and which are linearly proportional to distance (Marten and Marler, 1977; Marten *et al.*, 1977). In nature, the assumptions of frictionless atmospheric conditions and homogeneous environments are difficult to satisfy, but spherical spreading and excess attenuation effects do provide a valuable starting point for estimating hearing distance. The factors determining how far away a sound can be heard then are source intensity, inverse square attenuation, excess attenuation, spectrum level of the background noise, and masked threshold or critical ratio. An equation adopted from Marten and Marler (1977) considers these variables where I_0 = source intensity, ISA = inverse square attenuation, EA = excess attenuation (constant attenuation effects), I_{mat} = intensity of signal at a certain spectrum level of background noise, d_0 = distance at which source intensity is measured, and d_{mc} = maximum communication distance.

To estimate the maximum distance over which communication would be possible it is necessary to calculate the distance traveled by an acoustic signal as it drops from the original intensity (I_0) to an intensity just audible above the background noise (I_{mat}). The amount of signal attenuation, or drop, is given by $I_0 - I_{mat}$. To estimate maximum communication distance solve the following equation for d_{mc}:

$$\text{Drop} = 20 \log \frac{d_{mc}}{d_o} + \frac{\text{EA} \cdot d_{mc}}{100}$$

For each positive drop and positive EA there is a unique d. The relation among source intensity, excess attenuation, and background noise level as they affect maximum communication distance is shown in Figs. 11 and 12.

The level and spectral characteristics of the background noise are crucial to the

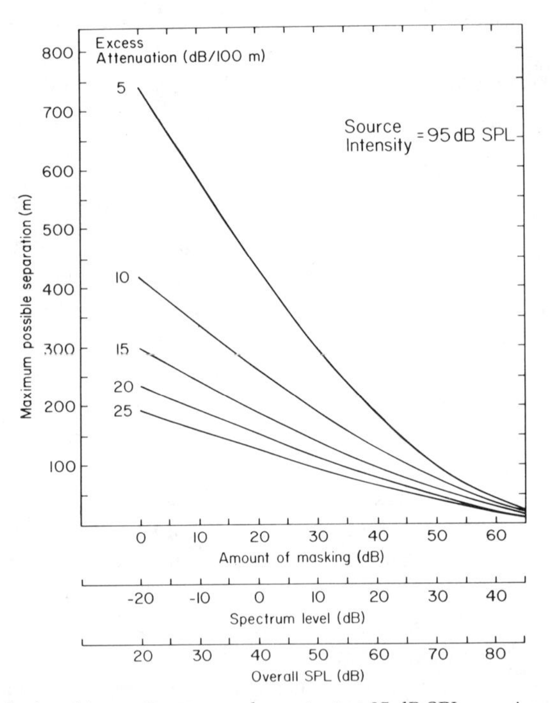

Fig. 11. With signal intensity assumed constant at 95 dB SPL, maximum transmission distance is calculated given background noise level at different values of excess attenuation (dB/100 m). Background noise level is given as overall SPL (i.e., reading on the C scale of a sound level meter), spectrum level (per cycle energy distribution over the entire band of noise), and the amount of masking assuming a critical ratio of 20 dB.

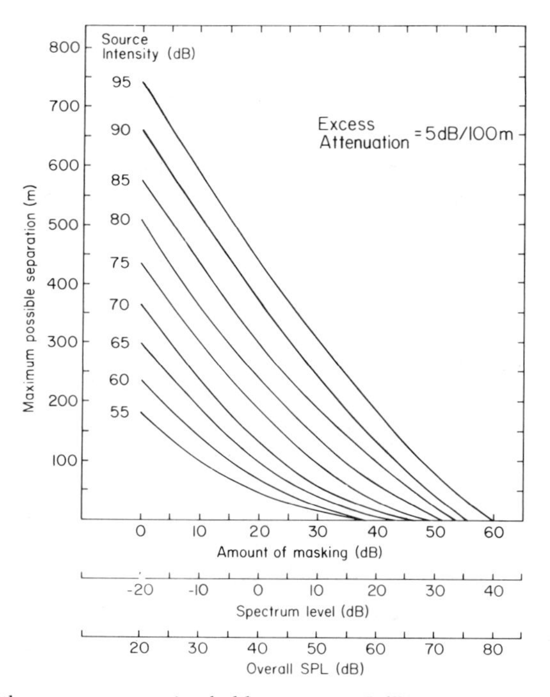

Fig. 12. With excess attentuation held constant at 5 dB/100 m, maximum transmission distance is given as a function of level of background noise for a number of different source intensities. Abscissa same as in Fig. 11.

determination of maximum communication distance. The threshold in the quiet for a 2.86-kHz tone is about 0 dB SPL for Budgerigars (Dooling and Saunders, 1975a). Critical ratio data for Budgerigars indicate that background noise spectrum level must be at least 18–20 dB below the power in a 2.86-kHz tone in order for the tone to be detected by this species. For a noise spectrum level of −20 dB then, masked threshold is the same as absolute threshold in the quiet and no masking occurs (Dooling, 1980; Dooling and Saunders, 1975a). If the spectrum level of the background noise is 0 dB SPL, however, the threshold of the signal is raised to 20 dB SPL, causing 20 dB of masking. Spectrum level can be computed from overall sound pressure level given the bandwidth of the noise and the assumption that energy is evenly distributed throughout the band:

$$\text{Spectrum level} = \text{overall SPL} - 10 \log (\text{bandwidth})$$

For a typical sound level meter reading of noise in the frequency band of 0–10 kHz, an overall SPL of 40 dB is equivalent to a spectrum level of about 0 dB (Hirsh, 1952). These relations are shown as three abscissas in Figs. 11 and 12.

In Fig. 11, source intensity is assumed constant at 95 dB SPL and the parameter shown is excess attenuation in decibels per 100 m ranging from 5 dB/100 m to 25 dB/100 m. A 5 dB/100 m excess attenuation value is appropriate in the situation where a bird is singing perched 10 m or more above the ground in an open field (Marten and Marler, 1977). On the other hand, an excess attenuation value of 20 dB/100 m is more appropriate to a bird singing at ground level in a coniferous forest (Marten and Marler, 1977).

Figure 12 shows the effect of varying source intensity with excess attenuation constant at 5 dB/100 m. For an overall background noise level of 45 dB SPL, there is approximately 25 dB of masking occurring and the maximum transmission distance of a vocalization given at 70 dB is about 100. By increasing its vocal output from 70 to 90 dB SPL, a bird can triple the maximum possible separation between signaler and receiver from 100 to about 300 m.

Since the above situation has been simplified and represents something of a best case under ideal conditions, several caveats deserve mention. First, the adverse acoustic conditions in natural environments can cause complex, nonstationary effects which exert a considerable influence on maximum transmission distance. Second, these data are based on experiments with Budgerigars and there are no masking data available for other birds. It may be that birds as a group show considerable variability in masked thresholds and hence in the maximum range at which weak signals can be detected against a background of noise. Third, the constant mobility of both signaler and receiver will exert tremendous influences on maximum transmission distances (Marten and Marler, 1977; Marten *et al.*, 1977). Fourth, the masking data presented for Budgerigars are only for situations in which the signal and masker are delivered from the same source. It is known for humans that a considerable improvement in audibility occurs when the signal and masker are delivered from different locations (Durlach and Colburn, 1978; Hirsh, 1971). The phenomenon known as the "cocktail party" effect relies in large part on the binaural auditory system's ability to separate certain parts of auditory space and thereby spatially separate the signal from the noise. In a bird's natural environment, it is probably the exception rather than the rule that the signal and noise masker are delivered from the same location in auditory space. All of the above caveats must be considered in any discussion relating maximum transmission distance to group dispersion, territory size, or other measures of organism spacing. The above discussion should not be taken to mean that bird song is always structured to maximize transmission distance. Recent studies of song characteristics of wood warblers (Parulidae) are not consistent with the notion that wood warblers maximize the distance over which their songs are transmitted. Rather, wood warblers may produce songs, which by virtue of their frequency and structure, carry mainly to other birds of biological relevance to the singers (Lemon *et al.*, 1981).

V. HEARING AND VOCALIZATIONS

A. Auditory Resolving Power and Vocal Precision

The stereotypy of songbird vocalizations has often served as a foundation for assertions about the resolving power of the avian ear (e.g., Pumphrey, 1961). A review of the psychophysical data on auditory discrimination in birds indicates a well-developed resolving power in the frequency and time domains and less sensitivity to intensity. Psychoacoustic data for birds indicate a frequency-resolving power of about 20 Hz at 2.0 kHz, and a duration resolving power of about 2–3 msec at 20 msec. This kind of precision by the avian ear easily rivals or surpasses that of the sonagraph, which to date has provided the firmest evidence for stereotypy in bird song.

Several studies of vocal production in birds have attempted to correlate the variability in sound production with the resolving power of the avian ear. From studies using period or zero crossing analyses, it is generally agreed that variation in the frequency of a note is about the same order of magnitude as frequency resolving power measured psychophysically (Dooling, 1980; Dooling and Saunders, 1975a; Greenewalt, 1968). Variation in internote interval and note duration estimated from studies of several species (Dooling, 1980; R. J. Dooling and M. H. Searcy, unpublished data; Wiley, 1976) corresponds well with the discrimination of auditory duration as measured in Budgerigars (Dooling and Haskell, 1978). Others (Greenewalt, 1968) estimate temporal variability in note duration and internote interval to be considerably smaller than that given above, however, and the reasons for this discrepancy are not clear. Less attention has generally been given to variation in amplitude across repeated occurrences of the same song element. In one study on the Canary, however, the coefficient of variation in note intensity was well matched to the intensity-resolving power of the ear measured psychophysically (Dooling, 1980).

In the above terms, then, the auditory system and the vocal production system of birds appear about equally matched in precision. This level of precision is quite close to the auditory discriminatory abilities of humans as measured on very similar acoustic tasks. For most students of vocal communication in birds, a nagging question remains. How is it that we are unable to "hear out" all of the subtle details seen in the sonagram of a complex vocalization when there is clear evidence from song-learning experiments that birds can learn to imitate these differences? There is no simple answer to this question, but some of the factors most likely involved in the phenomenon include the following: First, our inability to hear the fine details evident in the sonagram of a complex sound is certainly not unique to bird vocalizations. It is also the case that we would have equal difficulty in acoustically identifying the fine details of human speech sounds (e.g., formant transitions) in spite of the fact that we can learn to produce these transitions with little problem. Second, psychophysical studies of auditory discrimination, such as those reviewed in this chapter, usually involve considerable training with the stimuli and short interstimulus intervals over which a comparison must be made. Differences in fine detail which we perhaps cannot acoustically identify might still render two complex vocalizations discriminably different to our ear if we were first trained to the stimuli and then tested with the two sounds presented only a few hundred milliseconds apart. This implies the operation of a third factor involving the focus of attention. It is certainly plausible that when tested to the limits of their sensory capacity, birds and humans prove equally capable. But, when tested with complex species-specific vocalizations, variables such as prior experience, learning, and stimulus salience might serve to focus attention on certain critical aspects of the stimulus away from other irrelevant acoustic dimensions. The result of such an attentional mechanism would be that one species might learn a discrimination between two complex natural sounds much more easily than the other. This kind of interpretation gains considerable support from recent studies of discrimination of species-specific vocal signals in monkeys (Beecher *et al.*, 1979; Zoloth and Green, 1979; Zoloth *et al.*, 1979).

B. Hearing and Vocal Ontogeny

For many songbirds, hearing is critically involved in the development of normal vocal behavior as evidenced from experiments involving long-term noise exposure or deafening at various stages of vocal development (Konishi and Nottebohm, 1969; Marler *et al.*, 1973; Marler and Mundinger, 1971; Nottebohm, 1969). On a more subtle level, several songbirds show a selective bias toward learning only conspecific songs during the sensitive period for song learning (Konishi, 1978; Marler and Mundinger, 1971; Marler and Peters, 1977).

Conducting learning experiments is tedious. Whether or not a bird learns a song can only be determined later when the bird sings it. It is therefore an open question whether selective vocal learning is due to an early auditory perceptual process or whether later motor constraints play the primary role. Considerable attention has recently been focused on comparative aspects of vocal learning in two congeneric species of sparrow—the Swamp and the Song sparrows—in an attempt to gain understanding into the song learning process (Marler and Peters, 1977, 1981). When raised in isolation and tutored with a collection of Swamp and Song sparrow songs, both of these species have a marked tendency to learn only conspecific song over the song of the other species but with Swamp Sparrows showing the greatest selectivity. Using the cardiac orienting response, Dooling and Searcy (1980a) were able to show that, during the sensitive period for song learning, Swamp Sparrows respond differently to conspecific song than to songs of other species even when hearing these sounds for the first time. These results argue for an innate perceptual bias as a basis for selective learning in Swamp Sparrows.

C. Evidence for Species-Specific Perceptual Processes in Birds

The acoustic cues involved in species recognition can vary among species. The present review has tried to establish the limits of auditory resolving power which might be brought to bear on critical issues like species and individual recognition. Estimates of resolving power obtained from psychophysical studies are often better than estimates obtained from playback studies under natural conditions. Possible explanations for this discrepancy are that birds simply do not attend to discriminable changes in song even though they are capable of doing so or they are not motivated to respond to the differences they perceive. Are these issues testable using psychophysical procedures like those reviewed in this chapter? Results of a recent study of Red-winged Blackbirds (*Agelaius phoeniceus*) and Brown-headed Cowbirds (*Molothrus ater*) suggest that these issues can be dealt with in the laboratory.

Using Red-winged Blackbirds and Brown-headed Cowbirds, which have similar auditory capabilities (Hienz *et al.*, 1977; Sinnott *et al.*, 1980), Sinnott (1980) first trained both species to categorize conspecific song and songs of the other species. The song patterns of both species consist of a series of brief, low-amplitude ''introductory'' notes followed by a longer, more intense sustained portion (Fig. 13). In the Red-winged Blackbird, this terminal position is a trill and in the Brown-headed Cowbird a high-frequency ''whistle.''

Once the birds learned to categorize the full song themes shown in Fig. 13, they were tested again on elements derived from separating the introductory notes from the terminal portions of the songs. Clear species differences emerged in the ability to categorize these elements (Fig. 14). Both species were equally adept at categorizing the initial portions of conspecific song and the song of the other species. On the terminal portion of the song, however, the blackbirds outperformed the cowbirds on blackbird trills and the cowbirds outperformed the blackbirds on cowbird whistles. Sinnott (1980) argues for higher-level attentional mechanisms to account for these results. Birds categorizing conspecific themes may have a predisposition to attend to all of the information in the stimuli during training. But when categorizing themes of the other species, both blackbirds and cowbirds attended only to the bare minimum amount of the stimulus needed to categorize the song (i.e., introductory notes). Of course one component of such an attentional mechanism might be directly related to differential amounts of prior experience between conspecific song and song of the other

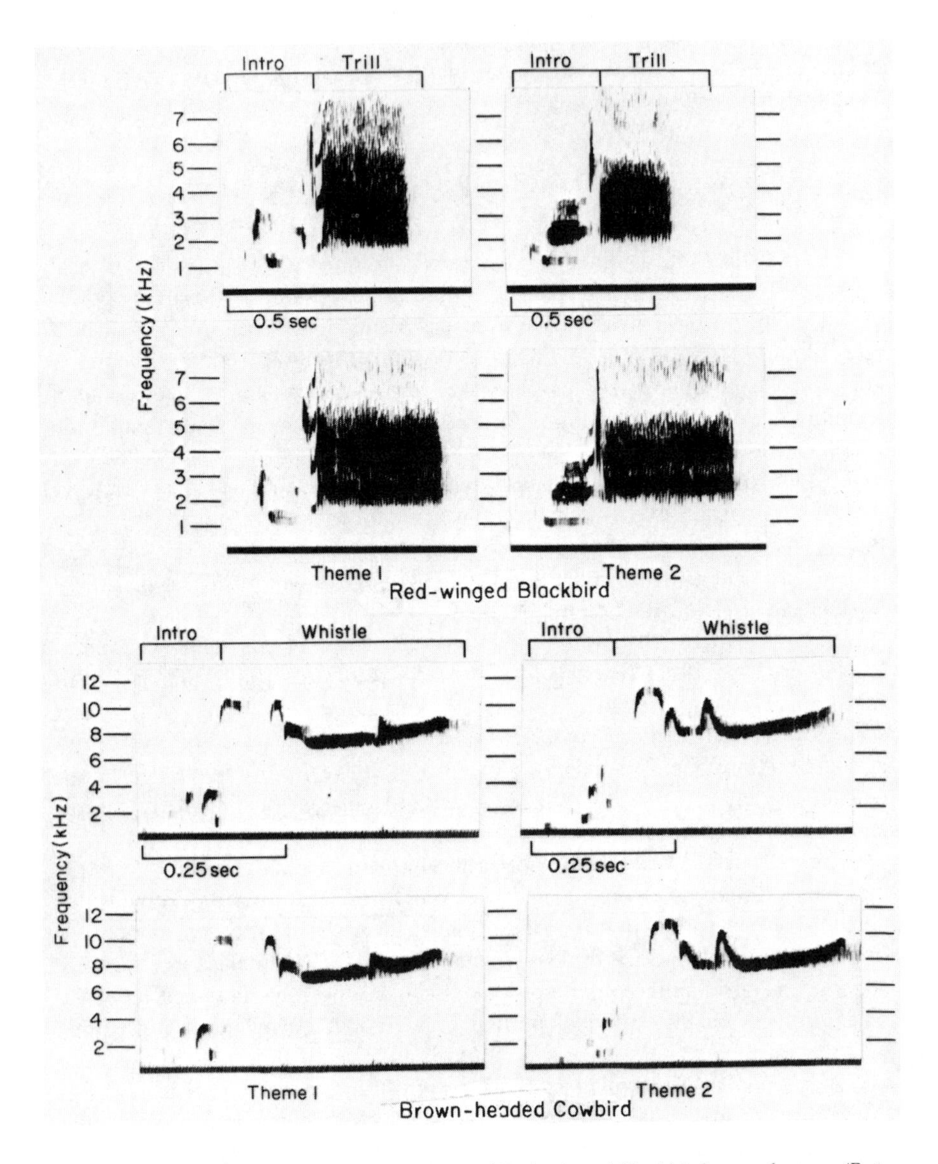

Fig. 13. (Top) Sonagrams of four tokens of Red-winged Blackbird song themes. (Bottom) Sonagrams of four tokens of Brown-headed Cowbird song themes. Brackets indicate the point where introductory notes were separated from trills during the tests with isolated elements. (From Sinnott, 1980.)

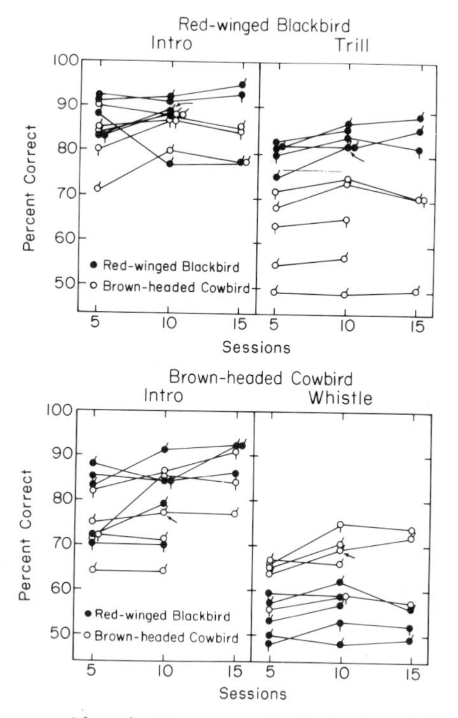

Fig. 14. Percent correct identification scores for Red-winged Blackbird (●) song elements (top) and Brown-headed Cowbird (○) song elements (bottom). Males are denoted by a mark at the upper right of circle and females with a mark at bottom center. (From Sinnott, 1980.)

species for both blackbirds and cowbirds. It would be interesting to test naive birds raised from the egg on this same stimulus set, thus ruling out any questions of prior experience with the stimuli.

VI. CONCLUSION

Audiograms from psychophysical studies of hearing in birds are available for 16 species. There are differences among species in absolute auditory sensitivity, but all birds hear best in the frequency region of 1–5 kHz and show a considerable lack of sensitivity for frequencies above 10 kHz. On measures of auditory discrimination—including temporal resolving power—birds appear similar to other vertebrates that have been tested, including man. Masked thresholds for Budgerigars, on the other hand, are considerably different from those obtained from other vertebrates and suggest a specialization for the detection of vocal signals in noise. Knowing the intensity with which a bird vocalizes and the signal-to-noise ratio the ear can obtain allows one to estimate the maximum possible communication distance between source and receiver for a number of hypothetical environments. Sound localization abilities of birds, with the exception of Barn Owls, have yet to be adequately measured. Preliminary evidence suggests poorer localization for pure tones than for noise and poorer localization for midfrequencies than for high or low frequencies. Interaural pathways in birds suggest an entirely different mechanism of spatial localization. There are hints from both young birds and adult birds of special perceptual processes which function in song learning and species recognition.

These above results indicate that the auditory world of birds is both rich and complex. Hearing plays an important role in a wide range of critical behaviors such as predator detection, mate selection, territorial defense, species recognition, and song learning. Important questions remain to be answered. What are the mechanisms of sound localization in birds with auditory systems not specialized for prey detection? Are the masked thresholds of other birds like those of Budgerigars? What is the role of auditory discriminatory abilities in the ontogeny of vocal categories and species recognition? Finally, to what extent are avian vocalizations, and indeed the avian ear, specialized for communication in different habitats? The answers to these questions will have far-reaching consequences for a "sensory" ecology and for our understanding of the evolution of acoustic communication systems.

ACKNOWLEDGMENTS

Preparation of this chapter was supported by NIMH Grant MH31165. I thank D. Kroodsma, P. Marler, W. Searcy, J. Sinnott, and W. Stebbins for thoughtful comments on earlier versions. I also thank M. Searcy and E. Arruzza for figure and manuscript preparations, respectively.

REFERENCES

Beecher, M. D., Peterson, M. R., Zoloth, S. R., Moody, D. B., and Stebbins, W. C. (1979). Perception of conspecific vocalizations by Japanese macaques. *Brain Behav. Evol.* **16,** 443–460.

Brackenbury, J. H. (1979). Power capabilities of the avian sound-producing system. *J. Exp. Biol.* **78,** 163–166.

Brémond, J. C. (1968). Recherches sur la semantique et les elements vecteurs d'information dans les signaux acoustiques du rouge-gorge (*Erithacus rubecula* L.). *Terre Vie* **2,** 109–220.

Brémond, J. C. (1975). Specific recognition in the song of Bonelli's warbler (*Phylloscopus bonelli*). *Behaviour* **58,** 99–116.

Brodkorb, P. (1971). Origin and evolution of birds. *In* "Avian Biology" (D. S. Farner, J. R. King, and K. C. Parkes, eds.), Vol. 1, pp. 19–55. Academic Press, New York.

Brooks, R. J., and Falls, J. B. (1975a). Individual recognition by song in White-throated Sparrows. I. Discrimination of songs by neighbors and strangers. *Can. J. Zool.* **53,** 879–888.

Brooks, R. J., and Falls, J. B. (1975b). Individual recognition by song in White-throated Sparrows. III. Song features used in individual recognition. *Can. J. Zool.* **53,** 1749–1761.

Brown, C. H., Beecher, M. D., Moody, D. B., and Stebbins, W. C. (1978a). Localization of pure tones by Old World monkeys. *J. Acoust. Soc. Am.* **63,** 1484–1492.

Brown, C. H., Beecher, M. D., Moody, D. B., and Stebbins, W. C. (1978b). Localization of primate calls by Old World monkeys. *Science* **201,** 753–754.

Bruns, V. (1976a). Peripheral auditory tuning for fine frequency analysis by the CF-FM bat, *Rhinolophus ferrumequinum.* I. Mechanical specializations of the cochlea. *J. Comp. Physiol.* **106,** 77–86.

Bruns, V. (1976b). Peripheral auditory tuning for fine frequency analysis by the CF-FM bat, *Rhinolophus ferrumequinum.* II. Frequency mapping in the cochlea. *J. Comp. Physiol.* **106,** 87–97.

Coles, R. B., Lewis, D. B., Hill, K. G., Hutchings, M. E., and Gower, D. M. (1980). Directional hearing in the Japanese Quail (*Coturnix coturnix japonica*). II. Cochlear physiology. *J. Exp. Biol.* **86,** 153–170.

Dooling, R. J. (1979). Temporal summation of pure tones in birds. *J. Acoust. Soc. Am.* **65,** 1058–1060.

Dooling, R. J. (1980). Behavior and psychophysics of hearing in birds. *In* "Comparative Studies of Hearing in Vertebrates" (A. N. Popper and R. R. Fay, eds.), pp. 261–288. Springer-Verlag, Berlin and New York.

Dooling, R. J., and Haskell, R. J. (1978). Auditory duration discrimination in the Parakeet (*Melopsittacus undulatus*). *J. Acoust. Soc. Am.* **63,** 1640–1642.

Dooling, R. J., and Saunders, J. C. (1975a). Hearing in the parakeet (*Melopsittacus undulatus*): Absolute thresholds, critical ratios, frequency difference limens, and vocalizations. *J. Comp. Physiol.* **88,** 1–20.

Dooling, R. J., and Saunders, J. C. (1975b). Auditory intensity discrimination in the parakeet (*Melopsittacus undulatus*). *J. Acoust. Soc. Am.* **58,** 1308–1310.

Dooling, R. J., and Searcy, M. H. (1979). The relation among critical ratios, critical bands, and intensity difference limens in the Parakeet (*Melopsittacus undulatus*). *Bull. Psycho. Soc.* **13,** 300–302.

Dooling, R. J., and Searcy, M. H. (1980a). Early perceptual selectivity in the Swamp Sparrow. *Dev. Psychobiol.* **13,** 499–506.

Dooling, R. J., and Searcy, M. H. (1980b). Forward and backward auditory masking in the Parakeet (*Melopsittacus undulatus*). *Hearing Res.* **3,** 279–284.

Dooling, R. J., and Searcy, M. H. (1981). Amplitude modulation thresholds for the Parakeet (*Melopsittacus undulatus*). *J. Comp. Physiol.* **143,** 383–388.

Dooling, R., Mulligan, J., and Miller, J. (1971). Auditory sensitivity and song spectrum of the common Canary (*Serinus canarius*). *J. Acoust. Soc. Am.* **50,** 700–709.

Dooling, R. J., Zoloth, S. R., and Baylis, J. R. (1978). Auditory sensitivity, equal loudness, temporal resolving power and vocalizations in the House Finch (*Carpodacus mexicanus*). *J. Comp. Physiol. Psychol.* **92,** 867–876.

Dooling, R. J., Peters, S. S., and Searcy, M. H. (1979). Auditory sensitivity and vocalizations of the Field Sparrow (*Spizella pusilla*). *Bull. Psychon. Soc.* **14,** 106–108.

Durlach, N., and Colburn, H. S. (1978). Binaural phenomena. *In* "Handbook of Perception. Vol. IV: Hearing" (C. Carterette and M. P. Friedman, eds.), pp. 365–466. Academic Press, New York.

Ehret, G. (1975a). Frequency and intensity difference limens and non linearities in the ear of the house mouse (*Mus musculus*). *J. Comp. Physiol.* **102,** 321–336.

Ehret, G. (1975b). Masked auditory thresholds, critical ratios, and scales of the basilar membrane of the house mouse (*Mus musculus*). *J. Comp. Physiol.* **103,** 329–341.

Ehret, G. (1976). Temporal auditory summation for pure tones and white noise in the house mouse (*Mus musculus*). *J. Acoust. Soc. Am.* **59,** 1421–1427.

Emlen, S. T. (1972). An experimental analysis of the parameters of bird song eliciting species recognition. *Behaviour* **41,** 130–171.

Falls, J. B. (1963). Properties of bird song eliciting responses from territorial males. *Proc. Int. Ornith. Congr.* **13,** 359–371.

Fay, R. R. (1974). Masking of tones by noise for the goldfish (*Carassius auratus*). *J. Comp. Physiol. Psychol.* **87,** 708–716.

Fletcher, L. E., and Smith, D. G. (1978). Some parameters of song important in conspecific recognition by Gray Catbirds. *Auk* **95,** 338–347.

Gabor, D. (1947). Acoustical quanta and the theory of hearing. *Nature (London)* **159,** 591–594.

Gatehouse, R. W., and Shelton, B. R. (1978). Sound localization of Bobwhite Quail (*Colinus virginianus*). *Behav. Biol.* **22,** 533–544.

Giraudi, D., Salvi, R., Henderson, D., and Hamernik, R. (1980). Gap detection by the chinchilla. *J. Acoust. Soc. Am.* **68,** 802–806.

Goldman, P. (1973). Song recognition by field sparrows. *Auk* **90,** 106–117.

Gottlieb, G. (1971). "Development of Species Identification in Birds." Univ. of Chicago Press, Chicago, Illinois.

Gourevitch, G. (1965). Auditory masking in the rat. *J. Acoust. Soc. Am.* **37,** 439–443.

Green, D. M. (1971). Temporal auditory acuity. *Psychol. Rev.* **78,** 540–551.

Greenewalt, C. H. (1968). "Bird Song: Acoustics and Physiology." Smithsonian Inst. Press, Washington, D.C.

Greenwood, D. D. (1961a). Auditory masking and the critical band. *J. Acoust. Soc. Am.* **33,** 484–502.

Greenwood, D. D. (1961b). Critical bandwidth and the frequency coordinates of the basilar membrane. *J. Acoust. Soc. Am.* **33,** 1344–1356.

Griffin, D. R. (1954). Acoustic orientation in the Oil Bird, *Steatornis*. *Proc. Natl. Acad. Sci. U.S.A.* **39,** 885–893.

Griffin, D. R., and Suthers, R. A. (1970). Sensitivity of echolocation in Cave Swiftlets. *Biol. Bull. (Woods Hole, Mass.)* **139,** 495–501.

Hack, M. (1971). Auditory intensity discrimination in the rat. *J. Comp. Physiol. Psychol.* **74,** 315–318.

Harris, G. G. (1960). Binaural interactions of impulsive stimuli and pure tones. *J. Acoust. Soc. Am.* **32,** 685–692.

Hawkins, A. D., and Chapman, C. J. (1975). Masked auditory thresholds in the cod, *Gadus morhua* L. *J. Comp. Physiol.* **103,** 209–226.

Hawkins, J. E., Jr., and Stevens, S. S. (1950). The masking of pure tones and of speech by white noise. *J. Acoust. Soc. Am.* **22,** 6–13.

Henderson, D. (1969). Temporal summation of acoustic signals by the chinchilla. *J. Acoust. Soc. Am.* **46,** 474–475.

Hienz, R. D., Sinott, J. M., and Sachs, M. B. (1977). Auditory sensitivity of the Redwing Blackbird and the Brown-headed Cowbird. *J. Comp. Physiol. Psychol.* **91,** 1365–1376.

Hill, K. G., Lewis, D. B., Hutchings, M. E., and Coles, R. B. (1980). Directional hearing in the Japanese Quail (*Coturnix coturnix japonica*). I. Acoustic properties of the auditory system. *J. Exp. Biol.* **86,** 135–151.

Hirsh, I. J. (1952). "The Measurement of Hearing." McGraw-Hill, New York.

Hirsh, I. J. (1971). Masking of speech and auditory localization. *Audiology* **10,** 110–114.

Jacobs, D., and Tavolga, W. (1967). Acoustic intensity limens in the gold fish. *Anim. Behav.* **15,** 324–335.

Jenkins, W. M., and Masterton, R. B. (1979). Sound localization in the pigeon (*Columba livia*). *J. Comp. Physiol. Psychol.* **93,** 403–413.

Jesteadt, W., Wier, C. C., and Green, D. M. (1977). Intensity discrimination as a function of frequency and sensation level. *J. Acoust. Soc. Am.* **61,** 169–177.

Johnson, C. S. (1968). Relation between absolute thresholds and duration-of-tone in the bottlenosed porpoise. *J. Acoust. Soc. Am.* **43,** 757–763.

Knudsen, E. I. (1980). Sound localization in birds. *In* "Comparative Studies of Hearing in Vertebrates" (A. N. Popper and R. R. Fay, eds.), pp. 289–322. Springer-Verlag, Berlin and New York.

Knudsen, E. I., and Konishi, M. (1978a). A neural map of auditory space in the owl. *Science* **200,** 795–797.

Knudsen, E. I., and Konishi, M. (1978b). Space and frequency are represented separately in the auditory midbrain of the owl. *J. Neurophysiol.* **41,** 870–884.

Knudsen, E. I., and Konishi, M. (1978c). Center–surround organization of auditory receptive fields in the owl. *Science* **202,** 778–780.

Knudsen, E. I., and Konishi, M. (1979). Mechanisms of sound localization in the Barn Owl (*Tyto alba*). *J. Comp. Physiol. A* **133,** 13–21.

Knudsen, E. I., Blasdel, G. G., and Konishi, M. (1979). Sound localization by the Barn Owl (*Tyto alba*) measured with the search coil technique. *J. Comp. Physiol.* **133,** 1–11.

Konishi, M. (1969). Time resolution by single auditory neurons in birds. *Nature (London)* **222,** 566–567.

Konishi, M. (1970). Comparative neurophysiological studies of hearing and vocalizations on song birds. *Z. Vgl. Physiol.* **66,** 257–272.

Konishi, M. (1973a). How the Barn Owl tracks its prey. *Am. Sci.* **61,** 414–424.

Konishi, M. (1973b). Locatable and nonlocatable acoustic signals for Barn Owls. *Am. Nat.* **107,** 775–785.

Konishi, M. (1978). Auditory environment and vocal development in birds. *In* "Perception and Experience" (R. D. Walk and H. L. Picks, eds.), pp. 105–118. Plenum, New York.

Konishi, M., and Knudsen, E. I. (1979). The Oilbird: Hearing and echolocation. *Science* **204,** 425–427.

Konishi, M., and Nottebohm, F. (1969). Experimental studies in the ontogeny of avian vocalizations. *In* "Bird Vocalizations: Their Relation to Current Problems in Biology and Psychology" (R. A. Hinde, ed.), pp. 29–48. Cambridge Univ. Press, London and New York.

Krebs, J. (1976). Birdsong and territory defense. *New Sci.* **70,** 534–536.

Kreithen, M. L., and Quine, D. M. (1979). Infrasound detection by the homing pigeon: A behavioral audiogram. *J. Comp. Physiol.* **129,** 1–4.

Kroodsma, D. E. (1976). Reproductive development in a female songbird: Differential stimulation by quality of male song. *Science* **192,** 574–575.

Kuhn, A., Leppelsack, H.-J., and Schwartzkopff, J. (1980). Measurement of frequency discrimination in the starling (*Sturnus vulgaris*) by conditioning of heart rate. *Naturwissenschaften* **67S,** 102.

Lemon, R. E., Struger, J., Lechowicz, M. J., and Norman, R. F. (1981). Song features and singing heights of American warblers: Maximization or optimization of distance? *J. Acoust. Soc. Am.* **69,** 1169–1176.

Lewis, B., and Coles, R. (1980). Sound localization in birds. *Trends Neurosci.* **3,** 102–105.

Long, G. R. (1977). Masked auditory thresholds from the bat, *Rhinolophus ferrumequinum*. *J. Comp. Physiol.* **116,** 247–255.

Lynn, G. K., and Small, A. M. (1977). Interactions of backward and forward masking. *J. Acoust. Soc. Am.* **61,** 185–189.

Marler, P. (1955). Characteristics of some animal calls. *Nature (London)* **176,** 6–7.

Marler, P. (1959). Developments in the study of animal communication. *In* "Darwin's Biological Work" (P. R. Bell, ed.), pp. 150–206. Cambridge Univ. Press, London and New York.

Marler, P. (1970a). Bird song and speech development: Could there be parallels? *Am. Sci.* **58,** 669–673.

Marler, P. (1970b). On the origin of speech from animal sounds. *In* "The Role of Speech in Language" (J. F. Kavanagh and J. E. Cutting, eds.), pp. 389–449. Academic Press, New York.

Marler, P., and Mundinger, P. (1971). Vocal learning in birds. *In* "Ontogeny of Vertebrate Behavior" (M. Moltz, ed.), pp. 389–450. Academic Press, New York.

Marler, P., and Peters, S. S. (1977). Selective vocal learning in a sparrow. *Science* **198,** 519–521.

Marler, P., and Peters, S. S. (1981). Birdsong and speech: Evidence for special processing. *In* "Perspectives on the Study of Speech" (P. Eimas and J. Miller, eds.), pp. 75–112. Erlbaum, Hillsdale, New Jersey.

Marler, P., Konishi, M., Lutjen, A., and Waser, M. S. (1973). Effects of continuous noise on avian hearing and vocal development. *Proc. Natl. Acad. Sci. U.S.A.* **70,** 1393–1396.

Marten, K., and Marler, P. (1977). Sound transmission and its significance for animal vocalization. I. Temperate habitats. *Behav. Ecol. Sociobiol.* **2,** 271–290.

Marten, K., Quine, D., and Marler, P. (1977). Sound transmission and its significance for animal vocalization. II. Tropical forest habitats. *Behav. Ecol. Sociobiol.* **2,** 291–302.

Masterton, B., Heffner, H., and Ravizza, R. (1969). The evolution of human hearing. *J. Acoust. Soc. Am.* **45,** 966–985.

Miller, J. D. (1964). Auditory sensitivity of the chinchilla in quiet and in noise. *J. Acoust. Soc. Am.* **36***(A),* 2010.

Mills, A. W. (1972). Auditory localization. *In* "Foundations of Modern Auditory Theory" (J. V. Tobias, ed.), Vol. 2, pp. 303–348. Academic Press, New York.

Moiseff, A., and Konishi, M. (1981). Neuronal and behavioral sensitivity to binaural time differences in the owl. *J. Neurosci.* **1,** 40–48.

Morton, E. S. (1970). Ecological sources of selection on avian sounds. Ph.D. Thesis, Yale Univ., New Haven, Connecticut.

Morton, E. S. (1975). Ecological sources of selection on avian sounds. *Am. Nat.* **109,** 17–34.

Nottebohm, F. (1969). The "critical period" for song learning. *Ibis* **111,** 386–387.

Novick, A. (1959). Acoustic orientation in the Cave Swiftlet. *Biol. Bull. (Woods Hole, Mass.)* **117,** 497–503.

Payne, R. S. (1962). How the Barn Owl locates its prey by hearing. *Living Bird* **1,** 151–159.

Payne, R. S. (1971). Acoustic location of prey by Barn Owls (*Tyto alba*). *J. Exp. Biol.* **54,** 535–573.

Peters, S. S., Searcy, W. A., and Marler, P. (1980). Species song discrimination in choice experiments with territorial male Swamp and Song Sparrows. *Anim. Behav.* **28,** 393–404.

Plomp, R. (1964). Rate decay of auditory sensation. *J. Acoust. Soc. Am.* **36,** 277–282.

Pumphrey, R. J. (1961). Sensory organs: hearing. *In* "Biology and Comparative Anatomy of Birds" (A. J. Marshall, ed.), pp. 69–86. Academic Press, New York.

Quine, D. B. (1978). Infrasound detection and ultra low frequency discrimination in the homing pigeon (*Columba livia*). *J. Acoust. Soc. Am.* **63,** S75.

Raab, D., and Ades, H. (1946). Cortical and midbrain mediation of a conditional discrimination of acoustic intensities. *Am. J. Psychol.* **59,** 59–83.

Richards, D. G., and Wiley, R. H. (1980). Reverberations and amplitude fluctuations in the propagation of sound in a forest: implications for animal communication. *Am. Nat.* **115,** 381–399.

Roberts, J. P., Kacelnik, A., and Hunter, M. J., Jr. (1980). Some consequences of sound interference patterns for birdsong. *Acoust. Lett.* **3,** 141–146.

Rooke, I. J., and Knight, T. A. (1977). Alarm calls of the honeyeaters with reference to locating sources of sound. *Emu* **77,** 193–198.

Rosowski, J. J., and Saunders, J. C. (1980). Sound transmission through the avian interaural pathways. *J. Comp. Physiol. A.* **136,** 183–190.

Sachs, M. B., Woolf, N. K., and Sinnott, J. M. (1980). Response properties of neurons in the avian auditory system: Comparisons with mammalian homologues and consideration of the neural encoding of complex stimuli. *In* "Comparative Studies of Hearing in Vertebrates" (A. N. Popper and R. R. Fay, eds.), pp. 323–353. Springer-Verlag, Berlin and New York.

Saunders, J. C., Denny, R. M., and Bock, G. R. (1978). Critical bands in the Parakeet (*Melopsittacus undulatus*). *J. Comp. Physiol.* **125,** 359–365.

Schwartzkopff, J. (1968). Structure and function of the ear and the auditory brain areas in birds. *In* "Hearing Mechanisms in Vertebrates" (A. V. S. DeReuck and J. Knight, eds.), pp. 41–59. Little, Brown, Boston, Massachusetts.

Schwartzkopff, J. (1973). Mechanoreception. *In* "Avian Biology" (D. S. Farner, J. R. King, and K. C. Parkes, eds.), Vol. 3, pp. 417–477. Academic Press, New York.

Shalter, M. D. (1978). Localization of passerine seeet and mobbing calls by Goshawks and Pygmy Owls. *Z. Tierpsychol.* **46,** 260–267.

Shalter, M. D., and Schleidt, W. M. (1977). The ability of Barn Owls, *Tyto alba,* to discriminate and localize avian alarm calls. *Ibis* **119,** 22–27.

Sinnott, J. M. (1980). Species-specific coding in bird song. *J. Acoust. Soc. Am.* **68,** 494–497.

Sinnott, J. M., Sachs, M. B., and Hienz, R. D. (1976). Differential sensitivity to frequency and intensity in songbirds. *J. Acoust. Soc. Am.* **60,** S87.

Sinnott, J. M., Sachs, M. B., and Hienz, R. D. (1980). Aspects of frequency discrimination in passerine birds and pigeons. *J. Comp. Physiol. Psychol.* **94,** 401–415.

Stebbins, W. C. (1970). Studies of hearing and hearing loss in the monkey. *In* "Animal Psychophysics: The Design and Conduct of Sensory Experiments" (W. C. Stebbins, ed.), pp. 41–66. Appleton, New York.

Stevens, S. S., and Neuman, E. B. (1936). The localization of actual sources of sound. *Am. Psychol.* **48,** 297–306.

Takasaka, T., and Smith, C. A. (1971). The structure and innervation of the pigeon's basilar papilla. *J. Ultrastruct. Res.* **35,** 20–65.

Tanaka, K., and Smith, C. A. (1978). Structure of the chicken's inner ear: SEM and TEM study. *Am. J. Anat.* **153,** 251–272.

Terhune, J. M., and Ronald, K. (1975). Masked hearing thresholds of ringed seals. *J. Acoust. Soc. Am.* **58,** 515–516.

Thorpe, W. H. (1961). "Bird-Song: The Biology of Vocal Communication and Expression in Birds," Cambridge Monographs in Experimental Biology, No. 12. Cambridge Univ. Press, London and New York.

Trainer, J. E. (1946). The auditory acuity of certain birds. Ph.D. Thesis, Cornell Univ., Ithaca, New York.

Waser, P. M., and Waser, M. S. (1977). Experimental studies of primate vocalization: specializations for long-distance propagation. *Z. Tierpsychol.* **43,** 239–263.

Watson, C. S. (1963). Masking of tones by noise for the cat. *J. Acoust. Soc. Am.* **35,** 167–172.

Watson, C. S., and Gengel, R. W. (1969). Signal duration and signal frequency in relation to auditory sensitivity. *J. Acoust. Soc. Am.* **46,** 989–997.

Wier, C. C., Jesteadt, W., and Green, D. M. (1977). Frequency discrimination as a function of frequency and sensation level. *J. Acoust. Soc. Am.* **61,** 177–184.

Wiley, R. H. (1976). Communication and spatial relationships in a colony of Common Grackles. *Anim. Behav.* **24,** 570–584

Wiley, R. H., and Richards, D. G. (1978). Physical constraints on acoustic communication in the atmosphere: implications for the evolution of animal vocalizations. *Behav. Ecol. Sociobiol.* **3,** 69–94.

Wilkinson, R., and Howse, P. E. (1975). Time resolution of acoustic signals by birds. *Nature (London)* **258,** 320–321.

Yodlowski, M. L. (1980). Infrasonic sensitivity in pigeons (*Columba livia*). Ph.D. Thesis, Rockefeller Univ., New York.

Yost, W. A., Wightman, F. L., and Green, D. M. (1971). Lateralization of filtered clicks. *J. Acoust. Soc. Am.* **50,** 957–962.

Zoloth, S. R., and Green, S. M. (1979). Monkey vocalizations and human speech: Parallels in perception? *Brain Behav. Evol.* **16,** 430–442.

Zoloth, S. R., Peterson, M. R., Beecher, M. D., Green, S., Marler, P., Moody, D. B., and Stebbins, W. (1979). Species-specific perceptual processing of vocal sounds by monkeys. *Science* **204,** 870–873.

5

Adaptations for Acoustic Communication in Birds: Sound Transmission and Signal Detection

R. HAVEN WILEY
DOUGLAS G. RICHARDS

ISBN 0-12-426801-3

I. INTRODUCTION

Birdsong typically serves for communication at long range, over distances of 50–200 m. In contrast, humans use their enormous capability for acoustic communication predominantly at short range. As a consequence, engineers have devoted little attention to the problems of long-range acoustic communication, and most people have little intuitive feel for its difficulties. Nevertheless, many distortions of acoustic signals during propagation through natural environments, such as frequency-dependent attenuation, boundary interference, reverberations, and irregular amplitude fluctuations, are easily detected by ear. Just in the last decade, as a result of initiatives for noise control in human environments, our understanding of the acoustics of sound transmission in the atmosphere has developed rapidly.

This review considers adaptations in acoustic signals that can improve the efficiency of transmitting information to a receiver. For this task, we first need to consider some general features of communication, in particular the concepts of information transfer and noise. Next, this review turns to the physical processes that affect sound as it propagates through the atmosphere. Our aim is to compare the optimal acoustic signals for communication in natural circumstances with the signals actually used by birds. We focus primarily on communication at long range, but consider some adaptations for short-range communication as well. Finally, a look at the theory of signal detection in a noisy environment identifies some basic compromises that any receiver must make and some adaptations in birdsong that could improve the performance of receivers.

II. INFORMATION TRANSFER IN COMMUNICATION

In seeking adaptations of acoustic signals for efficiency of information transfer, we presuppose that communicatory signals usually evolve to increase this efficiency, the ratio of the amount of information transmitted to energy or time necessary. This premise appears to contradict a suggestion that signals often evolve to manipulate receivers while transmitting as little information as possible about the sender (Dawkins and Krebs, 1978). The confusion arises over use of the term "information." This review, like much previous ethological literature

(Altmann, 1965; Hazlett and Bossert, 1965; Dingle, 1972; Hazlett and Estabrook, 1974a,b; Wilson, 1975), uses the term "amount of information," or "information" for short, as defined in information theory by Shannon's measure of uncertainty or entropy (Shannon and Weaver, 1949; Woodward, 1953; Quastler, 1958; Raisbeck, 1963; Pfeiffer, 1965; Cherry, 1966). For a set of discrete events, the average uncertainty of the next event, H, equals

$$-\sum_{\mathrm{i}} P(\mathrm{i}) \log_2 P(\mathrm{i})$$

where the $P(\mathrm{i})$ are the probabilities of each event. For continuously variable events, measurement of H requires an integration rather than a summation over the field of events, or H can be approximated by classifying events into discrete categories. Information received from a signal is then the reduction in the uncertainty of the receiver's responses after receiving the signal. This definition presupposes a nonparticipant observer, who records received signals and responses of the receiver. The *received* or transmitted *information,* H_T, equals $H_R - H_{R|S}$ where H_R is the uncertainty of the receiver's actions without regard to the occurrence of signals and $H_{R|S}$ is the uncertainty of a receiver's responses after a signal occurs.

The necessary condition for concluding that a signal transmits information to a receiver is $H_T > 0$. This condition is an exact statement of the usual criterion for communication: a change in the receiver's behavior after receiving a signal, where a signal is any pattern of energy or matter that reaches the receiver without providing the power to effect a response directly (Cherry, 1966; Altmann, 1967; Klopfer and Hatch, 1968; Hailman, 1977; Green and Marler, 1979). The measurement of any of these uncertainties involves a number of difficulties in classifying signals and in detecting subtle, delayed, or tonic responses (Schleidt, 1973; Smith, 1977), all of which lead to underestimates of the information transmitted by a signal (see Beer, Chapter 9, Volume 2). Nevertheless, to conclude that communication has occurred, an investigator must document some change in the receiver's behavior as a consequence of a signal.

Another measure of information in communication is the amount of *broadcast information,* the information included in a signal about a sender. Broadcast information, H_B, is the reduction in the uncertainty concerning the behavioral states and external circumstances of the sender as a result of the emission of a signal: $H_B = H_0 - H_{0|S}$, where H_0 is the uncertainty concerning the sender's states or external circumstances. Here the nonparticipant observer focuses on the broadcast signal and sender, rather than the received signal and receiver. Again, measures of the amount of broadcast information in a signal require judgments about the classification of variants of signals and the sender's states or circumstances. Ethologists, however, have long recognized that signals are often associated with particular states or circumstances of senders, such as species identity, individual identity, general motivational state, specific behavioral tendencies, and even events in the external environment, such as the presence of a predator or food (Marler, 1961, 1967; Smith, 1977; Green and Marler, 1979). Smith's (1963, 1969, 1977) concept of the *message* of a signal corresponds to a subset of the broadcast information, since he restricts the message to the association between signals and the sender's behavioral states and excludes associations between signals and external circumstances.

Although $H_T > 0$ is a necessary criterion for communication, $H_B > 0$ is not. Some communicatory signals could conceivably contain no information about a sender. In this case no association would exist between the sender's behavioral states or external circumstances and the occurrence of a signal. It is difficult to imagine that natural selection could favor responding to such signals, however.

Dawkins and Krebs (1978) emphasize that senders might include "misleading information" in signals. The distinction between "misleading" and "correct information" has no relation to the amount of transmitted or broadcast information. Instead, this distinction depends on changes in the fitnesses of the sender and receiver as a consequence of the receiver's responses. Signals contain "misleading information" when they evoke responses to the advantage of the sender and disadvantage of the receiver. The sender thus manipulates the receiver by exploiting the receiver's rules for interpreting signals. The separation of manipulative communication with "misleading information" from mutualistic communication with "correct information" depends entirely on whether or not the receiver's responses benefit the receiver, the sender, or both (Fig. 1).

Regardless of changes in the receiver's or sender's fitnesses, the criterion for communication is always $H_T > 0$. Furthermore, natural selection should favor senders that tend to maximize the efficiency of information transmission whenever interactions of individuals are either mutualistic or manipulative (Fig. 1). In either case, signals should evolve to produce responses by the recipient with the minimal commitment of the sender's time and energy. Receivers, on the other hand, should tend to minimize H_T in manipulative communication. As much as possible, receivers should ignore signals with "misleading information." Instead, receivers should tend to maximize H_T for mutualistic or altruistic communication (Fig. 1). In general, the evolution of manipulative and altruistic communication poses the same problems as the evolution of cheating in any mutualistic social interaction (Dawkins, 1976).

Owing to the attenuation and degradation of signals and the mixing of signals with background noise, there always exists some equivocation in the association between the properties of received (R) and broadcast (B) signals: $H_{S(B)|S(R)} > 0$ (Shannon and Weaver, 1949; Quastler, 1958; Wilson, 1975). This conditional uncertainty is noise. It is a property of the communication channel between the sender and the receiver; in fact, it defines the channel. Like H_T and H_B, it is available only to a nonparticipant observer. Noise as a result of degradation and

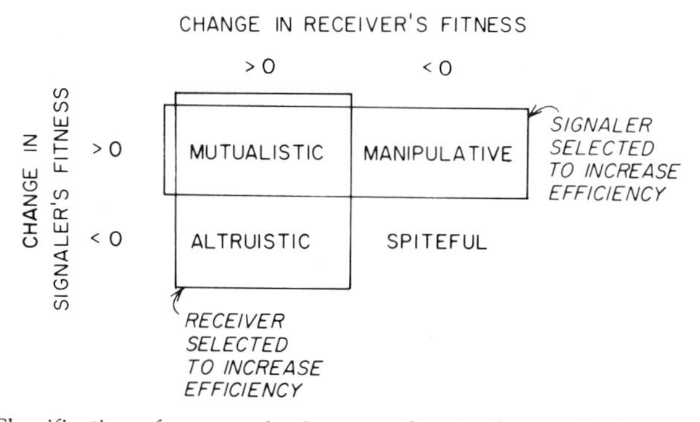

Fig. 1. Classification of communication according to changes in the sender's and the receiver's fitnesses. For senders, selection tends to increase the efficiency (see text) of mutualistic and manipulative communication, while for receivers selection tends to increase the efficiency of mutualistic and altruistic communication.

attenuation of signals and mixing with background sounds affects manipulative as well as mutualistic communication. All forms of communication must confront the difficulties of communication through a noisy channel.

Our primary concern in this chapter is the nature of noise in the channel for acoustic communication in terrestrial habitats. From a receiver's point of view, noise interferes with two tasks in the reception of signals, detection and recognition. *Detection* requires discrimination between the presence and absence of a signal, while *recognition* requires discrimination among classes and variants of signals as well as between signal and no signal. Detection and recognition are both affected by noise in the channel, by equivocation in the association between broadcast and received signals. Communication at long range strains a receiver's capabilities for both detection and recognition of signals, since attenuation of the intensity and degradation of the structure of signals increase with the distance between the sender and the receiver. Note that degradation of the structure of signals can affect a receiver's ability to discriminate among variants of signals even when detection is little problem, while attenuation always affects both detection and recognition.

III. FREQUENCY-DEPENDENT ATTENUATION

In the pioneering studies of sound transmission in natural environments in relation to properties of birdsong, one goal was to identify the optimal frequencies for long-range acoustic communication in different habitats (Morton, 1970, 1975; Chappuis, 1971). These studies showed greater attenuation of high frequencies in forests than in open habitats. Furthermore, a "sound window" existed for transmission near the ground. Frequencies around 1–2 kHz attenuated less than either lower or higher frequencies (Marten and Marler, 1977; Marten *et al.*, 1977).

To understand the attenuation of sound in the atmosphere, we need to consider four processes: spherical spread, atmospheric absorption, scattering, and boundary interference. The latter three produce frequency-dependent attenuation and thus determine the optimal frequencies for long-range acoustic communication. However, these processes should not produce any consistent differences in optimal frequencies among natural habitats. Furthermore, "sound windows" should vary in complex ways that depend on the exact locations of the sender and receiver with respect to the ground and on features of the ground and the atmosphere.

A. Spherical Spread

As a result of the spherical spread of sound energy radiating from a source, intensity decreases in proportion to $1/r^2$, where r is the distance between source and receiver, a decrease of 6 dB for a doubling of r. This spherical spread of sound energy holds only for a homogeneous medium. We shall see that scattering, vertical gradients in temperature and wind, and reflective layers such as the ground or strata of vegetation interfere with spherical spread of sound. Spherical spread thus provides a standard against which other forms of attenuation are compared. Studies of sound propagation in the atmosphere concentrate on excess attenuation, attenuation in excess of the 6 dB per doubling of distance from spherical spread of sound energy. Some authors also subtract attenuation from atmospheric absorption in calculating excess attenuation.

Sound does not attenuate with $1/r^2$ in the immediate vicinity of a sound source, even in a homogeneous medium. In the near field of a source, intensity goes through a series of maxima and nulls with increasing distance from the source. Only when $r > 2a^2/\lambda$, where a = the radius of a cylindrical piston source (approximately similar to a circular cone loudspeaker) and λ = wavelength, does intensity decrease inversely with the square of the distance. Thus the inverse square law of spherical spreading can be used to extrapolate sound intensity back to approximately $r = 2a^2/\lambda$ but not any closer to the source (Kinsler and Frey, 1962, pp. 175–177; Gaunaurd and Überall, 1978). For a small animal this distance is only a few centimeters, but for many speakers it is about one meter.

B. Atmospheric Absorption

The absorption of sound energy by the atmosphere increases with frequency as a result of two processes (see Kinsler and Frey, 1962; Evans *et al.*, 1971; Beranek, 1971a). Energy dissipated as heat owing to the viscosity of the atmosphere (the classical component of atmospheric absorption) is proportional to absolute temperature and the square of frequency. Energy dissipated in rotational and vibrational relaxation of oxygen molecules in the presence of water molecules (the molecular component) is strongly influenced by the percentage of water vapor in the atmosphere as well as by frequency. Harris (1966) presents detailed measurements of the absorption of sound in air as functions of temperature and relative humidity for frequencies between 125 Hz and 4 kHz, and Pöhlmann (1961) presents values for higher frequencies (see Griffin, 1971).

For any value of temperature and relative humidity, atmospheric absorption increases monotonically with frequency. For frequencies between 100 Hz and 10 kHz, except in very cold or dry air, absorption is proportional to frequency raised to a power between 1 and 2. In natural environments, measurements of sound attenuation invariably reveal greater attenuation of higher frequencies. At least some of this effect results from greater atmospheric absorption of higher frequencies. For instance, in Marten and Marler's (1977) measurements of excess attenuation over open fields at a height of 10 m, attenuation increased with frequency approximately as expected from Harris' data. Attenuation closer to the ground and in forested habitats is generally higher than expected for absorption alone, although increased absorption of higher frequencies affects the attenuation of sound in all environments.

Attenuation by absorption decreases as humidity increases, at least for conditions that apply to birdsong (frequencies below 8 kHz and percentage water vapor greater than 0.5%). In practice, this effect is small. A change in relative humidity from 50 to 90% reduces attenuation by absorption by less than 1 dB/100 m for frequencies below 2 kHz and temperatures greater than 5°C. For a frequency of 4 kHz, the effects of humidity are more appreciable in the range of temperature from 0° to 15°C, conditions that apply to songbirds in early spring. However, humidity and temperature have counteracting influences on attenuation by absorption. Since humidity often drops as temperature rises during a morning, attenuation by atmospheric absorption of sound might not vary in a regular pattern during a diurnal cycle.

C. Scattering

Scattering refers to the multiple reflection, diffraction, and refraction of sound by objects or heterogeneities in the atmosphere. It depends on the wavelength of sound in relation to the dimensions of objects or heterogeneities. As a consequence, scattering in natural environments is strongly dependent on frequency.

In acoustics the actual wave in a scattering medium is treated mathematically as if it were the result of a scattered wave interfering with the incident wave. The scattered wave is thus the difference between the actual wave and the undisturbed wave that would occur if the object were absent, and the total power of the scattered wave is the power removed from the incident wave by scattering from the object (Morse and Ingard, 1968, Chapter 8). The scattered wave normally has two components, a reflected portion consisting of energy radiating in all directions from the object, and an interfering portion producing a shadow along the line of propagation of the incident wave behind the object. When the object is large in relation to the wavelength of the incident sound, the interfering and reflected components are equal. In other words, the energy reflected from the object equals the energy removed from the shadow behind the object. For wavelengths that approximate the radius of the object (within a factor of two or three) the spatial distribution of the scattering wave becomes extremely complicated with lobes extending in many directions. When the wavelength of the incident sound is large in relation to the object, the total scattered power is small, and no distinct shadow is formed behind a relatively rigid object. Most calculations of scattered waves assume an incident plane wave rather than a spherical wave. However, for a small source, propagating sound approaches the conditions of a plane wave, at least for relatively small areas of the wave front and at long distances from the source.

Near a small source, such as an animal or a loudspeaker, sound often radiates in a beam. Under these conditions one effect of scattering is to reduce sound intensity along the axis of the beam by deflecting energy outward; in effect, the angular spread of the beam increases progressively. Since the ratio of circumference to cross-sectional area decreases with distance from the source, a beam of sound in a scattering environment should attenuate more over a given distance near the source, where a greater proportion of the scattered sound leaves the beam, than at a great distance (Givens *et al.*, 1946; Schilling *et al.*, 1947). As a consequence, description of excess attenuation in scattering environments requires measurements at several distances from the source.

To evaluate sound attenuation in natural environments as a result of scattering, we need to consider scattering from two sorts of heterogeneities, (1) vegetation and (2) variations in the velocity and temperature of the air as a result of turbulence.

1. Scattering from Vegetation

Vegetation presents surfaces that are approximately cylindrical or spherical, since roughly circular objects like leaves in random orientations have average effects similar to those of spheres. Scattering of sound by a single cylinder or sphere and by an array or cloud of cylinders or spheres has received considerable attention in the acoustic literature (Morse and Ingard, 1968, Chapter 8; Johnson, 1977; Embleton, 1966). In all of these cases, scattering is a function of the frequency of the incident sound, but scattering can either increase or decrease with frequency depending on the surface impedance of the objects. Acoustical impedance is analogous to electrical impedance in circuits with alternating current. It includes resistance and reactance that impede the propagation of sound.

Embleton has calculated the attenuation of a plane wave by an array of cylinders to illustrate the dependence of scattering on surface impedance. For rigid cylinders (impedance approaching infinity), attenuation by scattering increases as a sigmoid function of frequency. The inflection point occurs at the frequency corresponding to a ratio $d/\lambda = \pi$, where d is the diameter of the cylinders and λ is the wavelength. In contrast, soft cylinders (impedance approaches 0) scatter less sound as frequency increases. At very high frequencies ($d/\lambda > 10\pi$), regardless of the surface impedance of the cylinders, scattered power converges on a value of 8.8 dB(Na)/100 m, where N is the number of cylinders per unit area and a is the mean radius of the cylinders. In other words, the asymptotic attenuation at high frequencies is proportional to the density of objects times their average radius. Thus for approximately rigid objects, like the trunks and large limbs of trees, attenuation of sound by scattering depends primarily on the density of objects with radii substantially larger than 1λ. As frequency increases (shorter wavelengths), smaller objects become effective scatterers of sound. The density of effective scatterers thus increases with frequency in forests.

The only attempts to compare theoretical expectations for attenuation from scattering and actual measurements of excess attenuation in natural environments are Aylor's (1971) studies in hemlock and pine forests. For 8-kHz tones, he measured attenuation of 28 dB through 69 m of hemlock and 18.5 dB through 92 m of pine. These values agree nicely with Aylor's calculated values of 25 and 16 dB, respectively. However, his approximate expression for attenuation of high frequencies as a result of scattering from an array of rigid cylinders yields results somewhat different from those of Embleton's procedure, discussed above. The latter, including the corrections for the asymptotic attenuation at the high-frequency limit, yields expected values of about 15 dB for both of Aylor's measurements. Unlike Embleton's calculations, in which attenuation of sound by scattering is directly proportional to Na, at least for values such as those that apply to natural forests, Aylor's calculations do not yield values of attenuation directly proportional to Na (see also Morse and Ingard, 1968, p. 441). The resolution of these discrepancies must await further research. However, the available information suggests that we can expect attenuation by scattering in forests to reach at least 10 dB/100 m for frequencies above 1 kHz.

The effects of foliage on attenuation by scattering are clear in Aylor's (1971) measurements of excess attenuation in fields of corn with different densities of plants. Attenuation at 1 kHz increases less steeply as foliage density increases compared to attenuation at 4 kHz. Attenuation of sound by dense second-growth brush before and after the leaves had dropped in autumn showed similar effects. At 10 m above ground in a deciduous forest, Marten and Marler (1977) found that foliage increased attenuation of frequencies between 2 and 11 kHz by about 10 dB/100 m (Fig. 2). In this case, foliage had approximately the same effect on attenuation for all frequencies above 2 kHz. Deciduous forests have rather uniform leaf sizes, roughly 3–5 cm in radius. Frequencies above 2–3 kHz would

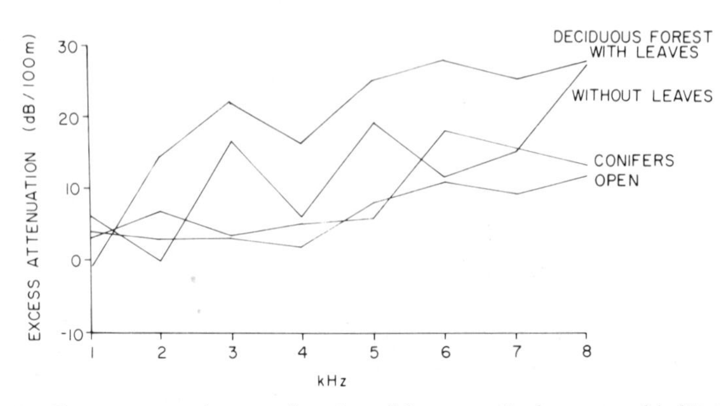

Fig. 2. Excess attenuation as a function of frequency in four natural habitats over a distance of 100 m at a height of 10 m above ground (after Marten and Marler, 1977): a deciduous forest with and without leaves; a coniferous forest; and an open field with little wind. Frequencies between 1 and 8 kHz are those used most for long-range communication by passerine birds.

thus have uniform attenuation by scattering. In coniferous forests at 10 m above ground, attenuation was lower than in a deciduous forest with leaves (Marten and Marler, 1977). Evidently conifer needles scatter little sound at frequencies below 10 kHz. Thus trunks, large limbs, and the broad leaves of dicotyledonous trees are important in attenuating sound above 1–2 kHz, well within the frequency range used by singing birds.

2. *Scattering from Turbulence*

Attenuation of sound by scattering from atmospheric turbulence results from the refraction of sound by cells of air that differ in temperature or velocity from the surrounding medium. Interference between the scattered and undisturbed waves can attenuate sound in the direction of propagation. Just as for scattering from solid objects, scattering from atmospheric heterogeneities in turbulent air becomes more pronounced at higher frequencies and thus shorter wavelengths in relation to the scale of the atmospheric heterogeneity. In addition, at higher frequencies the scattered wave is more narrowly concentrated in the forward direction, where it produces interference with the undisturbed wave, just as in scattering from solid objects.

The effects of turbulence in attenuating sound by scattering have not received experimental study. Nevertheless, theory suggests that it is worth serious consideration. For frequencies corresponding to wavelengths less than the scale of the turbulence, scattered power from a unit volume increases with the square of frequency (Lighthill, 1953):

$$\text{Attenuation} = 2k^2LM^2$$

where $k = 2\pi f/c$, L is the scale of the turbulence, and M is the Mach number of the fluctuations (RMS velocity in the direction of propagation divided by c, the velocity of sound). With $L = 1$ m and $v = 1$ m/sec, attenuation of a 1-kHz tone would amount to 2.6 dB/100 m. On the other hand, at 10 kHz attenuation would reach 3.8 dB/m, more than 100 times greater than at 1 kHz. Even moderate atmospheric turbulence should result in severe attenuation of the higher frequencies used by birds for singing. Scattering from atmospheric turbulence should thus have much the same effect on the attenuation of sound as scattering from vegetation and atmospheric absorption. All three processes in general produce increasing attenuation with increasing frequency, particularly in the range between 1 and 10 kHz. Attenuation by turbulence and vegetation should often exceed attenuation by absorption.

Atmospheric turbulence can result from wind shear over irregular surfaces and also from temperature gradients near the surface of the ground or the top of a forest canopy (Munn, 1966; Stringer, 1972, Chapter 3). Temperature gradients that exceed the adiabatic lapse rate for air (about −10°C/km) result in cells of warmer air rising from the surface and expanding. Over irregular surfaces wind shear produces vortices and thus fluctuations in wind velocity. Although any wind in a forest is likely to be highly turbulent, wind velocity is normally lower inside forests than in open areas. Thus heterogeneity in the atmosphere caused by turbulence primarily affects open habitats, where the effect is usually strongest during midday when the sun heats the surface. We can expect that on calm days open habitats attenuate sounds above 1–2 kHz less than forested habitats. Under windy conditions, scattering of sound should become much more comparable in open and forested habitats.

D. Boundary Interference

A reflecting boundary near the path of transmission affects sound in two ways: (1) as a result of interference between the direct wave from the sender to the receiver and the reflected wave from the boundary; and (2) as a result of additional waves propagating in and near the ground. In natural environments, this situation frequently arises for acoustic communication between individuals near the ground. Like all interference phenomena, this effect depends strongly on frequency. Thus it is the third major source of attenuation that varies with frequency.

1. *Reflected Waves*

The nature of interference between the direct and reflected waves depends on three factors: the difference in path length; the acoustical impedance of the surface (Embleton *et al.*, 1976; Donato, 1976; Chessell, 1977; Thomasson, 1977); and irregularities in the medium or the surface which affect the coherence of the direct and reflected waves (Ingard and Maling, 1963; Daigle *et al.*, 1978; Daigle, 1979).

The importance of the difference in path length is easily appreciated. For a perfectly homogeneous medium and a smooth rigid surface, the difference in the lengths of the direct and reflected paths in relation to the wavelength of the propagating sound completely determines the nature of interference at the receiver. When the difference in path lengths equals one-half wavelength (or an odd multiple of $\lambda/2$), destructive interference at the receiver would reach a maximum and intensity at the receiver would approach 0. When the path length difference equals one wave length (or an integer multiple of λ), constructive interference would double the intensity at the receiver.

The difference in path length, ΔR, between a sender and a receiver at the same height above ground equals

$$\sqrt{4h^2 + r^2} - r$$

where h is the height and r is the horizontal distance between sender and receiver. As height increases the difference in path length increases approximately proportionately. Thus for transmission over a given distance, the lowest frequency for maximal destructive interference decreases as height increases. As range increases from 0 to very large distances, the path length difference decreases from $2h$ to nearly 0. The lowest frequency for maximal destructive interference also increases, although very slowly when r is large in comparison to h.

For several reasons, these simple calculations of differences in path length do not permit an exact prediction of the frequencies for maximal destructive interference in natural habitats. First, reflection from a surface of finite impedance, like the ground, introduces a phase change in the reflected wave which shifts the frequencies for maximal interference. Second, measurements under field conditions usually do not reveal evidence of destructive interference at frequencies above 1 or 2 kHz (Marten and Marler, 1977; Marten *et al.*, 1977). This lack of clear interference patterns for higher frequencies results from irregularities in the medium and the surface, which scatter the propagating waves and thus destroy their coherence. Third, propagation of sound with angles of reflection close to grazing involves two additional modes of propagation, a ground wave in the ground itself and a surface wave coupled to the ground but in the air immediately above the surface. To understand the effects of boundary interference, we must consider these three features in turn (see Rudnick, 1947b; Ingard, 1951, 1953, 1969; Embleton *et al.*, 1976; Donato, 1976; Thomasson, 1977; Piercy *et al.*, 1977).

Comparisons of sound attenuation at 1.2 m above asphalt and closely mown grass show how a phase shift in the reflected wave can change the frequencies for maximal destructive interference (Embleton *et al.*, 1976). The grass introduces a large phase change in the reflected wave at frequencies above about 1 kHz. In

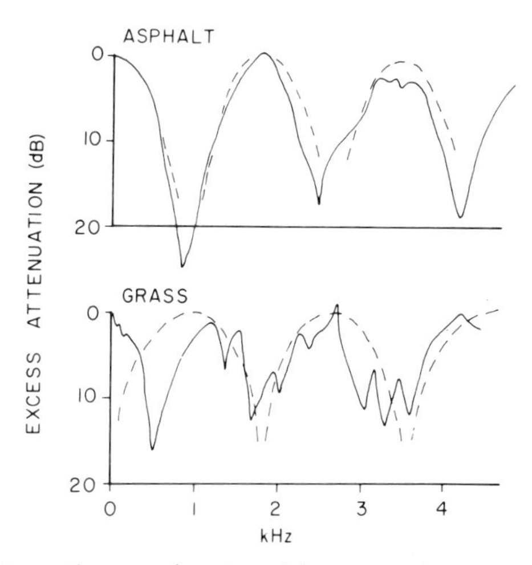

Fig. 3. Excess attenuation as a function of frequency above two surfaces over a distance of 15 m at a height of 1.2 m above ground (solid lines; after Embleton *et al.*, 1976). Dashed lines represent waves with no phase change on reflection (asphalt) or 180° phase change on reflection (grass).

contrast, the asphalt surface has an impedance large enough that phase changes on reflection are negligible. In fact, the phase change on reflection by the grass approaches 180° to judge from Embleton *et al.* (1976). The lowest frequency for destructive interference shifts from about 900 Hz for propagation over asphalt ($r = 15$ m, $h = 1.2$ m) to approximately 1800 Hz for propagation over grass (Fig. 3).

Note, however, that grass also attenuates frequencies around 500 Hz (Fig. 3). Below 1 kHz, the acoustic impedance of mown grass increases sharply. The impedance varies approximately inversely with the square root of frequency (Embleton *et al.*, 1976) and increases fivefold between 1 kHz and 100 Hz. This dependence of acoustic impedance on frequency introduces a major peak of attenuation at about 500 Hz, for a difference in path length (ΔR) about equal to $\frac{1}{4}\lambda$ (Fig. 3). A similar peak of attenuation at 300–800 Hz often appears in studies of sound propagation in natural habitats (Morton, 1975; Aylor, 1971; Marten and Marler, 1977; Marten *et al.*, 1977; Waser and Waser, 1977). The location of this peak is primarily determined by the frequency-dependent acoustic impedance of the ground below 1 kHz and the consequent frequency-dependent phase change on reflection. Changes in ΔR with changes in height or range have little effect on this attenuation.

We can see how this frequency-dependent change in impedance produces attenuation below 1 kHz as follows (Fig. 4). Recall that the surface changes from low to high impedance as frequency drops from 1 kHz toward 0. A low impedance surface, inducing a large phase change on reflection, produces *destructive*

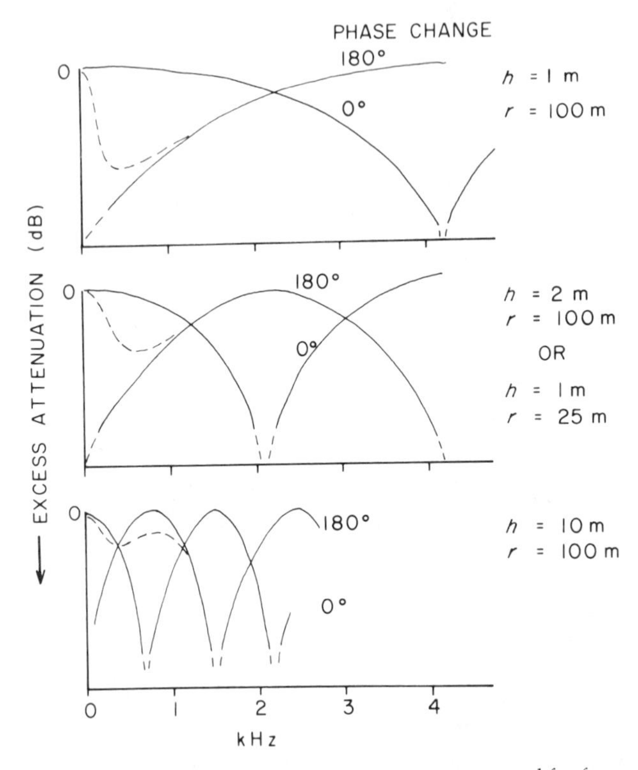

Fig. 4. Schematic representation of excess attenuation expected for frequencies below 1 kHz, when the acoustic impedance of the ground varies inversely with frequency below 1 kHz. For each combination of height (h) and horizontal range (r), two solid lines show the attenuation expected from interference between direct and reflected waves with 0° and 180° phase changes on reflection. The dashed line in each case indicates the attenuation expected when the phase change increases from 0° to 180° as frequency decreases from 1 to 0 kHz. Note that the peak in excess attenuation becomes less pronounced as height increases or range decreases.

interference for $\Delta R/\lambda$ near 0. In contrast, a surface of high impedance, inducing little phase change on reflection, produces *constructive* interference for $\Delta R/\lambda$ near 0. Thus as frequency decreases below 1 kHz (wavelength increases), attenuation passes through a maximum value and then decreases as frequency approaches 0 (Fig. 4). The maximal attenuation reached depends on the height of propagation above ground and the range, but the location of this peak changes little (see Piercy *et al.*, 1977, for more details).

In measurements of sound attenuation in natural environments, little evidence of interference appears for frequencies above 2 kHz (Marten and Marler, 1977; Morton, 1975; Waser and Waser, 1977). In the experiments by Embleton *et al.*

(1976) the interference maxima are irregular at higher frequencies. Also the maxima for destructive interference do not reach infinite attenuation as predicted from coherent interference between direct and reflected waves. This smoothing of the interference peaks particularly at higher frequencies results from turbulence in the atmosphere and irregularities of the surface which reduce the coherence of the propagating waves (Daigle *et al.*, 1978; Daigle, 1979). Recall that turbulence affects higher frequencies and correspondingly smaller wavelengths more than it does lower frequencies. Thus at long range under even mild conditions of turbulence and irregularity in the environment, the interference peaks at high frequencies largely disappear.

2. *Ground and Surface Waves*

When the wave reflected from a boundary travels at near grazing incidence, propagation of sound includes some additional properties (Embleton *et al.*, 1976; Piercy *et al.*, 1977; Chessell, 1977). Reflection at grazing incidence occurs when the height of the sender and receiver is small in relation to the range of propagation, so that the same considerations apply to insects or other small animals communicating over distances of a few meters just above the surface of the ground (Michelsen, 1978) and to larger animals communicating over longer distances at a height of several decimeters or even a meter.

For surfaces with finite impedance, such as grass or soil, reflection at grazing incidence involves essentially a 180° phase change. The reflection coefficient for a locally reacting surface equals

$$\frac{\sin\theta - \rho c/Z}{\sin\theta + \rho c/Z}$$

where ρc is the characteristic acoustic impedance of the air, Z is the normal specific impedance of the ground, and θ is the angle of incidence of the reflected wave. As θ becomes small, $\sin\theta$ becomes small in relation to $\rho c/Z$ and the reflection coefficient approaches -1. The consequence, in this first approximation of interference between direct and reflected waves at grazing incidence, is essentially complete destructive interference between direct and reflected waves; essentially no sound would propagate between the sender and the receiver. Actual measurements show that this first approximation is far from accurate. In fact, in this situation low frequencies propagate with little excess attenuation, while frequencies above some cutoff attenuate rapidly; the effect is rather like communication through a low-pass filter.

The propagation of low frequencies in spite of near cancellation between the direct and reflected waves results from two additional terms in the wave equations, usually termed a ground wave and a surface wave. These two terms describe the interaction of a curved wave front with a surface of finite impedance. For a spherically spreading wave, the ground wave and surface waves become negligible as the range of propagation becomes large. For transmission at grazing incidence above real surfaces for moderate distances, the ground wave in particular results in propagation of low frequencies. Under these conditions, the cutoff frequency above which sound is severely attenuated decreases as range increases. In practice, long-range propagation of sound close to the ground is effectively limited to frequencies below 1 kHz.

For a source near the ground, as the receiver moves higher above the ground, progressively more sound above the cutoff frequency reaches the receiver (Embleton *et al.*, 1976). If both source and receiver move upward away from the ground, the spectrum of received sound transforms into the pattern of interference peaks, discussed above.

3. *Applications*

In summary, boundary interference has complex effects on sound. Attenuation results from interference between direct and reflected waves from the sender to the receiver, which in turn depend on the height, range, surface impedance, and turbulence in the medium. In natural environments, destructive interference primarily affects a band of low frequencies, approximately 0.5–1 kHz, especially for propagation about 1 m above ground and ranges on the order of 100 m. Attenuation in this band of frequencies results primarily from the sharp increase in the acoustic impedance of the ground as frequency decreases below 1 kHz. Attenuation by destructive interference between direct and reflected waves is much less pronounced at frequencies above 1–2 kHz, since scattering of sound destroys the coherence of direct and reflected waves. In addition, for very low height/range ratios sound propagation is effectively restricted to low frequencies, with the cutoff frequency increasing as the h/r ratio increases.

The acoustic impedance of the ground, critical for understanding boundary interference in natural environments, has not received extensive study. Technical difficulties make precise measurements difficult. Apparently the surface impedance of closely mown grass and that of forest floor do not differ greatly (Piercy *et al.*, 1977). The porosity of the soil as a result of root action seems to determine the acoustic impedance (Aylor, 1971; Piercy *et al.*, 1977). Other surfaces, snow, water, and sand for instance, might differ more. In addition, a surface covered with relatively deep vegetation, such as an old field or a forest canopy, would act like an absorbing baffle and reflect little sound, at least at frequencies above 0.5–1 kHz. It seems possible that birds singing from shrubs projecting from an old field might encounter little boundary interference. In fact, in some measurements of sound propagation above tall grass, there is little evidence of boundary interference (Morton, 1975; Marten and Marler, 1977).

Interference between direct and reflected waves is unlikely to have much effect on frequencies above 1–2 kHz at moderate or long distances from the source. The coherence of these waves is lost during transmission as a result of scattering. The suggestion that such interference could create pronounced patterns of frequency-dependent attenuation at distances of 50–200 m from the source (Roberts *et al.*, 1979) needs further verification.

Most measurements of attenuation in natural environments are not easily compared with predictions concerning the effects of boundary interference. It has become standard procedure to use a microphone close to the speaker to determine the output of the speaker. The response from a more distant microphone is then compared to the response from this reference microphone. However, it is clear that both microphones are affected by interference between direct and reflected waves. Owing to the differences in the distances of the two microphones from the speaker, interference differs at the two microphones. To avoid this problem, Embleton *et al.* (1976) used a single microphone and determined the frequency response of their speaker independently.

Furthermore, attenuation by boundary interference, like attenuation of a beam of sound by scattering and attenuation by spherical spread, is not a linear function of the distance between source and receiver. A determination of excess attenuation at a single distance from the source does not permit conclusions about excess attenuation at other distances, whenever scattering or boundary effects are appreciable. In these circumstances, it is necessary to determine excess attenuation at two or more distances from the source in order to extrapolate attenuation at other distances (Michelsen, 1978; Wiley and Richards, 1978).

E. Differences between Habitats

The preceding review leads to no clear prediction for differences among natural habitats in the optimal frequency for long-range acoustic communication. Consider the possibilities for such differences from each of the sources of attenuation. Atmospheric absorption results in increasing attenuation as frequency increases, regardless of habitat. Scattering also increases attenuation as frequency increases. In forests, scattering from vegetation has the greatest effect, while in open habitats, scattering from atmospheric turbulence predominates. Thus attenuation from scattering is probably lower in open habitats under conditions of low atmospheric turbulence, such as at night or in early morning with little wind. The absolute level of attenuation from absorption or scattering could differ between open and forested habitats, depending on the humidity and turbulence of the atmosphere, but both processes have a similar dependence on frequency under all conditions. To reduce attenuation by absorption and scattering, it is always best to use the lowest frequencies possible.

Boundary interference has more complicated effects. It primarily influences possibilities for acoustic communication within 1–2 m above ground. Again, these effects should not differ with habitat in any systematic way. For sound transmission over deep vegetation, such as old fields or marshes, boundary interference would decrease as a result of absorption of the reflected wave by the vegetation. In this case there is no restriction on the use of the 0.5- to 1-kHz band close to the surface. To summarize, lower frequencies attenuate less in all habitats, with the exception that sound propagation near the ground is subject to destructive interference. For maximum efficiency, long-range acoustic communication in any habitat should employ the lowest frequencies possible.

In spite of these considerations, several studies have reported that the average frequencies of songs differ in forested and open habitats. In particular, birds in tropical forests average lower frequencies in their long-range songs than do birds of open habitats (Morton, 1970, 1975; Chappuis, 1971). On the other hand, in a survey of birds in open habitats and forests in North Carolina, we could find no differences in the frequencies of songs (Richards and Wiley, 1980). We thus confront two problems: no clear reason why birds in forests and open habitats should emphasize different frequencies; and a possible difference between tropical and temperate habitats.

These problems are perhaps resolved in part by one striking feature of birds in tropical forests: some species habitually sing very close to or actually on the ground. Singing on or near the ground rarely occurs in temperate forests, as indicated by our information for a North Carolina deciduous forest (Fig. 5). Even those species that feed on the ground rarely sing any lower than 3 m. The Wood Thrush (*Hylocichla mustelina*) and Ovenbird (*Seiurus aurocapillus*), which feed almost exclusively on the ground, usually sing about 10–20 m above ground. This avoidance of long-range communication close to the ground reduces the effects of boundary interference.

In contrast, in tropical forests some species sing while actually walking on the ground. In neotropical forests, species in several genera of the family Formicariidae (antbirds) regularly sing either on or within 1 m of the ground (particularly genera like *Formicarius* and *Grallaria;* Slud, 1960; Ridgely, 1976). These songs are long-range signals, as individuals of these species are often widely separated in the forest. They have the striking features that we associate with birdsong in neotropical forests: steady tones of unusually low frequencies for a passerine (1–2 kHz). Because propagation of sound at grazing incidence with respect to the ground is subject to stringent low-pass filtering, frequencies above 1–2 kHz are effectively useless. It is not possible to predict the exact cutoff frequencies for these antbirds without further study. Simply standing on a log would make a substantial difference in the attenuation of an antbird song.

A number of non-passerine birds produce remarkably low frequency sounds for long-range communication from positions near the ground. The booming calls of many grouse (Tetraonidae) are prime examples, often clearly audible more than a kilometer away (Greenewalt, 1968; Hjorth, 1970; Wiley, 1973). Ground-dwelling curassows (Cracidae), a neotropical gallinaceous group, also produce sounds of extremely low frequency (100–200 Hz) (Delacour and

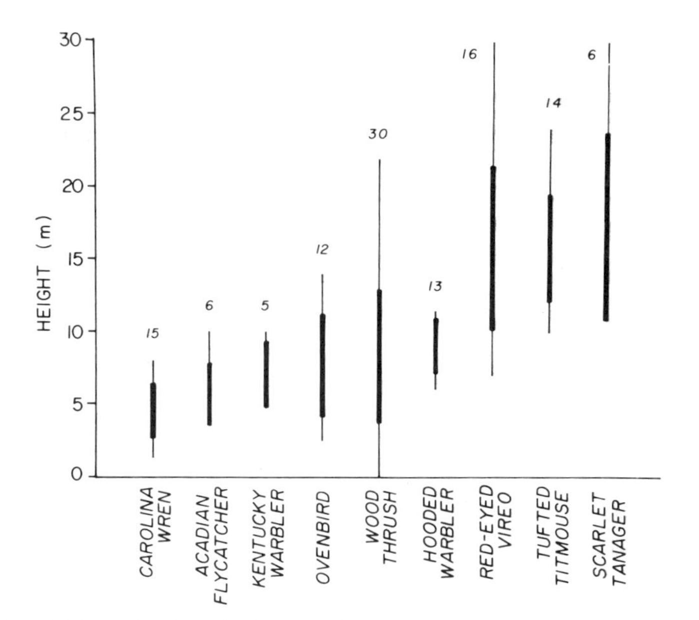

Fig. 5. Heights of undisturbed singing birds in a deciduous forest in North Carolina during May. For each species, the plot shows the mean, 1 SD on either side of the mean, and the range of observed heights. Above each bar is the sample size (number of separate perches on which uninterrupted singing continued for at least 30 sec); samples included two to eight individuals of each species. Wood Thrushes and Ovenbirds feed on the ground; Carolina Wrens feed primarily within 2 m of the ground. The remaining species forage at approximately the same heights at which they sing. Species listed are *Thryothorus ludovicianus, Empidonax virescens, Oporornis formosus, Seiurus aurocapillus, Hylocichla mustelia, Wilsonia citrina, Vireo olivaceus, Parus bicolor,* and *Piranga olivacea,* respectively.

Amadon, 1973). The Ruffed Grouse (*Bonasa umbellus*) uses its wings to produce mechanical sounds of very low frequency (Hjorth, 1970). All of these apparently bizarre acoustic signals seem ideally adapted for long-range communication at grazing incidence with the ground.

Aside from interference near the ground, patterns of attenuation in all habitats favor low frequencies for long-range communication. Many owls, doves, and cuckoos employ frequencies below 1 kHz for long-range acoustic signals (Greenewalt, 1968). Why, then, do passerines rely on frequencies above 1 kHz for long-range communication (Fig. 6)? The answer probably lies partly in the small sizes of most songbirds. Sound sources with small dimensions, regardless of the exact method of sound production, cannot efficiently produce low-frequency sounds (Kinsler and Frey, 1962). In general, the frequencies emphasized in loud vocalizations vary inversely with a bird's size (Konishi, 1970a; Bowman, 1979).

One technique for increasing the efficiency of a sound radiator at low frequencies is to couple it with an exponential horn (Kinsler and Frey, 1962). At low frequencies, a horn serves to match the radiation impedance to the relatively large impedance of the large volume of air outside. In addition, a resonating air sac, crop, esophagus, gular pouch, or tracheal sac could increase the radiation efficiency of low-frequency sounds. However, most passerines have no anatomical structures large enough to serve as effective horns for frequencies below 5–10 kHz nor large enough to resonate at frequencies below about 2 kHz. The tropical antbirds that sing on the ground are undoubtedly under the strongest selection to lower the frequencies of their songs; perhaps they have gone as far in this direction as a small bird can. The grouse and curassows already noted for the extremely low frequencies in their long-range signals could use resonating esophageal sacs of trachea for producing these sounds, although acoustical studies have not established this point. Woodpeckers, even small ones, can use hollow trees for this purpose.

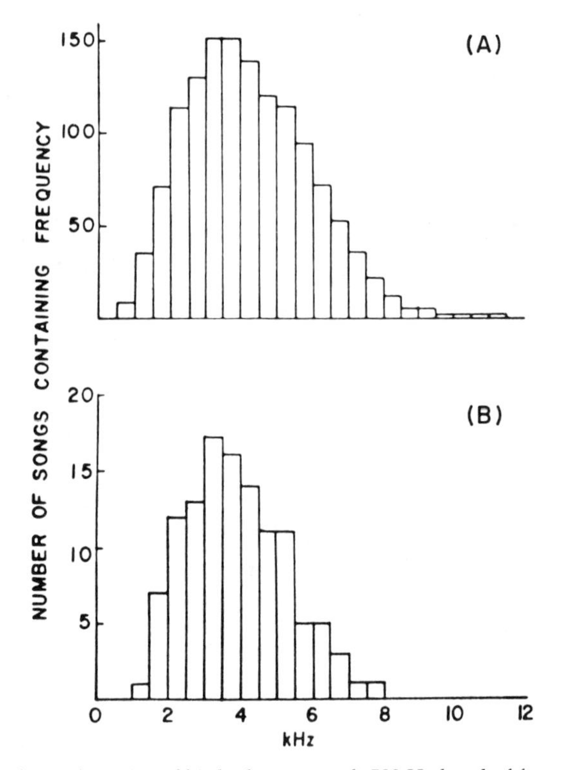

Fig. 6. Numbers of species of birds that use each 500 Hz band of frequencies in their songs. (A) All North American species; (B) species that breed in a deciduous forest in North Carolina. (Both based on measurements of maximum and minimum frequencies on the sonagrams in Robbins *et al.*, 1966.)

An additional limitation on the use of low-frequency acoustic signals for long-range communication is the threshold for hearing, which in passerines rises steeply below 2 or 3 kHz (Konishi, 1970b, 1973b; Dooling *et al.*, 1971, 1978; Dooling, Chapter 4, this volume). The frequencies in the long-range songs of passerines (Fig. 6) are roughly the mirror image of hearing thresholds for passerines.

At present, we are left with questions. Do grouse and curassows have unusual thresholds for hearing at low frequencies? Or does the reduced attenuation of low frequencies compensate for the higher thresholds of hearing? What limits the evolution of greater sensitivity to low frequencies in those passerines that use long-range acoustic communication? Or do the constraints on radiation of low-frequency sounds alone explain the reliance on frequencies above 1 kHz?

Although the best frequencies for long-range propagation of sound probably do not differ much among natural habitats, there still might be differences in the bands of acceptable frequencies. In open habitats, at least during periods of calm weather, frequencies above 3–4 kHz attenuate less than in forests, so that in the open these frequencies might prove acceptable, even though not optimal. In forests, scattering from foliage would always attenuate these frequencies regardless of weather. Perhaps a difference in the acceptable bandwidth for long-range songs explains why songs of Great Tits (*Parus major*) average lower frequencies in forests than in open woodland (Hunter and Krebs, 1979). In any habitat, the upper limit of the band of acceptable frequencies is also constrained by auditory thresholds. Since auditory thresholds of passerines rise 25 dB between 4 and 8 kHz (Konishi, 1970b; Dooling *et al.*, 1971, 1978), frequencies much higher than 4 kHz would not permit long-range communication in any case. Bowman (1979) presents some suggestive examples of Darwin's finches (Emberizidae) with bandwidths of song that match the patterns of attenuation in natural habitats. It is not yet clear how general these patterns are, nor how they relate to differences in vegetation.

In studies of attenuation in natural environments, measurements are often obtained for only one location within each habitat and even then not replicated. Furthermore, most such studies have intentionally measured attenuation in open habitats at times when atmospheric turbulence is slight, such as early in the morning in the absence of wind. Birds of open habitats as well as forests sing more at such times than later in the day or during windy weather, but nevertheless long-range communication in open habitats must often contend with turbulence, which as we have seen would attenuate higher frequencies more than lower ones. To compare attenuation of sound in open and forested habitats definitively, we need repeated measurements within each habitat, at different locations, heights above ground, distances, times of day, and weather conditions.

IV. DEGRADATION OF ACOUSTIC SIGNALS

So far we have focused on attenuation, the progressive reduction in the intensity of sound in the course of propagation through natural environments. In most cases, communication requires more than simple detection by the receiver of the presence or absence of sound of a given frequency. The receiver must discriminate among signals with different acoustic structures, in other words different patterns of frequency and intensity in time. Consequently, degradation of these patterns during the propagation of a signal is another major limitation on long-range communication, in addition to any attenuation of the signal.

If we turn to two processes that degrade the structure of acoustic signals, a clear difference between habitats emerges. Communication through dense vegetation, for instance, between birds in a forest, must contend with strong reverberations, while communication through a turbulent atmosphere, as between birds in an open field at midday, must contend with strong fluctuations in the amplitude of signals. Reverberations and irregular amplitude fluctuations have different consequences for communication.

A. Reverberation

Reverberations, which result from multiple reflections and scattering of sound during propagation, are often apparent in sonagrams of sounds recorded in a forest (Fig. 7). When a source produces a square pulse of sound at a given frequency, a receiver some distance away in a forest detects a pulse that rises to its maximal amplitude and then decays gradually. In environments with many reflecting surfaces, like a forest, the decay of received energy results from sound

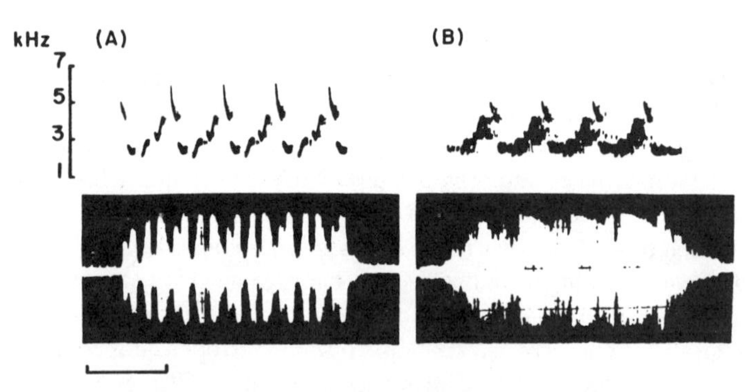

Fig. 7. Songs of Carolina Wrens (*Thryothorus ludovicianus*) recorded 10 m (A) and 50 m (B) from the singing bird to show the effects of reverberation. Upper, sonagram; lower, oscillogram. Time mark, 1.0 sec. Note that the basic pattern of frequency is preserved at a distance of 50 m although the pattern of amplitude is completely lost.

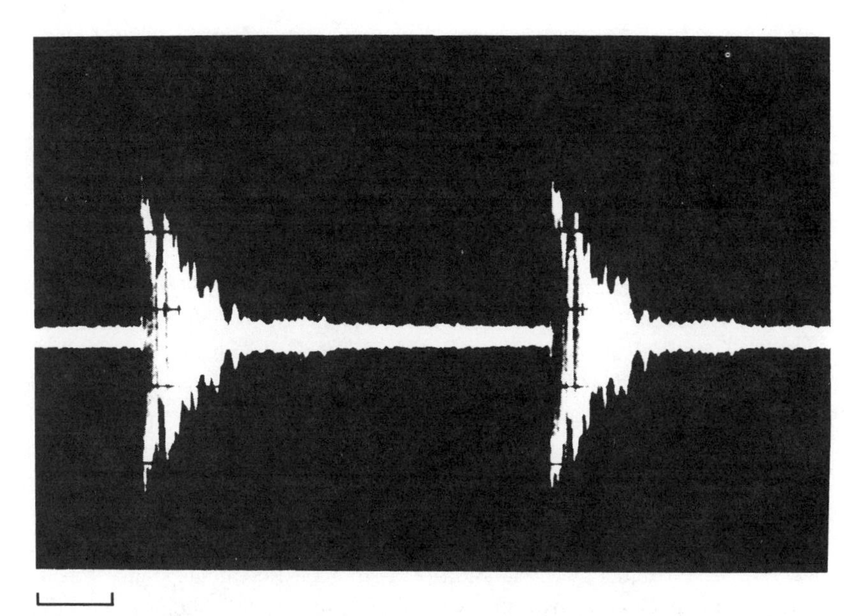

Fig. 8. Reverberations of two 25-msec square pulses of 8 kHz sound recorded 25 m from a speaker in a deciduous forest. Time mark, 100 msec. (After Richards and Wiley, 1980.)

that has traveled over progressively longer paths in the course of multiple reflections and scatterings. For environments with numerous reflecting surfaces in complex arrangement, the decay is roughly exponential (Kinsler and Frey, 1962; Beranek, 1971b; Embleton, 1971). The increase in sound intensity on the leading edge depends on the relative intensity of the direct wave in relation to reflected and scattered waves. At a distance of 25 m in a mixed deciduous forest, a pulse of 8 kHz has a relatively sharp leading edge but decays slowly (Fig. 8) (Richards and Wiley, 1980).

Since reverberations depend on multiple reflections and scattering, they should vary in intensity with frequency and the presence of foliage. In a mixed deciduous forest, reverberations for frequencies above 4 kHz are stronger with leaves on the trees than without leaves. Reverberations are essentially absent in an open field at a distance from any forest edge on a calm day. In addition, at least at moderate distances between the source and the receiver, reverberations should depend on the directivity of the source. Reverberations are substantially greater for an omnidirectional speaker than for a much more directional horn speaker (Richards and Wiley, 1980). Reverberations should also increase markedly as the range of communication increases, although this effect has not yet been measured. The dependence of reverberations on frequency provides another reason for long-range acoustic communication in forests to avoid signals with high frequencies.

Reverberations create difficulties for a receiver attempting to resolve rapid amplitude modulations at a given frequency. Reverberations decay rapidly (>3 dB/50 msec) in a forest even at high frequencies. Consequently, reverberations primarily interfere with the separation of rapidly repeated pulses of sound. For instance, in the presence of reverberation, the trills in some birds' songs blend into a sound of almost constant intensity.

In contrast, reverberations have little effect on recognition of frequency modulation. For receivers that can discriminate frequencies, like all birds and mammals, reverberations at one frequency do not interfere with the reception of a pulse of sound at another frequency (outside the masking bandwidth of the first frequency). Signals that avoided rapid repetition of any one frequency would minimize interference from reverberations. The song of the Carolina Wren (*Thryothorus ludovicianus*) provides a good example of a complex acoustic structure that avoids rapid repetitions of a given frequency (Fig. 7).

Reverberations thus primarily mask rapid amplitude modulation (AM) and rapid, repetitive frequency modulation (FM) in acoustic signals. Since forested and open habitats clearly differ in reverberation, forest birds should, in general, avoid rapid, repetitive FM in comparison to birds of open habitats. Morton (1970, 1975) and Chappuis (1971) reported just this trend for birds in tropical forests and open habitats. Even populations of the same species differ in this way: Rufous-collared Sparrows (*Zonotrichia capensis*) sing slower trills in wooded than in open habitats (Nottebohm, 1975). Among passerines breeding in North Carolina, more than 40% of forest birds lack rapid, repetitive FM (minimum time between repetitions of the same frequency greater than 0.1 sec) while only 15% of birds in open habitats lack such rapid repetitions (Richards and Wiley, 1980).

Nevertheless, most birds in forests (60%) did include some rapid, repetitive FM in their territorial songs. Properties of signals that degrade rapidly with distance can serve for communication between individuals nearby. In addition, acoustic signals can incorporate some degradable features in order to permit a receiver to judge the source's distance (see below).

B. Amplitude Fluctuations

Sound propagating through turbulence acquires irregular amplitude fluctuations. While reflections from stationary objects always produce standing patterns of amplitude in space, moving or nonstationary heterogeneities in the medium produce nonstationary amplitude patterns at any one locus. A constant tone at the source thus acquires irregular amplitude fluctuations after propagating through a turbulent atmosphere or in the presence of irregularly moving objects (Fig. 9)

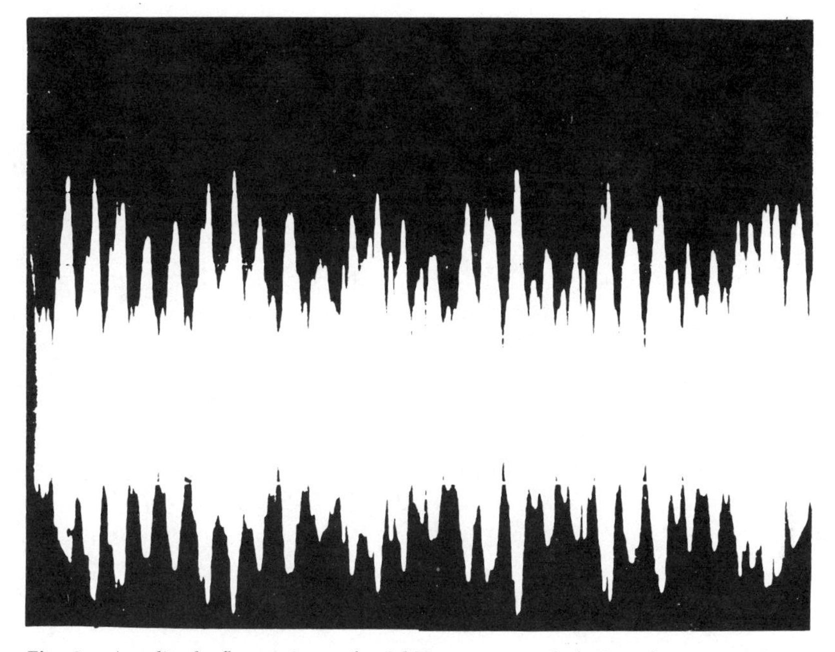

Fig. 9. Amplitude fluctuations of a 2-kHz tone recorded 60 m from a speaker in a deciduous forest. Total duration, 1 sec. (After Richards and Wiley, 1980.)

(Knudsen, 1946; Rudnick, 1947a; Richardson, 1950; Wiener and Keast, 1959; Kriebel, 1972; Embleton *et al.*, 1974; Marten and Marler, 1977; Marten *et al.*, 1977; Waser and Waser, 1977).

The theory of wave propagation through a randomly varying, isotropic medium is well developed (Chernov, 1960; Tatarski, 1961; Tolstoy and Clay, 1966; Ishimaru, 1978). In general, amplitude fluctuations increase in direct proportion to the square of frequency, the distance of propagation, and two parameters of the turbulence, the mean square fluctuations of refractive index and the dimensions of the largest scale of eddies or heterogeneities. Amplitude fluctuations of sound in a forest are appreciable even during light breezes (Richards and Wiley, 1980). Peak values of amplitude fluctuations for pure tones exceed 36 dB under all conditions except barely perceptible breezes. These fluctuations generally increase with frequency and distance, but turbulence as a result of even a light wind near the ground is the principal determinant of the intensity of the fluctuations.

These amplitude fluctuations mask amplitude modulation in a signal. To know which rates of AM are masked, we need to know the spectrum of the amplitude fluctuations caused by turbulence. Fluctuations recorded under various conditions of wind speed all have similar spectra (Richards and Wiley, 1980). In all cases, spectral density decreases approximately exponentially with frequency more than 30 dB between 1 and 50 Hz. Thus, random amplitude fluctuations induced by turbulence mask primarily low rates of AM (less than 10–20 Hz). High rates of AM, including high-frequency components in AM as a result of sharp onsets and terminations of pulses of sound, should thus propagate through turbulence with little degradation.

Irregular amplitude fluctuations have no appreciable effect on the frequencies of propagating sound. Amplitude modulation changes the frequency spectrum of a signal by adding side bands above and below the carrier frequency, with the difference in frequency between the carrier and the side band equalling the frequency of amplitude modulation. Thus random amplitude fluctuations from turbulence produce an exponentially decreasing envelope of side bands on either side of the carrier frequency. This envelope drops to more than 30 dB below the carrier frequency for frequencies 50 Hz on either side of the carrier and thus would have no appreciable effect on perception of frequency. Owing to the temporal resolution of hearing by birds and mammals, irregular amplitude fluctuations from turbulence are primarily perceived as fluctuations in intensity.

C. Frequency-Modulated Tones in Long-Range Communication

The combined effects of reverberations and irregular amplitude fluctuations suggest that long-range acoustic signals should not encode information in patterns of amplitude modulation. Reverberations and amplitude fluctuations have complementary effects in degrading the structure of AM in acoustic signals. Reverberations primarily interfere with reception of rapid AM; amplitude fluctuations primarily mask low rates of AM. In contrast, patterns of frequencies should prove much less susceptible to degradation during propagation.

Information can be encoded in frequency patterns in two ways: either as simultaneous combinations or as sequences of independently selected frequencies. In practice, the vocal tracts of animals cannot produce more than two or three simultaneous, independently variable frequency components. Differentiation of vowels in human speech by formants is probably the closest approach to this sort of system among animals. Any one human language differentiates less than 10 sounds by this means. In some birds, the syrinx can produce two independent frequencies (Greenewalt, 1968). Encoding information in more complex patterns of frequencies usually requires frequency modulation, the use of successive tones of different frequencies, or tones of continuously varying frequency.

Tonal signals have an additional advantage in long-range communication. When receivers can discriminate frequencies, the concentration of power in a single tone increases the receiver's signal/noise ratio. Thus, for animals that can analyze or discriminate frequencies, tonal frequency-modulated signals provide the optimal arrangement for long-range acoustic communication.

Long-range advertising songs of birds almost invariably consist of frequency-modulated tones (for an exception, see Wiley and Cruz, 1980). In addition, some long-range signals of mammals also fit the expectation of frequency-modulated tones. Prime examples include the howls of wolves (Harrington and Mech, 1978), the whistling of bull elks (Struhsaker, 1967), the songs of gibbons (Tembrock, 1974; Tenaza, 1976), and the whoop component of the mangabey's whoop-gobble (Waser, 1977a).

For effective use of amplitude modulation in long-range signals, the amplitude patterns must incorporate enough redundancy, usually simple repetition, to counteract degradation during propagation through the environment. The trills of many passerine birds and the long-range vocalizations of some primates show this sort of repetitive amplitude modulation (Marler, 1969, 1973). For a variety of passerine birds, playbacks in the field of artificially modified recordings of songs have shown that recognition of conspecific songs does not depend on amplitude patterns (Falls, 1963; M. Schubert, 1971; G. Schubert, 1971). The North American Ovenbird is particularly striking in this respect, since human observers standardly use the pattern of increasing amplitude to identify this species' songs, although the Ovenbirds themselves appear unaffected by this feature (Falls, 1963).

V. STRATIFIED ENVIRONMENTS

Since the velocity of sound depends on the temperature and velocity of air, vertical gradients of temperature and wind velocity refract sound (Wood, 1966; Eyring, 1946; Ingard, 1953; Pridmore-Brown and Ingard, 1955; Wiener and Keast, 1959; Piercy *et al.*, 1977). In warmer air, sound travels faster. In moving air, the velocity of sound equals the resultant of the velocity of sound in stationary air of the same composition and temperature and the velocity of the air itself. When temperature decreases with height above the ground, a shadow zone exists for horizontal propagation of sound above the ground as the wave front is refracted upward into zones of lower sound velocity. Likewise, when wind velocity increases with distance above ground, a shadow zone exists for upwind propagation of sound horizontally above the ground.

The exclusion of sound energy from the shadow zone is not absolute. Some energy is refracted into the shadow zone as a result of turbulence in the atmosphere. In addition, sound propagating near the ground includes ground and surface waves (Pridmore-Brown and Ingard, 1955), which transmit energy into the shadow zone. Nevertheless, a shadow zone has a marked effect on horizontal propagation of sound. Attenuation can increase by 20–30 dB at the boundary of the shadow zone (Wiener and Keast, 1959; Piercy *et al.*, 1977). Piercy *et al.* (1977) report that excess attenuation in the shadow zone is largely independent of frequency or distance. This finding suggests that most of the energy in the shadow zone derives from energy scattered by turbulence.

The horizontal distance from the source to the shadow zone equals

$$-4(hT/\gamma)^{1/2}$$

where T = temperature, h = height, and $\gamma = dT/dh$, the temperature gradient. Thus one way that a bird can avoid the effects of sound shadows on long-range communication in open habitats is to sing at a height above the ground, since the distance along the ground to the shadow zone increases with the square root of the height of the singer. In fact many birds of open country either select isolated high perches for singing or perform song flights high above ground (see Morton, 1975).

Low frequencies of sound are much less affected by atmospheric temperature gradients than are higher frequencies, since the scale of the strong gradients in wind and temperature close to the ground becomes smaller than the wavelength of sound for low frequencies (Piercy *et al.*, 1977). Experimental measurements suggest that shadow zones do not form for frequencies below 200–400 Hz.

When sound velocity increases with distance above ground, rather than decreases, refraction deflects sound downward toward the ground. At night, as the ground cools by radiating heat to the sky, the layer of air immediately adjacent to the ground cools faster than the air higher above ground. Consequently, temperature often increases with distance above ground, rather than decreases as it does during midday (Munn, 1966; Geiger, 1965). When temperature increases with height, termed an inversion, sound becomes effectively trapped in a layer next to the ground. Under these conditions, sound propagates horizontally with much less attenuation than otherwise. Even at moderate distances attenuation would vary with the first power of range, rather than the square of range as in spherical spreading.

Long-range acoustic communication horizontally above the ground should thus avoid conditions that tend to produce shadow zones and should take as much advantage as possible of temperature inversions. Above open ground or dense vegetation, shadow zones are most likely during midday in sunny weather, when temperature usually decreases with height above the surface and winds tend to be strong. At night temperature inversions would favor long-range acoustic communication.

Temperature gradients inside forests might produce some complex patterns of shadow zones for acoustic communication. The forest canopy absorbs more radiation during the daytime and radiates more at night compared to the lower levels of vegetation. Temperatures below the canopy fluctuate much less from day to night and often vary only slightly with height above ground. Consequently, during midmorning there is a major maximum in temperature in the upper part of the canopy. At night this peak in temperature in the canopy is reduced (Evans, 1939; Schilling *et al.*, 1946; Eyring, 1946; Hales, 1949; Christy, 1952; Heckert, 1959; Geiger, 1965).

A temperature maximum in the upper canopy tends to refract sound upward above the canopy and downward below the canopy. Under these conditions, a position just below the canopy might have advantages for long-range acoustic communication as a result of the temperature inversion immediately above (Fig. 10) (Wiley and Richards, 1978). In measurements of sound transmission in a tropical forest, Waser and Waser (1977) found evidence for just such an effect. Birds in forests might take advantage of such temperature gradients, to judge from the heights of singing birds in relation to the main canopy of a mixed

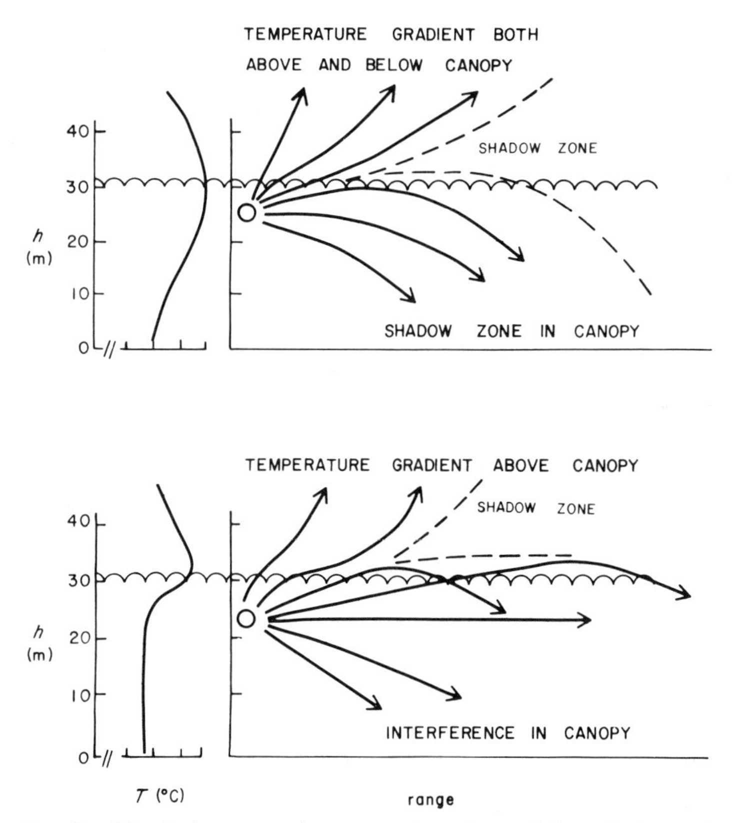

Fig. 10. Possible shadow zones in propagation of sound through the canopy of a forest. Left, representative temperature gradients; right, rays perpendicular to wave fronts for the propagation of sound from a source (circle) in the canopy. Scalloped line, top of canopy. (After Wiley and Richards, 1978.)

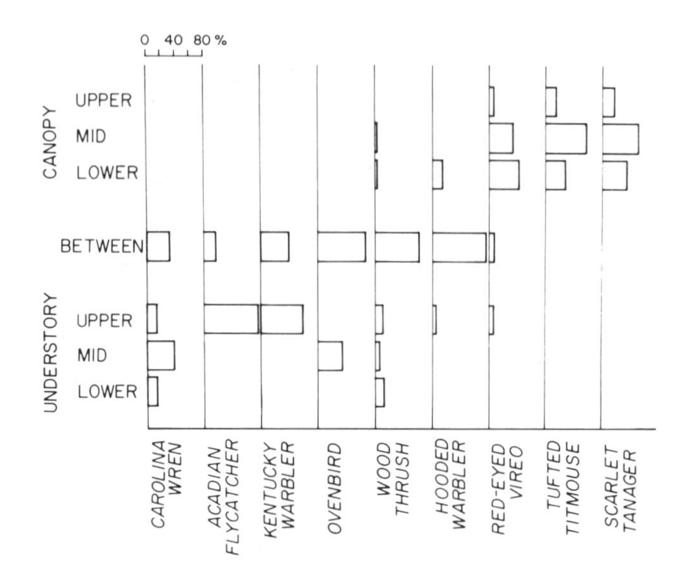

Fig. 11. Locations of undisturbed singing birds in relation to strata of foliage in a deciduous forest. Upper, mid, and lower designate thirds of the range of dense foliage in the canopy and understory. Between designates a zone of less dense foliage between understory and canopy. Ovenbirds and Wood Thrushes (see Fig. 5) actually favored locations just below the canopy in this open stratum.

deciduous forest in North Carolina (Fig. 11). Several species preferentially select singing positions near the lower edge of the canopy.

Another possibility arises in forests that have layers of dense foliage above and below a more open layer, an arrangement that might form a wave guide for sound propagation in the open layer. To evaluate this possibility, we need measurements of excess attenuation at heights corresponding to maxima and minima in density of foliage. For layers of foliage to create this effect, they must reflect a substantial fraction of incident sound.

VI. OTHER CONSIDERATIONS IN LONG-RANGE COMMUNICATION

A. Time of Day

At midday, particularly during sunny weather, the acoustic properties of the atmosphere are particularly unfavorable for long-range acoustic communication. On sunny days, large temperature and wind gradients develop above the ground on layers of dense vegetation. The temperature gradients often exceed the adiabatic lapse rate. Under these conditions turbulence results from cells of rising air. At night, temperature gradients are much more favorable for sound transmission near the ground. In the absence of strong winds, there should also be little turbulence to scatter sound.

While many nocturnal animals (including birds) call throughout most of the night, diurnal species tend to concentrate their long-range communication in the first hours of day and to a lesser extent in the evening (Henwood and Fabrick, 1979). Since the temperature cycles tend to lag the solar cycle somewhat, atmospheric conditions for long-range sound propagation should usually be more favorable at sunrise than at sunset. It is significant that a dawn chorus is less apparent near the ground inside tall tropical forests. In this environment, winds are rarely strong at any time, and temperature gradients are unusually small.

B. Communication across Water

Acoustic communication across water encounters some conditions that differ from those for communication above the ground or vegetation. The acoustic impedance of water is high in comparison to the relatively porous ground. As a consequence, a still surface of water should create boundary interference for sound propagating near the surface approximating that above asphalt (Embleton *et al.*, 1976). In particular, the prominent peak of attenuation below 1 kHz for propagation over porous surfaces like grass should not appear over water. If waves on the surface are large in relation to the wavelength of sound, they should create a strong spatial variation in the intensity of reflected sound much as a defraction grating does for reflected light.

Another important difference between sound transmission above water and above land would result from differences in temperature gradients above the surface. Water has such a high heat capacity that temperature changes at the surface as a result of solar radiation or radiation to the sky at night are normally much less than the changes in the surface temperatures of ground or vegetation. Consequently, temperature gradients above water probably diverge less from the adiabatic lapse rate than gradients above soil or vegetation. As a result, shadow zones resulting from temperature gradients should have less effect on horizontal communication above water than above soil. In addition, gradients of wind velocity near the surface might be less over water than over land. It is also possible that less turbulence would develop above water at least in the absence of large waves, so that sound would attenuate less by scattering from turbulence than above ground.

C. Directivity of the Source

The angular spread of sound radiating from a source, termed its directivity, influences several of the processes of attenuation and degradation considered above. As a result of scattering, a narrowly beamed sound attenuates more rapidly with distance than expected for spherical spread alone. On the other hand, by focusing energy into a narrow beam, a source increases the intensity at every position along the axis of propagation. A beamed sound also encounters less reverberation.

All of these considerations suggest that directivity increases the detection and recognition of signals at the distance. However, this advantage accrues only when the sender knows the location of the receiver so that it can aim the beam of sound in the correct direction. Otherwise, a sender would need to produce repeated signals aimed successively in different directions in order to cover all possible locations for receivers. In view of the advantages of directivity, it might well pay for a bird to produce beamed acoustic signals for long-range communication but to aim them in different directions whenever the location of potential receivers was unknown (Richards and Wiley, 1980). There is little information available on the directivity of avian vocalizations (Witkin, 1977).

D. Background Noise

The distance at which an acoustic signal is detectable depends ultimately on the levels of masking background sounds, including sounds generated by physical processes, like wind and rain, and sounds used for acoustic communication by other species.

A receiver confronts somewhat different problems in detecting or recognizing signals in the presence of more or less continuous background sounds in comparison to relatively brief background sounds. As a result of wind and rain, for instance, more or less continuous energy over a wide band of frequencies is mixed with signals from the source. The intensity of this more or less continuous background sound, relatively easily measured in natural environments (Morton, 1970; Waser and Waser, 1977), is highest at midday and early afternoon. In addition, some animals create nearly continuous background sounds for other species, like the cicadas (*Tibicen*) in North American forests in late summer.

For receivers of any one species, brief sounds that serve for communication by other species are also important in masking. These extraneous sounds are more difficult to measure than the continuous sounds. They usually occur too intermittently to affect the mean level of background sound. During the hour or so of dawn chorus, however, the cacophony of singing birds amounts to nearly continuous background sound.

In some communities of frogs, species appear to use distinct frequency bands (Littlejohn and Martin, 1969; Straughn, 1973). In communities of birds, species or individuals can avoid singing simultaneously (Cody and Brown, 1969; Ficken *et al.*, 1974; Wasserman, 1977) but no clear allocations of the frequency dimension occur. In fact, most passerine birds use approximately the same frequency band for their long-range acoustic signals.

VII. ESTIMATING A SIGNALER'S LOCATION

The physical properties of sound in the atmosphere suggest several other adaptations in addition to long-range communication. Consider first those adaptations for indicating the signaler's location with respect to the receiver. A singing bird might find it advantageous to let receivers judge its location. For instance, consider a song that discourages conspecific males from settling near the singer. If the receiver could judge where the singer was, it could settle at a distance that would reduce the chances of competition with a previously established resident. On the other hand, if the song included no cues for locating the signaler, the receiver would have no choice but to approach the singing bird in order to ascertain its location or simply to settle in a location without regard to the singer's location. Including cues for location in a song could thus prove advantageous for both the singer and the receiver, since both could avoid unnecessary confrontations. A singer might find it advantageous to appear closer to a receiver than it actually was. Like all forms of deception in communication, this technique would require some gullibility in the receiver.

To judge the signaler's location, a receiver must estimate its direction and range. To estimate direction, larger vertebrates, whose ears act as pressure receptors, can use binaural comparisons of intensity, phase, or time of arrival of a sound. Smaller animals, including some birds, have ears that appear to behave like pressure-gradient receptors (Lewis and Coles, 1980). Determination of direction by each of these methods depends on differences in the simultaneous amplitudes at the two ears. Consequently, propagation of acoustic signals through scattering environments or turbulence degrades cues for estimating the direction of the signaler. Nevertheless, man and other primates are remarkably good at estimating the directions of sounds in a forest (Eyring, 1946; Waser, 1977b). One way to enhance the locatability of an acoustic signal in the presence of scattering or turbulence is to include highly repetitive amplitude and frequency modulation. With a highly predictable signal, a receiver could extract the precise relationship of the signals received by its two ears by cross-correlating sufficient samples from each ear. Birds can locate wide-spectrum noise better than any pure tone (Konishi, 1973a, 1977; Shalter and Schleidt, 1977; Knudsen and Konishi, 1979) but, as we have seen, such signals have disadvantages in long-range acoustic communication for other reasons.

To judge the signaler's range, a receiver could use cues from most processes of attenuation and degradation considered above (Coleman, 1963; Schleidt, 1973; Griffin and Hopkins, 1974; Wiley and Richards, 1978). Frequency-dependent attenuation can serve to indicate the distance to the source when a receiver can compare the relative intensities in different frequency bands of the signal. The receiver would need to have some knowledge of the intensities in different frequency bands of the signal at the source, or it would need to hear the same signal at different distances. Reverberations and amplitude fluctuations also increase with range and thus can be used to judge the distance to the signaler. Estimates of range based on reverberations or amplitude fluctuations would not require detailed information about the signal at the source or at different distances. The receiver could make at least a crude estimate on the basis of its own information about the transmission path. Use of reverberations for ranging is easiest if a signal includes a variety of repetition rates of a given frequency. Examples include trills that change continuously in repetition rate or trills of different rates in the same song. Many songs incorporate frequency–time patterns that look like an inverted chevron, a pattern that would provide a continuous range of repetition rates within a limited time period.

It is difficult to obtain experimental evidence that birds use the attenuation or degradation of songs for estimating distance. In most cases, the natural response of a bird to a signal at a greater distance is simply less consistent or intense than to a closer one. An experimenter playing back recorded songs to birds at different distances cannot judge whether a subject fails to hear a song or chooses to ignore it. Carolina Wrens, however, provide an opportunity for differentiating these two possibilities. In natural circumstances, Carolina Wrens respond to songs broadcast within their own territory by silent approach and agonistic calls rather than song. In contrast, a song from an adjacent territory stimulates a wren to sing.

Richards (1981a) compared responses to test tapes of songs recorded at distances of 10 and 50 m from wrens in forests. When played at the same intensity within the territories of Carolina Wrens at distances of approximately 25 m from the subjects, these two tapes evoked different responses from the wrens. The subjects stopped singing and approached the speaker in response to the clean recordings but increased their singing in response to the degraded recordings, a demonstration that Carolina Wrens can use the natural degradation of song to judge a singer's distance.

The assumption that birdsong should evolve to maximize range of communication is unreasonable for many species. In many cases acoustic signals probably evolve for communication with receivers at some optimal distance corresponding to the usual spacing of individuals rather than simply for communication at maximal distances (Schleidt, 1973; Jilka and Leisler, 1974; Wiley and Richards, 1978). Communication at longer distances would waste time and energy, attract predators unnecessarily, and require a territorial resident to respond to individuals answering their songs from unnecessarily great distances. The advantages of restricting communication to an optimal distance are of course closely related to the advantages of including cues for estimating distance in a signal.

To insure effective communication at a distance of approximately one territorial diameter, signals should certainly not attenuate or degrade completely at that distance. Instead a song should reach the limits of detectability at a distance substantially greater than that required for effective communication. Nevertheless, the properties of songs that resist attenuation and degradation should correlate with the usual spacing of individuals. Evidence along these lines is available for *Cercopithecus* monkeys in Africa (Waser and Waser, 1977). Among Darwin's finches, population density, and thus presumably the usual spacing of individuals, correlates with average frequencies in songs and body size (Bowman, 1979). Small species are more abundant and have higher pitched songs.

VIII. SHORT-RANGE COMMUNICATION

Much vocal communication in birds, particularly between parents and young and between mates, occurs over distances of a few meters at most. In order to avoid attracting predators or sexual competitors, natural selection in many cases favors limitation of such signals to the minimal distance necessary. The characteristics of many such calls seemed designed to increase rather than decrease attenuation. Calls used in parent–offspring communication are not only low in intensity, but they often employ high frequencies (above 6 kHz) which attenuate rapidly. Precopulatory calls also tend to use high frequencies.

In colonial species, short-range communication between mates encounters some special problems. Attraction of predators is not likely. On the other hand, sexual competitors during the days of sexual receptivity pose a greater problem for colonial birds than for isolated pairs. The most pervasive problem for mates in a dense colony, however, is the difficulty of recognizing the mates' vocalizations in the continuous background sounds of other pairs. This problem is analogous to the difficulty of recognizing words in a continuous background of irrelevant speech, termed the "cocktail party effect" in psychophysics (Cherry, 1966).

One way to improve the detectability of signals in this situation is to increase their locatability. In psychophysical experiments, detection of words in irrelevant speech is improved if the two signals have different phase relationships in dichotic presentations, an arrangement analogous to signals from sources located in different directions from the receiver. For communication between mates among colonial birds, selection should thus favor acoustic signals that maximize the precision of locating the sender. Wide-spectrum sounds with sharp changes in amplitude, such as sudden onsets and terminations, result in maximal locatability (Konishi, 1977). In fact this sort of structure recurs in the vocalizations of many colonial birds, both passerine and non-passerine (White and White, 1970; Wiley, 1976).

Mates in dense colonies usually have visual and sometimes tactile signals available as well as many short-range calls for communication of information about the sender's internal states and external circumstances. In this case a complex stereotyped vocalization serves primarily to establish the signaler's identity and to attract the mate's attention, so that communication with the rich diversity of subtle visual and acoustic signals can begin (Wiley, 1976). Primary requirements for these signals are individuality and ease of locating.

The use of wide-spectrum sounds with sharp amplitude modulation and limited frequency modulation in colonies contrasts with the use of tonal sounds with complex frequency modulation in long-range communication. These two types of acoustic structure are suited for nearly opposite purposes, one for maximal locatability at short-range, the other for maximal resistance to degradation by reverberations and amplitude fluctuations in long-range communication.

IX. DETECTION AND RECOGNITION: MAXIMIZING THE RECEIVER'S PERFORMANCE

So far the distinction between detection, a receiver's judgment concerning the occurrence of a signal without classifying it, and recognition, a classification of a signal by its parameters, has not affected our conclusions. We have noted that attenuation and degradation affect both sorts of tasks. However, recognition of a signal places greater demands on the performance of a receiver than does pure detection. For effective communication in natural circumstances, a receiving individual must usually distinguish among a large number of different signals from the same and other species. In other words, recognition, not simple detection, of signals is crucial.

In the following discussion, we shall see that both detection and recognition of signals by any receiver are subject to some general constraints in a noisy environment. In terms of a receiver's performance, recognition is analogous to detection of signals with unknown parameters. It will develop that complex signals, by proper design, can improve the reliability of detection and recognition by a receiver. To see these points, we need to start with some basics.

A. Signal Detection

The theory of signal detection, developed primarily for electronic applications, like radar, characterizes the properties of the optimal receiver for detecting signals in noise (Woodward, 1953; Selin, 1965; Hancock and Wintz, 1966; Helstrom, 1968; Egan, 1978). The characteristics of an optimal receiver provide a standard against which real receivers are compared. The theorems require a number of simplifying assumptions that make exact application to animals in natural situations difficult. Nevertheless, they establish some very general limitations of signal detection, which have found wide application in psychophysics (Green and Swets, 1974; see also references cited below). Experiments that utilize this theory often require hundreds of repetitions with subjects that can follow precise instructions and whose motivation can be partially controlled with rewards. Thus methodological difficulties partly explain why this theory has never been applied to animal communication.

Nevertheless, two fundamental concepts from the theory of signal detection have such general application that they suggest some ways that animals might increase the efficiency of information transfer: (1) the distinction between the inherent detectability of a signal in noise and the receiver's criterion for a positive response; and (2) the necessary trade-off for any receiver between the probability of correct detections and the probability of false alarms.

The first step in the theory of signal detection is to establish a basic criterion that any receiver can use to decide whether or not it has detected a particular signal (see Swets *et al.*, 1961; Blachman, 1966). The general form of this criterion is a likelihood ratio. To see this point, we first need to define some probabilities: let $P(SB_1)$ and $P(SB_2)$ equal the prior probabilities that each of two signals are broadcast by a source, where $P(SB_1) + P(SB_2) = 1$. The source does one of two things; for instance, it either emits a signal (SB_1) or does not (SB_2). Alternatively, if we assume that detection of a signal is no problem, SB_1 and SB_2 can represent two different signals that constitute the entire repertoire of the sender. Now suppose that the receiver records an input (SR). The receiver's task then is to decide whether SR indicates that the source emitted SB_1 or SB_2. Thus $P(SB_1|SR)$ and $P(SB_2|SR)$ are the respective posterior probabilities of each broadcast signal given the received signal, with $P(SB_1|SR) + P(SB_2|SR) = 1$. The posterior odds in favor of SB_1 are $P(SB_1|SR){:}P(SB_2|SR)$.

The posterior probabilities are related to the *prior* probabilities as follows (by application of Bayes' formula):

$$\frac{P(SB_1|SR)}{P(SB_2|SR)} = \frac{P(SB_1)}{P(SB_2)} \cdot \frac{P(SR|SB_1)}{P(SR|SB_2)}$$

The posterior odds in favor of SB_1 thus depend on two terms: (1) the prior odds in favor of SB_1 (the odds that the receiver could guess the broadcast signal without even paying attention to the received input); and (2) a term, called the likelihood ratio, which summarizes the information provided by the input (SR) concerning the broadcast signal. If the two signals degrade so much that they result in identical stimuli for the receiver, then $P(SR|SB_1) = P(SR|SB_2)$, the likelihood ratio equals 1, and the posterior odds in favor of signal SB_1 simply equal the prior odds. At the other extreme, if each received signal is associated with one and only one broadcast signal, the likelihood ratio goes to infinity and consequently the posterior odds go to infinity; in other words the receiver can be almost certain of the broadcast signal upon receiving SR.

The likelihood ratio can thus provide a basis for a receiver's decisions. When the likelihood ratio is large enough, then the receiver should conclude that SB_1 occurred. The receiver must thus select a threshold value which separates the set of all possible received signals into two subsets, one that leads to the conclusion that SB_1 occurred and the other to the conclusion that SB_2 occurred.

In the simple situation when broadcast signals fall into two exclusive classes, there are exactly four possibilities each time the receiver samples its input and makes a decision: it can correctly decide that SB_1 occurred (a correct detection of SB_1): it can decide that SB_1 occurred when in fact SB_2 did (a false alarm); it can conclude correctly that SB_2 occurred; or finally it can conclude that SB_2 occurred when in fact SB_1 did (a missed detection of SB_1). The second and fourth of these possibilities are errors by the receiver, like errors of the first and second type in statistical tests. To select an appropriate criterion for decisions, a receiver must not only adopt an acceptable probability of false alarm but must also weigh the values (costs or benefits) of detections and each of the two kinds of errors.

B. Receiver Operating Characteristics

A receiver attempting to detect signals in the presence of noise cannot increase its probability of correct detections without inevitably increasing its probability of false alarms. Suppose a signal is simply a pulse of constant frequency, which must then propagate through a channel with the same frequency present as masking noise and with random amplitude fluctuations. The receiver then samples this channel for a specified time at the frequency of the signal and records the varying intensity of the received signal. When the sender is silent, we assume the received signal is a randomly fluctuating intensity with a Gaussian probability density function (PDF; Fig. 12). When the sender emits a signal, the received signal is also a randomly fluctuating intensity with a Gaussian PDF that has a standard deviation equal to that of the previous case but now has a greater mean intensity. The receiver must then select a threshold for deciding whether or not the sender emitted a signal. In the present case, selecting a threshold value of intensity is equivalent to selecting a threshold value of the likelihood ratio.

For any such threshold, the receiver achieves a certain probability of correct detections (P_{CD}), which equals the integral of the PDF of intensities for signal plus noise from the threshold to infinity. In addition, the receiver faces a probability of false alarm (P_{FA}), which equals the integral of the PDF of intensities for noise only from the threshold to infinity (Fig. 12). Now if the receiver selects different values of the threshold in succession, we can construct a graph of P_{CD} versus P_{FA} as the threshold varies. This function is the receiver operating characteristic (ROC).

If the signal attenuates by different amounts during propagation from the sender to the receiver, we can also study ROCs for different signal/noise ratios (Fig. 12). Note several things about these ROCs. First, if the PDF for signal plus

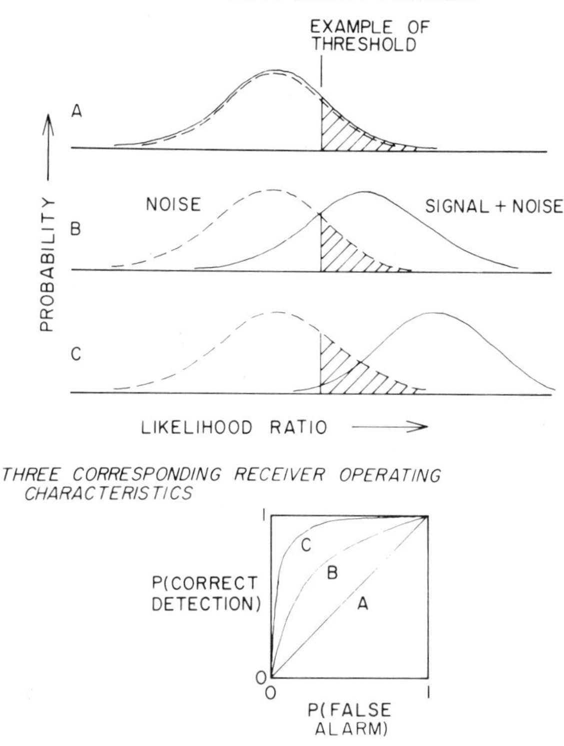

Fig. 12. Schematic demonstration of receiver operating characteristics for simple cases of signal detection. Cases A, B, and C represent increasing signal/noise ratios and hence increasing separation of the probability density functions (PDF) of amplitudes for noise only and noise plus signal. In each case, the probability of a correct detection is the integral of the PDF for signal plus noise from the threshold to infinity; the probability of false alarm is the corresponding integral for noise only. In each case, one can construct the probability of correct detection as a function of the probability of false alarm (the receiver operating characteristic) by letting the threshold vary from $-\infty$ to $+\infty$.

noise and the PDF for noise only are identical, then the probability of correct detection always equals the probability of false alarm, regardless of the threshold. Second, the greater the signal/noise ratio, the more the ROC is displaced away from the positive diagonal and toward the optimal point ($P_{FA} = 0$, $P_{CD} = 1$). Third, when the two PDFs have identical shapes, as in the present case, the ROCs are symmetrical with respect to the negative diagonal. Fourth, for any signal/noise ratio, the probability of correct detection increases monotonically with the probability of false alarm. On the other hand, given a threshold, the probability of false alarm is fixed at all signal/noise ratios, while the probability of correct detections increases as signal/noise ratio increases. Thus we reach the major conclusion that the receiver's performance depends on two factors: the criterion (threshold) it adopts for deciding that a particular signal has occurred; and the inherent detectability (signal/noise ratio) of the signal.

Although the preceding discussion has dealt with an extremely simplified case, the final conclusion has great generality. A receiver cannot adjust its probability of correct detections and its probability of false alarms independently. The performance of a receiver depends on two factors: its criterion for the presence of a particular signal; and the inherent detectability of signals. These conclusions apply to all communication in noisy environments, ones in which the broadcast signals result in overlapping distributions of inputs sensed by the receiver. Birds attempting to differentiate complex acoustic structures at long range in natural environments provide a good example.

C. More Complex Tasks for a Receiver

A receiver reaches maximal performance when it knows in advance the precise form of the signal, including the exact time of onset. In this case, theory shows that the likelihood ratio for the presence of a signal is a monotonic function of the cross-correlation of the received input and the signal (see general references above). This situation of course occurs rarely even in engineering applications.

Uncertainty on the part of the receiver concerning the exact form of the signal has an effect equivalent to decreasing the signal/noise ratio for a known signal. Likewise, the problem of detecting several different signals also results in an ROC similar to those for lower signal/noise ratios with one signal. Finally, recognition of several different signals, in addition to detection, once again resembles a decrease in the signal/noise ratio for detection alone. All three of these comparisons—detection of a single signal with unknown parameters versus detection of a single completely known signal, detection of several different signals versus detection of a single signal, and recognition versus detection of several signals—follow the same pattern: the ROC moves toward the positive diagonal, analogous to a decrease in signal/noise ratio or detectability.

In contrast, repetition of the same signal has the opposite effect, namely, an increase in signal/noise ratio; the ROC shifts away from the positive diagonal. This result conforms with predictions from information theory that increased redundancy in signals improves the probability of receiving a message in a noisy channel (Shannon and Weaver, 1949; Cherry, 1966). Of course, repetition and other forms of sequential redundancy increase the time required for communication.

These effects of unknown parameters of signals, multiple signals, recognition, and repeated signals have all been established in psychophysical studies on humans (unknown parameters: Creelman, 1960; Egan *et al.*, 1961a,b; Watson and Nichols, 1976; multiple signals: Veniar, 1958; Pollack, 1959a; Green, 1960; Pollack, 1964; Moray and O'Brien, 1967; Ahroon *et al.*, 1977; recognition: Tanner, 1956; Clarke *et al.*, 1959; Lindner, 1968; Green *et al.*, 1977; Swets *et al.*, 1978; repetition: Pollack, 1959b; Swets and Birdsall, 1978). We can expect that all of these results apply to animal communication just as to human or electronic communication.

D. Adaptations to Improve a Receiver's Performance

The effects of multiple signals, repeated signals, and uncertainty in the time of arrival of signals on the probability of correct detection suggest three adaptations that birds could use to improve a receiver's probability of detecting an acoustic signal: reduction in the diversity of alternative signals, repetition, and alerted detection.

The effect of multiple signals might pertain most clearly to problems of species recognition of songs. One would expect that in ecological communities with greater species diversity, species should use smaller repertoires of song types in order to permit more reliable detection of conspecific songs in the midst of a more complex background of heterospecific signals. Such an effect has been reported for North American wrens (Kroodsma, 1977). This effect should be most apparent among the rare species in diverse communities or species represented by widely scattered individuals, which confront the most extreme signal/noise ratios in long-range communication.

Repetition of an identical signal also improves a receiver's performance. Once again, this technique would have the greatest advantages for communication in ecologically diverse communities, particularly for the rarer species, and for communication at especially long ranges between widely spaced individuals. Exact repetition of a phrase occurs in the songs of many species. In addition, many species repeat the same song pattern a number of times before switching to a new one. The Carolina Wren, for instance, does both.

Alerted detection requires that some easily detectable signal precede, in a defined temporal relationship, a signal that carries important information (Raisbeck, 1963). Easy detection of the alerting component requires a high signal/noise ratio, as a result of low attenuation and degradation, and a simple predictable structure. This component need not carry any essential information, in the sense that the receiver need not make decisions on the basis of classifying (recognizing) this component. In the case of birdsong for instance, the alerting component need not include information on species identity or individual identity. Instead the alerting component serves to define for the receiver the period of time in which it can expect the message component. Since this latter component conveys information to the receiver, it must have enough complexity of structure to permit the receiver to distinguish among alternatives. In summary, the alerting component permits easy detection; the message component permits classification (recognition).

A variety of species of birds begin their territorial songs with one or a series of relatively simple tones. The songs then become progressively more complex in acoustic structure during their course. Shiovitz (1975) has noted that the song of the Indigo Bunting (*Passerina cyanea*) fits this pattern. He feels that the introductory portions serve as a signature to identify the species while later portions convey more detailed information. Theory of signal detection suggests that the initial portions of a signal should in fact convey less information than the final portions, but this consideration could apply to species identification as well as other kinds of information.

Experiments with Rufous-sided Towhees (*Pipilo erythrophthalmus*) have provided evidence that the initial tones in this species' characteristic songs have an alerting effect, while the final trill carries most of the information that permits territorial towhees to recognize a conspecific song (Richards, 1981b). Realizing that the effect of an alerting note would resemble an increase in signal/noise ratio, Richards compared the responses of territorial towhees to a series of tape recordings of complete songs, introductory notes, and final trills, which in each case were either artificially degraded with reverberations or not (Fig. 13). All recordings came from towhees beyond hearing of the subjects. He found that either clean or degraded recordings of the entire song (introduction plus trill) evoked strong responses. The introductory note alone, either clean or degraded, evoked little response. In contrast, when he tried the final trill alone, he found a dramatic difference between clean and degraded recordings (Fig. 14). Degraded trills evoked little response while clean recordings of trills evoked nearly a full response. Thus the trill provides the information that territorial towhees use to

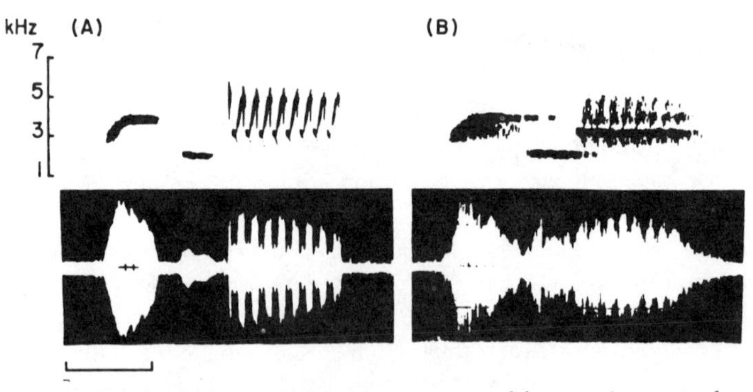

Fig. 13. Example of Rufous-sided Towhee songs used for experiments to demonstrate alerted detection. (A) Clean recording of an entire song without significant reverberation; (B) artificially reverberated version of the same song. Time mark, 1 sec. Above, sonagrams; below, oscillograms. (After Richards, 1981b.)

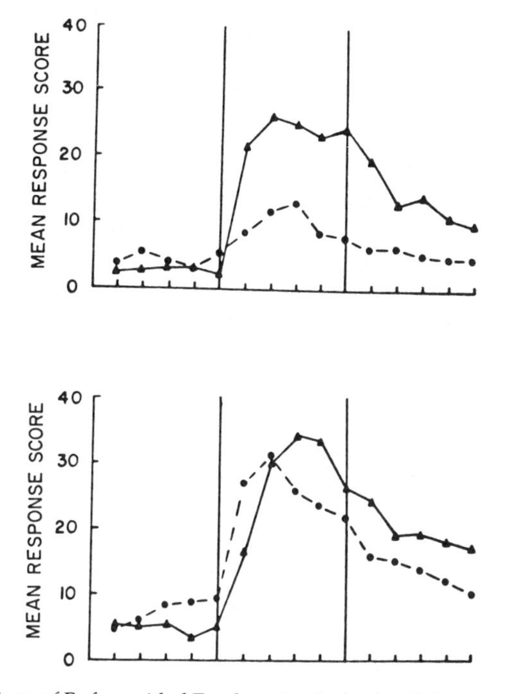

Fig. 14. Reactions of Rufous-sided Towhees to playbacks of clean and artificially reverberated songs and trills (songs without the initial notes). Ordinate, composite response score averaged over all subjects. Abscissa, 1 min blocks before (first five), during (middle five), and after (last five) playback. Dashed lines, response to reverberated recordings; solid lines, response to clean recordings. Above, playback of trills without initial notes; below, playback of entire songs (trills with initial notes). Initial notes alone, both clean and degraded, evoked little response. Thus two degraded components, initial notes and trills, each ineffective alone, combine to make an effective stimulus. (After Richards, 1981b.)

identify a conspecific song. On the other hand, the simple introductory notes, which had little effect alone, had a major effect in increasing the response of towhees to a degraded trill, inasmuch as the degraded trill alone had little effect, while the degraded entire song, complete with introductory notes, had a strong effect. The introductory notes thus allow towhees to recognize degraded trills.

E. Other Applications of Signal Detection Theory

The theory of signal detection has implications for the interpretation of playback experiments with birds under natural conditions. In some circumstances, a reduced response by subjects does not allow the investigator to decide whether the subjects have changed their criterion for a response or whether the signals differ in detectability. The large number of experiments on species recognition confront this problem. When subjects respond less intensely to artificially modified songs, the inference is usually drawn that the birds no longer recognize these songs as conspecific. In other words, the inference is usually that the songs now fall below the subject's criterion. An alternative possibility, however, is that the modified songs are inherently less detectable.

Detailed application of the theory of signal detection to animal communication requires measurements of ROC curves, including false alarm rates in birds' responses. A problem arises in distinguishing the birds' spontaneous responses from reactions to an experimental stimulus like a playback. Unless one can arrange a response that occurs only in an experiment, in which a signal is either presented or not within a defined time interval, it will prove impossible to separate false alarms from spontaneous response. This caveat, however, need not preclude the application of ROCs to studies of birdsong in the field. Even though probabilities of correct detection and probabilities of false alarm both include a constant probability of spontaneous response, ROCs should still permit experimenters to decide whether or not responses to different conditions of playback result from changes in the detectability of the signals or changes in the subjects' criterion for a response. For methods of determining ROCs that might apply to field studies of birds, see Egan *et al.* (1961c), Pollack *et al.* (1964), Swets and Kristofferson (1970), and Watson and Nichols (1976).

X. CONCLUSIONS

Signals evolve for efficient transfer of information whether the interaction between sender and receiver is mutualistic or manipulative. One way to increase the efficiency of acoustic communication is to minimize attenuation and degradation of signals during propagation. In all habitats, both forested and open, attenuation from absorption and scattering increases monotonically with increasing frequency. Scattering from vegetation and from atmospheric turbulence primarily affects frequencies above about 2 kHz. Thus, provided there is little wind, these frequencies attenuate less in open habitats than in forests. Nevertheless, except near the ground, lower frequencies always attenuate less. Near the ground in any habitat, interference primarily attenuates frequencies between 0.5 and 1 kHz. Thus, there is no clear reason to expect systematic differences in the optimal frequencies for long-range communication in different natural environments.

Degradation of acoustic signals by reverberations and amplitude fluctuations, on the other hand, differs systematically between environments with dense vegetation, such as forests, and open environments. Reverberations are usually more severe in the former and amplitude fluctuations in the latter. In forests, birds tend to avoid rapid repetition of any one frequency, as one would expect in reverberant environments. The general character of long-range signals in birds, frequency-modulated tones, avoids the effects of reverberations and amplitude fluctuations on acoustic signals.

Attenuation and degradation of acoustic signals can serve to encode information about the distance to the sender. Attenuation can also help to limit acoustic signals to nearby partners, for instance in communication between mates and between parents and young, when attraction of predators or sexual competitors has disadvantages. In the noisy environment of dense colonies, the principal problem in acoustic communication by mates is the "cocktail party effect." Here, use of wide-spectrum sounds with sharp amplitude modulation, in spite of minimal frequency modulation, has the advantage of increased locatability, which in turn helps a receiver to detect a signal in the high levels of background noise.

The theory of signal detection reaches two general conclusions that apply to birds as well as all other receivers: a distinction between the receiver's criterion (threshold) for a positive response and the inherent detectability of a signal; and an inevitable covariation between a receiver's probability of a correct detection and its probability of a false alarm. These features of signal detection are best summarized in a receiver operating characteristic (ROC), a plot of the probability of correct detection versus the probability of false alarm at different thresholds for a positive response. Certain complexities in a receiver's task, when compared to the task of detecting a single known signal, all have effects that resemble a decreased signal/noise ratio: detection of signals with unknown parameters; detection of one-of-several different signals; and recognition (classification) of signals in addition to detection. On the other hand, repetition of a signal has the opposite effect, namely increased signal/noise ratio.

As a result, three kinds of adaptations in acoustic signals could increase correct reception of signals by receivers. Where noise levels are high, as in ecological communities with a high diversity of species and for communication at extremely long ranges, birds could improve the detectability of their signals by limiting their repertoire and by employing repeated phrases or song patterns. In addition, an alerting signal increases the reliability of detecting and recognizing subsequent signals. Thus, repetition of phrases in a song and alerting notes are complementary techniques for achieving the same end: improved performance of receivers.

ACKNOWLEDGMENTS

This study, a contribution from the Behavioral Research Station in the North Carolina Botanical Garden, was supported in part by the Frank M. Chapman Memorial Fund of the American Museum of Natural History, Sigma Xi, and the National Institute of Mental Health (MH22316). We thank P. M. Waser, H. C. Gerhardt, and B. S. Simpson for their helpful comments on the manuscript.

REFERENCES

Ahroon, W. A., Jr., Pastore, R. E., and Wolz, J. P. (1977). Selective attention I. Two-channel simultaneous frequency difference limen. *J. Acoust. Soc. Am.* **61,** 811–815.

Altmann, S. A. (1965). Sociobiology of rhesus monkeys: II. Stochastics of social communication. *J. Theor. Biol.* **8,** 490–522.

Altmann, S. A. (1967). The structure of primate social communication. *In* "Social Communication among Primates" (S. A. Altmann, ed.), pp. 325–362. Univ. of Chicago Press, Chicago, Illinois.

Aylor, D. (1971). Noise reduction by vegetation and ground. *J. Acoust. Soc. Am.* **51,** 197–205.

Beranek, L. L., ed. (1971a). "Noise and Vibration Control." McGraw-Hill, New York.

Beranek, L. L. (1971b). Sound in small spaces. *In* "Noise and Vibration Control" (L. L. Beranek, ed.), pp. 194–218. McGraw-Hill, New York.

Blachman, N. M. (1966). "Noise and its Effect on Communication." McGraw-Hill, New York.

Bowman, R. I. (1979). Adaptive morphology of song dialects in Darwin's finches. *J. Ornithol.* **120,** 353–380.

Chappuis, C. (1971). Un exemple de l'influence du milieu sur les émissions vocales des oiseaux: L'évolution des chants en forêt équatoriale. *Terre Vie* **118,** 183–202.

Chernov, L. A. (1960). "Wave Propagation in a Random Medium." McGraw-Hill, New York.

Cherry, C. (1966). "On Human Communication," 2nd ed. MIT Press, Cambridge, Massachusetts.

Chessell, C. I. (1977). The propagation of noise along a finite impedance boundary. *J. Acoust. Soc. Am.* **62,** 825–834.

Christy, H. R. (1952). Vertical temperature gradients in a beech forest in central Ohio. *Ohio J. Sci.* **52,** 199–209.

Clarke, F. R., Birdsall, T. G., and Tanner, W. P., Jr. (1959). Two types of ROC curves and definitions of parameters. *J. Acoust. Soc. Am.* **31,** 629.

Cody, M. L., and Brown, J. H. (1969). Song asynchrony in neighboring bird species. *Nature (London)* **222,** 778–780.

Coleman, P. D. (1963). An analysis of cues to auditory depth perception in free space. *Psychol. Bull.* **60,** 302–315.

Creelman, C. D. (1960). Detection of signals of uncertain frequency. *J. Acoust. Soc. Am.* **32,** 805–810.

Daigle, G. A. (1979). Effects of atmospheric turbulence on the interference of sound waves above a finite impedance boundary. *J. Acoust. Soc. Am.* **65,** 45–49.

Daigle, G. A., Piercy, J. E., and Embleton, T. F. W. (1978). Effect of atmospheric turbulence on the interference of sound waves near a hard boundary. *J. Acoust. Soc. Am.* **64,** 622–630.

Dawkins, R. (1976). "The Selfish Gene." Oxford Univ. Press, London and New York.

Dawkins, R., and Krebs, J. R. (1978). Animal signals: Information or manipulation? *In* "Behavioral Ecology" (J. R. Krebs and N. B. Davies, eds.), pp. 282–309. Blackwell, Oxford.

Delacour, J., and Amadon, D. (1973). "Curassows and Related Birds." Am. Mus. Nat. Hist., New York.

Dingle, H. (1972). Aggressive behavior in stomatopods and the use of information theory in the analysis of animal communication. *In* "Behavior of Marine Animals: Current Perspectives in Research. Vol. 1: Invertebrates" (H. E. Winn and B. L. Olla, eds.), pp. 126–156. Plenum, New York.

Donato, R. J. (1976). Propagation of a spherical wave near a plane boundary with complex impedance. *J. Acoust. Soc. Am.* **60,** 34–39.

Dooling, R. J., Mulligan, J. A., and Miller, J. D. (1971). Auditory sensitivity and song spectrum in the Common Canary (*Serinus canarius*). *J. Acoust. Soc. Am.* **50,** 700–709.

Dooling, R. J., Zoloth, S. R., and Baylis, J. R. (1978). Auditory sensitivity, equal loudness, temporal resolving power and vocalizations in the House Finch (*Carpodacus mexicanus*). *J. Comp. Physiol. Psychol.* **92,** 867–876.

Egan, J. P. (1978). "Signal Detection Theory and ROC Analysis." Academic Press, New York.

Egan, J. P., Greenberg, G. Z., and Schulman, A. I. (1961a). Interval of time uncertainty in auditory detection. *J. Acoust. Soc. Am.* **33,** 771–778.

Egan, J. P., Schulman, A. I., and Greenberg, G. Z. (1961b). Memory for waveform and time uncertainty in auditory detection. *J. Acoust. Soc. Am.* **33,** 779–781.

Egan, J. P., Greenberg, G. Z., and Schulman, A. I. (1961c). Operating characteristics, signal detectability, and the method of free response. *J. Acoust. Soc. Am.* **33,** 993–1007.

Embleton, T. F. W. (1966). Scattering by an array of cylinders as a function of surface impedance. *J. Acoust. Soc. Am.* **40,** 667–670.

Embleton, T. F. W. (1971). Sound in large rooms. *In* "Noise and Vibration Control" (L. L. Beranek, ed.), pp. 219–244. McGraw-Hill, New York.

Embleton, T. F. W., Olson, N., Piercy, J. E., and Rollin, D. (1974). Fluctuations in the propagation of sound near the ground. *J. Acoust. Soc. Am.* **55,** 485 (A).

Embleton, T. F. W., Piercy, J. E., and Olson, N. (1976). Outdoor sound propagation over ground of finite impedance. *J. Acoust. Soc. Am.* **59,** 267–277.

Evans, G. C. (1939). Ecological studies on the rain forest of southern Nigeria. II. The atmospheric environmental conditions. *J. Ecol.* **27,** 436–482.

Evans, L. B., Bass, H. E., and Sutherland, L. C. (1971). Atmospheric absorption of sound: theoretical predictions. *J. Acoust. Soc. Am.* **51,** 1565–1575.

Eyring, D. F. (1946). Jungle acoustics. *J. Acoust. Soc. Am.* **18,** 257–270.

Falls, J. B. (1963). Properties of bird song eliciting responses from territorial males. *Proc. Int. Ornithol. Congr.* **13,** 259–271.

Ficken, R. W., Ficken, M. S., and Hailman, J. P. (1974). Temporal pattern shifts to avoid acoustic interference in singing birds. *Science* **183,** 762–763.

Gaunaurd, G. C., and Uberall, H. (1978). Acoustics of finite beams. *J. Acoust. Soc. Am.* **63,** 5–16.

Geiger, R. (1965). "The Climate Near the Ground." Harvard Univ. Press, Cambridge, Massachusetts.

Givens, M. P., Nyborg, W. L., and Schilling, H. K. (1946). Theory of the propagation of sound in scattering and absorbing media. *J. Acoust. Soc. Am.* **18,** 284–295.

Green, D. M. (1960). Psychoacoustics and detection theory. *J. Acoust. Soc. Am.* **32,** 1189–1203.

Green, D. M., and Swets, J. A. (1974). "Signal Detection Theory and Psychophysics." Krieger, New York.

Green, D. M., Weber, D. L., and Duncan, J. E. (1977). Detection and recognition of pure tones in noise. *J. Acoust. Soc. Am.* **63,** 948–954.

Green, S., and Marler, P. (1979). The analysis of animal communication. *In* "Handbook of Behavioral Neurobiology. Vol. 3: Social Behavior and Communication" (P. Marler and J. G. Vandenbergh, eds.), pp. 73–158. Plenum, New York.

Greenewalt, C. H. (1968). "Bird Song: Acoustics and Physiology." Smithsonian Inst. Press, Washington, D.C.

Griffin, D. R. (1971). The importance of atmospheric attenuation for the echolocation of bats (Chiroptera). *Anim. Behav.* **19,** 55–61.

Griffin, D. R., and Hopkins, C. D. (1974). Sounds audible to migrating birds. *Anim. Behav.* **22,** 672–678.

Hailman, J. P. (1977). "Optical Signals: Animal Communication and Light." Indiana Univ. Press, Bloomington.

Hales, W. B. (1949). Micrometeorology in the tropics. *Bull. Am. Meteorol. Soc.* **30,** 124–137.

Hancock, J. C., and Wintz, P. A. (1966). "Signal Detection Theory." McGraw-Hill, New York.

Harrington, F. H., and Mech, L. D. (1978). Wolf vocalization. *In* "Wolf and Man: Evolution in Parallel" (R. L. Hall and H. S. Sharp, eds.), pp. 109–132. Academic Press, New York.

Harris, C. M. (1966). Absorption of sound in air versus humidity and temperature. *J. Acoust. Soc. Am.* **40,** 148–159.

Hazlett, B. A., and Bossert, W. H. (1965). A statistical analysis of the aggressive communication systems of some hermit crabs. *Anim. Behav.* **13,** 357–373.

Hazlett, B. A., and Estabrook, G. F. (1974a). Examination of agonistic behavior by character analysis. I. The spider crab *Microphrys bicornutus*. *Behaviour* **48,** 131–144.

Hazlett, B. A., and Estabrook, G. F. (1974b). Examination of agonistic behavior by character analysis. II. Hermit crabs. *Behaviour* **49,** 88–110.

Heckert, L. (1959). Die klimatischen Verhältnisse in Laubwäldern. *Z. Meteorol.* **13,** 211–223.

Helstrom, C. W. (1968). "Statistical Theory of Signal Detection," 2nd ed. Pergamon, New York.

Henwood, K., and Fabrick, A. (1979). A quantitative analysis of the dawn chorus: temporal selection for communicatory optimization. *Am. Nat.* **114,** 260–274.

Hjorth, I. (1970). Reproductive behavior in Tetraonidae. *Viltrevy* 7, 183–596.

Hunter, M. L., Jr., and Krebs, J. R. (1979). Geographical variation in the song of the Great Tit (*Parus major*) in relation to ecological factors. *J. Anim. Ecol.* **48,** 759–785.

Ingard, U. (1951). On the reflection of a spherical sound wave from an infinite plane. *J. Acoust. Soc. Am.* **23,** 329–335.

Ingard, U. (1953). A review of the influence of meteorological conditions on sound propagation. *J. Acoust. Soc. Am.* **25,** 405–411.

Ingard, U. (1969). On sound-transmission anomalies in the atmosphere. *J. Acoust. Soc. Am.* **45,** 1038–1039.

Ingard, U., and Maling, G. C., Jr. (1963). On the effect of atmospheric turbulence on sound propagated over ground. *J. Acoust. Soc. Am.* **35,** 1056–1058.

Ishimaru, A. (1978). "Wave Propagation and Scattering in Random Media. Vol. 2: Multiple Scattering, Turbulence, Rough Surfaces and Remote Sensing." Academic Press, New York.

Jilka, A., and Leisler, B. (1974). Die Einpassung dreier Rohrsängerarten (*Acrocephalus schoenobaemus, A. scirpaceus, A. arundinaceous*) in ihrer Lebensräume in bezug auf das Frequenzspecktrum ihrer Reviergesänge. *J. Ornithol.* **115,** 192–212.

Johnson, R. K. (1977). Sound scattering from a fluid sphere revisited. *J. Acoust. Soc. Am.* **61,** 375–377.

Kinsler, L. E., and Frey, A. R. (1962). "Fundamentals of Acoustics," 2nd ed. Wiley, New York.

Klopfer, P., and Hatch, J. J. (1968). Experimental considerations. *In* "Animal Communication" (T. A. Sebeok, ed.), pp. 31–43. Indiana Univ. Press, Bloomington.

Knudsen, E. I., and Konishi, M. (1979). Mechanisms of sound localization in the Barn Owl (*Tyto alba*). *J. Comp. Physiol.* **133,** 12–21.

Knudsen, V. O. (1946). The propagation of sound in the atmosphere—attenuation and fluctuations. *J. Acoust. Soc. Am.* **18,** 90–96.

Konishi, M. (1970a). Evolution of design features in the coding of species-specificity. *Am. Zool.* **10,** 67–72.

Konishi, M. (1970b). Comparative neurophysiological studies of hearing and vocalization in songbirds. *Z. Vgl. Physiol.* **66,** 257–272.

Konishi, M. (1973a). Locatable and nonlocatable acoustic signals for Barn Owls. *Am. Nat.* **107,** 775–785.

Konishi, M. (1973b). How the owl tracks its prey. *Am. Sci.* **61,** 411–424.

Konishi, M. (1977). Spatial localization of sound. *In* "Dahlem Workshop on Recognition of Complex Acoustic Signals" (T. Bullock, ed.), pp. 127–143. Dahlem Konf., Berlin.

Kriebel, A. R. (1972). Refraction and attenuation of sound by wind and thermal profiles over a ground plane. *J. Acoust. Soc. Am.* **51,** 19–23.

Kroodsma, D. E. (1977). Correlates of song organization among North American wrens. *Am. Nat.* **111,** 995–1008.

Lewis, B., and Coles, R. (1980). Sound localization in birds. *Trends Neurosci.* **3,** 102–105.

Lighthill, M. J. (1953). On the energy scattered from the interaction of turbulence with sound or shock wave. *Proc. Cambridge Philos. Soc.* **49,** 531–551.

Lindner, W. A. (1968). Recognition performance as a function of detection criterion in a simultaneous detection–recognition task. *J. Acoust. Soc. Am.* **44,** 204–211.

Littlejohn, M. J., and Martin, A. A. (1969). Acoustic interaction between two species of leptodactylid frogs. *Anim. Behav.* **17,** 785–791.

Marler, P. (1961). The logical analysis of animal communication. *J. Theor. Biol.* **1,** 295–317.

Marler, P. (1967). Animal communication signals. *Science* **157,** 769–774.

Marler, P. (1969). *Colobus guereza:* Territoriality and group composition. *Science* **163,** 93–95.

Marler, P. (1973). A comparison of vocalizations of red-tailed monkeys and blue monkeys, *Cercopithecus ascanius* and *C. mitis,* in Uganda. *Z. Tierpsychol.* **33,** 223–247.

Marten, K., and Marler, P. (1977). Sound transmission and its significance for animal vocalization. I. Temperate habitats. *Behav. Ecol. Sociobiol.* **2,** 271–290.

Marten, K., Quine, D., and Marler, P. (1977). Sound transmission and its significance for animal vocalization. II. Tropical forest habitats. *Behav. Ecol. Sociobiol.* **2,** 291–302.

Michelsen, A. (1978). Sound reception in different environments. *In* "Perspectives in Sensory Ecology" (B. A. Ali, ed.), pp. 345–373. Plenum, New York.

Moray, N., and O'Brien, T. (1967). Signal-detection theory applied to selective listening. *J. Acoust. Soc. Am.* **42,** 765–772.

Morse, P. M., and Ingard, K. U. (1968). "Theoretical Acoustics." McGraw-Hill, New York.

Morton, E. S. (1970). Ecological sources of selection on avian sounds. Ph.D. Thesis, Yale Univ., New Haven, Connecticut.

Morton, E. S. (1975). Ecological sources of selection on avian sounds. *Am. Nat.* **109,** 17–34.

Munn, R. E. (1966). "Descriptive Micrometeorology." Academic Press, New York.

Nottebohm, F. (1975). Continental patterns of song variability in *Zonotrichia capensis:* some possible ecological correlates. *Am. Nat.* **109,** 605–624.

Pfeiffer, P. E. (1965). "Concepts of Probability Theory." McGraw-Hill, New York.

Piercy, J. E., Embleton, T. F. W., and Sutherland, L. C. (1977). Review of noise propagation in the atmosphere. *J. Acoust. Soc. Am.* **61,** 1402–1418.

Pöhlmann, W. (1961). Die Schallabsorption von Luft verschiedener Feuchtigkeit zwischen 10 und 100 kHz. *Proc. Int. Congr. Acoust., 3rd* **1,** 532–535.

Pollack, I. (1959a). Message uncertainty and message reception. *J. Acoust. Soc. Am.* **31,** 1500–1508.

Pollack, I. (1959b). Message repetition and message reception. *J. Acoust. Soc. Am.* **31,** 1509–1515.

Pollack, I. (1964). Message probability and message reception. *J. Acoust. Soc. Am.* **36,** 937–945.

Pollack, I., Galanter, E., and Norman, D. (1964). An efficient non-parametric analysis of recognition memory. *Psychon. Sci.* **1,** 327–328.

Pridmore-Brown, D. C., and Ungard, U. (1955). Sound propagation into the shadow zone in a temperature-stratified atmosphere above a plane boundary. *J. Acoust. Soc. Am.* **27,** 36–42.

Quastler, H. (1958). A primer on information theory. *In* "Symposium on Information Theory in Biology" (H. P. Yockey, R. L. Platzman, and H. Quastler, eds.), pp. 3–49. Pergamon, New York.

Raisbeck, G. (1963). "Information Theory." MIT Press, Cambridge, Massachusetts.

Richards, D. G. (1981a). Estimation of distance of singing conspecifics by the Carolina Wren. *Auk,* **98,** 127–133.

Richards, D. G. (1981b). Alerting and message components in songs of Rufous-sided Towhees. *Behaviour* **76,** 223–249.

Richards, D. G., and Wiley, R. H. (1980). Reverberations and amplitude fluctuations in the propagation of sound in a forest: Implications for animal communication. *Am. Nat.* **115,** 381–399.

Richardson, E. G. (1950). The fine structure of atmospheric turbulence in relation to the propagation of sound over the ground. *Proc. R. Soc. London, Ser. A* **203,** 149–164.

Ridgely, R. S. (1976). "A Guide to the Birds of Panama." Princeton Univ. Press, Princeton, New Jersey.

Robbins, C. S., Bruun, B., and Zim, H. S. (1966). "Birds of North America." Golden Press, New York.

Roberts, J., Kacelnik, A., and Hunter, M. C., Jr. (1979). A model of sound interference in relation to acoustic communication. *Anim. Behav.* **27,** 1271–1273.

Rudnick, I. (1947a). Fluctuations in intensity of an acoustic wave transmitted through a turbulent heated lamina. *J. Acoust. Soc. Am.* **19,** 202–204.

Rudnick, I. (1947b). The propagation of an acoustic wave along a boundary. *J. Acoust. Soc. Am.* **19,** 348–356.

Schilling, H. K., Drumheller, W., Nyborg, W. L., and Thorpe, H. A. (1946). On micrometeorology. *Am. J. Phys.* **14,** 343.

Schilling, H. K., Givens, M. P., Nyborg, W. L., Pielemeier, W. A., and Thorpe, H. A. (1947). Ultrasonic propagation in open air. *J. Acoust. Soc. Am.* **19,** 222–234.

Schleidt, W. M. (1973). Tonic communication: continual effects of discrete signs in animal communication systems. *J. Theor. Biol.* **42,** 359–386.

Schubert, G. (1971). Experimentelle Untersuchungen über die artkennzeichnenden Parameter in Gesang des Zilpzalps, *Phylloscopus c. collybita* (Vieillot). *Behaviour* **38,** 289–314.

Schubert, M. (1971). Untersuchungen über die reaktions-aulösenden Signalstrukturen des Fitisgesanges, *Phylloscopus t. trochilus* (L.), und das verhalten gegenüber arteigenen Rufen. *Behaviour* **38,** 250–288.

Selin, I. (1965). "Detection Theory." Princeton Univ. Press, Princeton, New Jersey.

Shalter, M. D., and Schleidt, W. M. (1977). The ability of Barn Owls *Tyto alba* to discriminate and localize avian alarm calls. *Ibis* **119,** 22–27.

Shannon, C. E., and Weaver, W. (1949). "The Mathematical Theory of Communication." Univ. of Illinois Press, Urbana.

Shiovitz, K. A. (1975). The process of species-specific song recognition by the Indigo Bunting, *Passerina cyanea,* and its relationship to the organization of avian acoustical behavior. *Behaviour* **55,** 128–179.

Slud, P. (1960). The birds of Finca "La Selva," Costa Rica: A tropical wet forest locality. *Bull. Am. Mus. Nat. Nist.* **121,** 49–148.

Smith, W. J. (1963). Vocal communication in birds. *Am. Nat.* **97,** 117–125.

Smith, W. J. (1969). Messages of vertebrate communication. *Science* **165,** 145–150.

Smith, W. J. (1977). "The Behavior of Communicating." Harvard Univ. Press, Cambridge, Massachusetts.

Straughn, I. R. (1973). Evolution of anuran mating calls. Bioacoustical aspects. *In* "Evolutionary Biology of the Anurans" (J. L. Vial, ed.), pp. 321–327. Univ. of Missouri Press, Columbia.

Stringer, E. T. (1972). "Foundations of Climatology." Freeman, San Francisco, California.

Struhsaker, T. T. (1967). Behavior of elk (*Cervus canadensis*) during the rut. *Z. Tierpsychol.* **24,** 80–114.

Swets, J. A., and Birdsall, T. G. (1978). Repeated observation of an uncertain signal. *Percept. Psychophys.* **23,** 269–274.

Swets, J. A., and Kristofferson, A. B. (1970). Attention. *Annu. Rev. Psychol.* **21,** 339–366.

Swets, J. A., Tanner, W. P., and Birdsall, T. G. (1961). Decision process in perception. *Psychol. Rev.* **68,** 301–340.

Swets, J. A., Green, D. M., Getty, D. J., and Swets, J. B. (1978). Signal detection and identification at successive stages of observation. *Percep. Psychophys.* **23,** 275–289.

Tanner, W. P. (1956). Theory of recognition. *J. Acoust. Soc. Am.* **28,** 882–888.

Tatarski, V. I. (1961). "Wave Propagation in a Turbulent Medium." McGraw-Hill, New York.

Tembrock, G. (1974). Sound production of *Hylobates* and *Symphalangus*. *Gibbon Siamang* **3,** 176–205.

Tenaza, R. R. (1976). Songs, choruses, and countersinging of Kloss' gibbons (*Hylobates klossi*) in Siberut Islands, Indonesia. *Z. Tierpsychol.* **40,** 37–52.

Thomasson, S. I. (1977). Sound propagation above a layer with a large refraction index. *J. Acoust. Soc. Am.* **61,** 659–674.

Tolstoy, I., and Clay, C. S. (1966). "Ocean Acoustics." McGraw-Hill, New York.

Veniar, F. A. (1958). Signal detection as a function of frequency ensemble, I & II. *J. Acoust. Soc. Am.* **30,** 1020–1024, 1075–1078.

Waser, P. M. (1977a). Individual recognition, intragroup cohesion and intergroup spacing: evidence from sound playback to forest monkeys. *Behaviour* **60,** 28–74.

Waser, P. M. (1977b). Sound localization by monkeys: a field experiment. *Behav. Ecol. Sociobiol.* **2,** 427–431.

Waser, P. M., and Waser, M. S. (1977). Experimental studies of primate vocalization: Specializations for long distance propagation. *Z. Tierpsychol.* **43,** 239–263.

Wasserman, F. E., (1977). Intraspecific acoustical interference in the White-throated Sparrow (*Zonotrichia albicollis*). *Anim. Behav.* **25,** 949–952.

Watson, C. S., and Nichols, T. L. (1976). Detectability of auditory signals presented without defined observation intervals. *J. Acoust. Soc. Am.* **59,** 655–668.

White, S. J., and White, R. E. C. (1970). Individual voice production in gannets. *Behaviour* **37,** 40–54.

Wiener, F. M., and Keast, D. N. (1959). Experimental study of the propagation of sound over ground. *J. Acoust. Soc. Am.* **31,** 724–733.

Wiley, R. H. (1973). The strut display of male Sage Grouse: a "fixed" action pattern. *Behaviour* **47,** 129–152.

Wiley, R. H. (1976). Communication and spatial relationships in a colony of Common Grackles. *Anim. Behav.* **24,** 570–584.

Wiley, R. H., and Cruz, A. (1980). The Jamaican Blackbird: A "natural experiment" for hypotheses in socioecology. *In* "Evolutionary Biology" (M. K. Hecht, W. C. Steere, and B. Wallace, eds.), Vol. 13, pp. 261–293. Plenum, New York.

Wiley, R. H., and Richards, D. B. (1978). Physical constraints on acoustic communication in the atmosphere: Implications for the evolution of animal vocalizations. *Behav. Ecol. Sociobiol.* **3,** 69–94.

Wilson, E. O. (1975). "Sociobiology." Harvard Univ. Press, Cambridge, Massachusetts.

Witkin, S. R. (1977). The importance of directional sound radiation in avian vocalization. *Condor* **79,** 490–493.

Wood, A. B. (1966). "Acoustics." Dover, New York.

Woodward, P. M. (1953). "Probability and Information Theory, with Applications to Radar." Pergamon, Oxford.

6

Grading, Discreteness, Redundancy, and Motivation-Structural Rules

EUGENE S. MORTON

I. INTRODUCTION

A. Differentiating between "How," "Why," and "What" Birds Communicate

Interest in the evolutionary origins and adaptiveness of avian vocalizations has increased in the last decade. This interest is forcing a reexamination of old concepts and assumptions so that descriptions of how birds communicate can be

ACOUSTIC COMMUNICATION IN BIRDS
VOLUME 1

ISBN 0-12-426801-3

tied to the ultimate question of why they vocalize. We sometimes make assumptions when describing how birds are communicating that make it difficult or impossible to interpret these data from an evolutionary perspective. MacKay (1972, p. 3) sensed this when he stated "how can we avoid foreclosing empirical issues, and missing essential points. . . ., by our choice of conceptual apparatus and working distinctions? . . . in a new field, we must expect our concepts to need constant refinement, and must be alert for signs that something important is escaping us because our customary ways of looking at the phenomena have a significant 'blind spot'."

The ultimate question, "why," can be answered when we know the fitness benefits gained by the sender. Traditionally, the information in and function of vocalizations were assessed by studying the reactions of receivers and to do otherwise was considered anthropomorphic or nonoperational (Marler, 1961, 1967; but see Smith, 1977, p. 287). Communication involves the change in the sender and receiver during and immediately after their interaction, but the change of interest to the evolutionary biologist is the average increase in inclusive fitness the interaction brings to the sender. Selection does not favor a bird "sending" unless there is, on the average, a fitness benefit. Granted, there is value in studying "how" birds communicate but there is danger that, in the interpretation of the data, only fitness benefits shared by the sender and receiver will be brought out. How, what, and why questions should not be separated if the evolutionary history of communication is to be a focus.

Smith (1974, p. 1018) suggests that communication should be mutually beneficial to sender and receiver because:

> In the evolution of communication within a species, one normal constraint usually is that the exchange of information must be useful to both the communicator and the recipient. The behavioral patterns evolved by the communicator enable him to transmit information that increases the probability of a social response suited to his needs. A recipient evolves the tendency to respond to this information only when the response suits his needs, which often differ, at least superficially, from those of the communicator. When their needs are not compatible, lack of selection pressure for the recipient to respond appropriately usually removes any advantages for the communicator, or at best yields an evolutionarily unstable situation of misinforming, to which recipients are always counter-adapting.

Smith emphasizes the mutual benefits of communication in his own work (see Smith, 1977) but, nonetheless, suggests the framework for an evolutionary approach in the quote above. What are the sender's needs, how do they differ from the receiver's, is there misinforming, and how are recipients counteradapting? An evolutionary approach involves only a change in emphasis: from downplaying these questions to focusing on them when interpreting data.

However, more than a change in emphasis is needed before communication studies conform to an evolutionary perspective more consistently. We must critically examine some assumptions. I do this by presenting them in their historical perspective below before discussing discreteness, redundancy, and motivation-structural rules in communication.

B. Assumptions Derived from Ethology, Linguistics, and Information Theory

In ethology, displays are defined as behaviors specially adapted for social signal functions (Moynihan, 1956). These adaptations included stereotypy and exaggeration, characteristics serving to make them conspicuous against background noise. Displays could be named, counted, and compared among species (Moynihan, 1970; Wilson, 1972), and variations in them were considered unimportant until recently (Barlow, 1977). Displays are said to appear and disappear over evolutionary time depending on their effectiveness in providing information in interactions (Moynihan, 1970). The form and number of displays characteristic of a species were often compared between species in making taxonomic judgements. Displays were viewed as species-typical characteristics with little attention given to how selection acting at the individual or genic level promoted particular displays. There was an interest in the evolutionary origins of displays from nondisplay precursors. Some visual displays and their precursors were extensively studied (e.g., Morris, 1956) but the precursors for particular vocalizations were not obvious and the ethological display concept did not provide an explanation for the origins of particular vertebrate vocalizations (Smith, 1977, p. 325).

The display concept was used as the logical background for hundreds of "vocalizations of" papers that listed vocalizations as stereotyped and named entities. Often the vocalization names were based on the apparent function of the sound or were given onomatopoetic names. Sounds were named so that they could be discussed, usually at the level of the species (e.g., the *caw*) but no adaptive significance was attached to their physical structure: a *caw* was considered to be arbitrary in structure like words are. I believe that people are uncritically comfortable with the concept of arbitrariness in animal sound structures due to our use of arbitrary sounds in speech. In Section II, I present some ideas that suggest that sound structures are not arbitrary.

Marler (1955) first countered the view that the physical structure of bird sounds is arbitrary by arguing that some sounds are structurally designed to provide clues to the sender's location while others are not. Recent studies on the adaptations of sounds for long-distance transmission have further reduced the veracity of the assumption that vocalizations have arbitrary physical structures (Morton, 1975; Marten *et al.*, 1977; Waser and Waser, 1977; Wiley and Richards, 1978).

Ethology has not produced a conceptual model or unifying theoretical framework for analyzing animal communication (Smith, 1977, p. 18). Ethologists used concepts derived from linguistics and information theory that were not based on evolutionary theory. From linguistics, zoosemiotics has, inadvertently, come to be used as a one-word equivalent of the study of animal communication (Sebeok, 1977, p. 1055). Information theory provided a mathematical framework within which a series of terms (e.g., channel, entropy, bits) were rigorously definable; even "information," a term that might better have remained ambiguous, was defined as the reduction of uncertainty. Game theory may replace information theory, even in the field of man–computer speech interfacing (Thomas, 1978). These nonevolutionary concepts led ethologists to separate the process of communication from other aspects of the biology of animals, aspects that provide the sources of natural selection acting on vocalizations and communication. The concept that senders provide and share information, and that displays contribute to the "orderliness" of interactions, pervades notable works on communication (e.g., Smith, 1977) with little consideration that these assumptions need verification or, at least, explanation.

An example will clarify what I have just said. Chickadees (*Parus* spp.) utter *chip* sounds that are termed "contact notes" when the birds forage in flocks during the winter. Contact note is a name given to the sound ascribing its function, to keep flock members together. We have an explanation for the sound: flock members share information about their locations and thereby stay in proximity, and receivers and senders are benefited because hawks cannot easily approach a flock undetected (Morse, 1973). But this explanation is a series of untested assumptions and is too general and vague to produce testable hypotheses. The *chip* is chevron-shaped when analyzed on a spectrograph, and when this sound structure is compared across many species the information conveyed is general: something of interest has been perceived by the sender (Morton, 1977). Therefore, another interpretation of contact notes is possible. The sender may *chip* when it discovers food or when it is moving rapidly from perch to perch, but the sender is benefited in either case if the movement of others away from the sender is slowed or if the *chip* attracts them toward the sender. For receivers, if food resources are sufficiently rich, responding to *chips* is advantageous in a trade-off between loss of foraging time and the predator detection benefit to being a flock member. This approach, emphasizing the importance of the sound's structure and differences in fitness benefits between sender and receiver, provides us with a series of testable hypotheses about the function of chips. This approach also places the communication within an ecological setting: food richness becomes important to our understanding of receivers' positive responses to these "contact notes." The study of communication thus becomes integrated with other aspects of biology.

C. Is Communication a Sharing of Information, Manipulation, or What?

Why is it that cooperation among communicators, sharing beneficial information, is assumed by so many researchers? Above, I suggested that this might be due to a sort of "cryptic anthropomorphism" wherein human speech has influenced our thinking about animal communication. However, it is also likely that communicating birds do have the appearance of sharing information or cooperatively communicating. But, the evolutionary process leading to this appearance may have little to do with mutually beneficial information between sender and receiver. The appearance of cooperation in communication may be analogous to the appearance of cooperation in pairbonding. What used to be thought of as a cooperative, mutually beneficial series of behaviors resulting in reproduction "for the good of the species" is now viewed more as a contest between the sexes, and terms such as mate testing and anticuckoldry behavior are used to describe the "cooperating" effort of mating (e.g., Trivers, 1972).

Wallace (1973) suggested that vocalizations might be used to misinform. His model showed how signals that successfully misinform receivers may improve the signaler's relative fitness to the point where the entire population eventually consists of misinformers. Now, the formerly manipulative signal is used by all and has become a convention. The human observer may not find the original manipulative function and the signal's origins may be undetectable. However, if Wallace is correct, we should be able to detect new cases of misinformation for new forms should continue to arise.

Dawkins and Krebs (1978) suggested that the word "communication" as used by ethologists should be abandoned in favor of "manipulation." This may be an overemphasis of the fitness benefits derived from misinforming. Is a convention, in the sense described above, still a form of misinforming? Conventions and manipulative signals are vague terms and probably not readily distinguishable empirically but have conceptual value if they are viewed as components of communication. But perhaps Darwin's (1872) term, "vocal expression," is less biased than communication and might better indicate that the researcher is concerned with the evolutionary history together with the immediate process of communicating.

II. THE MOTIVATION-STRUCTURAL RULES MODEL

A. Background

My suggestion that the physical structures of vocalizations have significance for understanding vocal expression in vertebrates was published earlier (Morton, 1977). There I suggested that vocalizations given by living species, particularly those used in close range, follow a structural code that was established long ago in the evolutionary history of terrestrial vertebrates. I suggested that in evolutionarily older vertebrates, such as amphibians and reptiles, which continue to grow after sexual maturity, call frequency may directly reflect body size. Thus large individuals tend to have dominant frequencies lower than those of small individuals (e.g., Martin, 1972; Zweifel, 1968). Vocalizations have subsequently been shown to operate as effective size indicators in *Bufo bufo* (Davies and Halliday, 1978) and *Physalaemus pustulosus* (Ryan, 1980). This is also likely in crocodilians (J. Lang, unpublished data). In the amphibians just mentioned, the lower-frequency calls of larger males function to repel smaller males or form the basis upon which females choose larger mates. In several *Hyla* species smaller males remain silent but position themselves near larger males that are calling to attract females (Fellers, 1979). These smaller males are, in effect, parasitizing the sound of large size to obtain matings.

The sounds used by aggressive birds and mammals are low in frequency, whereas fearful or appeasing individuals use high-frequency sounds (Morton, 1977). Size symbolism provides an explanation for the evolutionary origin of these vocalizations. But since size in homeothermic vertebrates is constant at maturity, relative to amphibians and reptiles, the use of low- or high-frequency sounds does not directly reflect differences in size between communicating individuals. Instead, the sounds are generally thought to reflect differences in motivation. I term the code suggesting the relationship between sound frequency and motivation, motivation-structural rules.

B. The Motivation-Structural Rules Model

The model predicts that vocalization structures follow a simple code with two physical dimensions, one dimension pertains to sound quality and ranges from harsh or atonal to pure-tone-like, the second dimension pertains to sound frequency. The two dimensions may vary independently but are often synchronized such that the lowest-frequency sounds are harsh while higher-frequency sounds are tonal.

Figure 1 illustrates the relationship between sound structures and motivation showing the "endpoint" (lowest harsh and highest tonal) sounds with the intermediate sounds grading between the endpoints in structure and depicting a variety of intermediate vocal expressions. An aggressive motivation is expressed through a low, usually harsh, vocalization and a fearful or appeasing motivation with a high, tonal sound. If a sound increases in frequency, it expresses lowered aggressive tendencies; if it falls, it expresses increased aggressive tendencies.

I emphasize that Fig. 1 simply shows a sampling of sound structures at various motivational stages from aggressive to fearful, to illustrate the structural coding

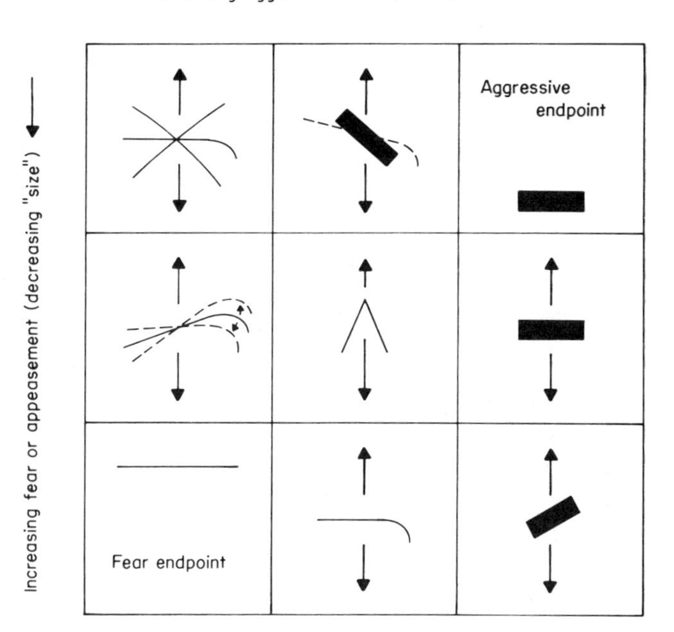

Fig. 1. A diagrammatic representation of sound structures to illustrate the motivation-structural code. Each block shows a hypothetical sound spectrogram (vertical scale, frequency; horizontal scale, time), with thin lines depicting a tonal sound and thick lines, harsh or broad band sounds. The arrows mean that the frequency of the depicted sound may vary up or down, approaching either the low- or high-frequency endpoints.

In the upper left block, motivation is weakly tending toward fear if the thin line slopes upward (its frequency rises) or weakly toward aggression if the slope is downward. In the middle left block, closer to the fear endpoint, sounds rise variously upward, between the dashed lines, and are tonal. The three blocks on the aggressive (right) side of the diagram are all broadband but the frequency is rising in the "distress" call, where fear and aggression are interacting. The central block depicts a chevron since the motivation and its sound structure are not nearer one endpoint than the other. (From Morton, 1977.)

of these moods. The potential combination of sound qualities and frequency changes is enormous.

The bark appears in the center and is characterized by a rise and fall (or fall and rise) in frequency. The rise and fall means that neither endpoint is being approached; therefore, unlike nonchevron sounds that tend toward one of the motivation-structural endpoints, barks symbolize neutral or adaptively indecisive motivation. Barks code arousal and interest but not necessarily motivational tendencies if they have an average frequency intermediate to the low and high endpoints. However, bark structures may also vary in frequency and sound quality representing motivational tendencies toward either endpoint (e.g., Morton and Shalter, 1977). Barks function in a great diversity of contexts, for example, from precopulation and contact calling to mobbing, territorial defense, and alarm. A vocalizer may either attract or repel others by barking, depending on its species and context.

C. Motivation-Structural Code and the Functions of Vocalizations

Through evolutionary time, various sound structures from the motivation-structural code may be favored by selection and become characteristic of the vocalizations of species, populations, or individuals. Some species might evolve to use only the aggressive end of the code while others use the entire code and others replace vocalizations with other communication modes. The selection pressures are derived from functional benefits to vocalizer fitness. Functions determine which particular sound structures will be used, what modifications may be adaptive (e.g., to enhance locatability or species distinctiveness), the contexts in which it is beneficial to use them, the amount of structural variation, and the timing of use or disuse during ontogeny. Functions can only be determined by studying living animals, of course, but comparative studies of closely related species can give some insight to the evolutionary time scale in which selection operates (Thielcke, 1976, p. 142).

However, the study of function apart from signal structure, as has been prevalent in the past, may not lead to an understanding of the evolution of vocal expression, for there is little basis for generalization (e.g., the *caw* of a crow and the *chip* of a sparrow would not be identified as barks with motivation-structural code similarity).

Selection may wed the motivation-structure code to function in ways that are subject to misinterpretation if only function and receiver reactions are considered. For example, a vocalization of the Herring Gull (*Larus argentatus*) termed the "food finding" call is reported to function in attracting other gulls to food (Frings *et al.*, 1955). In fact, single gulls remain silent when they locate food; it is only when potentially competing gulls are near that the call is given (J. Hand and E. Morton, unpublished data). Furthermore, a small single food item that can be swallowed immediately does not elicit a call from a gull finding it, while large items which a single gull is unable to swallow quickly do elicit the call. It seems unlikely that the "food finding" call functions to attract others since it is not given simply when food is found but only when other gulls are close enough to compete for the food. The call's structure is tonal and high in frequency, which the motivation-structural code predicts would be given by a frightened gull. The context, with other gulls coming in to chase the sender, seems consistent with the idea that the sender is frightened but this does not explain why selection has favored the sender giving the call. What benefit does the sender receive? One hypothesis is that the call sometimes causes hesitation in the receivers, they may look for a predator, permitting the sender to grasp the food.

Another example is provided by the Purple Martin (*Progne subis*), a cavity-nesting colonially breeding swallow. Competition, often involving vicious fights, occurs over nest sites in "martin houses" provided by humans. The plumage of first-year males resembles that of females and they arrive at colonies several weeks after adult males have usurped nest sites (Rohwer and Niles, 1979). Young males encounter intense aggression by adult males when the former try to take nest sites (Johnston, 1966). During attempts to land on the martin house, it is common for young males to utter a series of high-frequency barks, identical to those elicited by dangerous predators (E. S. Morton, unpubl. data). This causes members to vacate, leaving time for the young male to enter a nesting cavity. A relatively high frequency bark is given by alarmed martins. In this case again, selection has favored the vocal expression of fear in a context where it is only of benefit to the sender.

What selection has favored in both gull and martin is the *vocal expression of mood*. The gull and martin probably are fearful: they experience aggression in the contexts described. There is no deceit involved on their part; their use of fear sounds in the two contexts may be manipulative in effect, but the sounds appear to truly reflect their motivational states. Selection has favored their vocal expression of this mood since reactions to it are often beneficial to them. In the gull and martin examples, receivers respond to the senders' calls because the average consequences of doing so are also of benefit to them, even if not beneficial in the specific contexts described. If this were not so, selection would favor receivers who ignore these sounds.

The motivation-structural code thus provides a basis to study the significance of sound structures in communicating birds. It provides an explanation of the historical origin of the physical structure of vocalizations that should be useful to researchers studying the functions of vocalizations.

III. GRADING, DISCRETENESS, AND REDUNDANCY

A. Distance and the Motivation-Structural Code

Marler (1967) and others have noted that long distance calls of birds tend to form discrete classes more frequently than do those used over short distances. As the distance between sender and receiver increases, small changes in sound structures will increasingly tend to be obscured by sound attenuation to the point that there is no selection pressure favoring the use of graded vocalizations. With sounds used only in long-distance communication, such as many passerine bird songs, selection may favor the use of sound structures adapted for propagation across long distances that also code for species specificity. Thus selection favoring motivation coding may take a back seat for these distance-adapted sounds. When the sender is usually highly visible to receivers, long-distance calls may also contain the motivation-structural code and exhibit grading. Payne (1979), for example, reported the lowest and harshest songs of the Senegal Indigo-bird (*Vidua chalybeata*), a bird of open habitat, to be correlated with sender aggressiveness.

Vocalizations of the Carolina Wren (*Thryothorus ludovicianus*) show the relationship between variability of sounds within one class and the general distance between sender and receiver(s) (Table I). The sound classes are given onomatopoetic names (except song) with sound quality to the human ear indicated. The physical structures of sounds within each class are more variable in those sound classes used by birds close to recipients. The harsh quality sounds are more variable than sounds with a tonal quality. *Chirts* and *pi-zeets* may grade between classes as may *pees* and *scees;* grading occurs within but not between other classes. The two long-distance sounds, individual male songs and male *cheer,* are discrete, but each male has from 27 to 41 different songs.

B. Grading and Discreteness

1. Grading

More attention is being given to the significance of graded sound structures to vocal expression as equipment to analyze long sequences of vocalizations has become available. Statistical techniques to analyze graded vocalizations are also being developed (see Miller, 1979, and references therein).

The motivation-structural code is followed both within and between vocalization classes. The vocalization classes in Table I show relatively discrete sounds that would have been called displays by earlier ethologists. The *growl, pee,* and *dit,* for example, are sound classes whose structures fit the aggressive endpoint, fear endpoint, and chevron-shaped bark, respectively, as shown in Fig. 1. Within-class variation is illustrated in Fig. 2. Here, *scees* (Table I) showing structural variation are being uttered by a male Carolina Wren losing in a fight with another male, who remained silent. During the first 12 sec of the fight (see Fig. 2) the losing male continually attempted to flee but was held by the strong feet of the winning male. The sounds during this time have a rising frequency and harsh quality. The sounds are similar to the structure diagrammed in the lower right block of Fig. 1, and show a high level of both fear and aggressive motivation. At 20 sec, the sounds do not consistently rise in frequency and at 22.5 sec the losing male pecked back. This was accompanied by a change in sound structure: the sound became greater in bandwidth and lower in frequency. The structure of the losing male's vocalizations followed his overt behavior in the ways predicted by

TABLE I

Sounds of the Carolina Wren (*Thryothorus ludovicianus*) Arranged in Order of Decreasing Sender to Receiver Distance

Sound	Distance from sender to receiver	Sound structure	Structural variation within sound class
♂ Song	Long	Tonal	None
♂ *Cheer*	Long to medium	Tonal	None
♀ *Chatter*	Medium to short	Harsh	Little
♀ *Dit*	Medium to short	Tonal	Little
♂ *Dit*	Medium to short	Tonal	Little
Rasp	Short	Harsh	Great
Chirt	Short	Tonal to harsh	Great
Pi-zeet	Short	Tonal to harsh	Great
Tsuck	Short	Tonal	Little
Nyerk	Short	Harsh	Great
Scee	Short	Tonal to harsh	Great
Pee	Short	Tonal to harsh	Great
Growl	Short	Harsh	Moderate

the motivation-structural code. Morton (1977) and Morton and Shalter (1977) discuss grading in other species and vocalization classes of the Carolina Wren as examples of the motivation-structural code. Miller and Baker (1980) describe graded calls of the Magellanic Oystercatcher (*Haematopus leucopodus*) and suggest that grading provides a "high information content." In the Common Loon (*Gavia immer*) gradiation in the tremolo call indicated the probability that the calling bird would attack. Higher frequencies indicated a greater reluctance to attack (Barklow, 1979). Perhaps the most intensely studied avian graded vocalization system is that of the Northern Jacana (*Jacana spinosa*) with five of six vocal classes grading into one another (Jenni *et al.*, 1974). Jenni *et al.* discuss the problems of subjectivity encountered when attempting to categorize such an extensively graded vocal system.

2. *Discreteness*

When classes of vocalizations do not grade they are termed discrete, but this designation is a relative term and should not imply that no variation exists. Often birds are said to have discrete vocalizations relative to mammals (Marler, 1967). This apparent difference between the two groups could be related to differences in vocal tracts: mammals have flexible pharynges with a high potential for resonating sounds; birds have relatively rigid tracheas with little resonating potential (Greenewalt, 1968). Therefore, mammals are able to vary sound structures through resonance changes without changing tension in the vocal membranes.

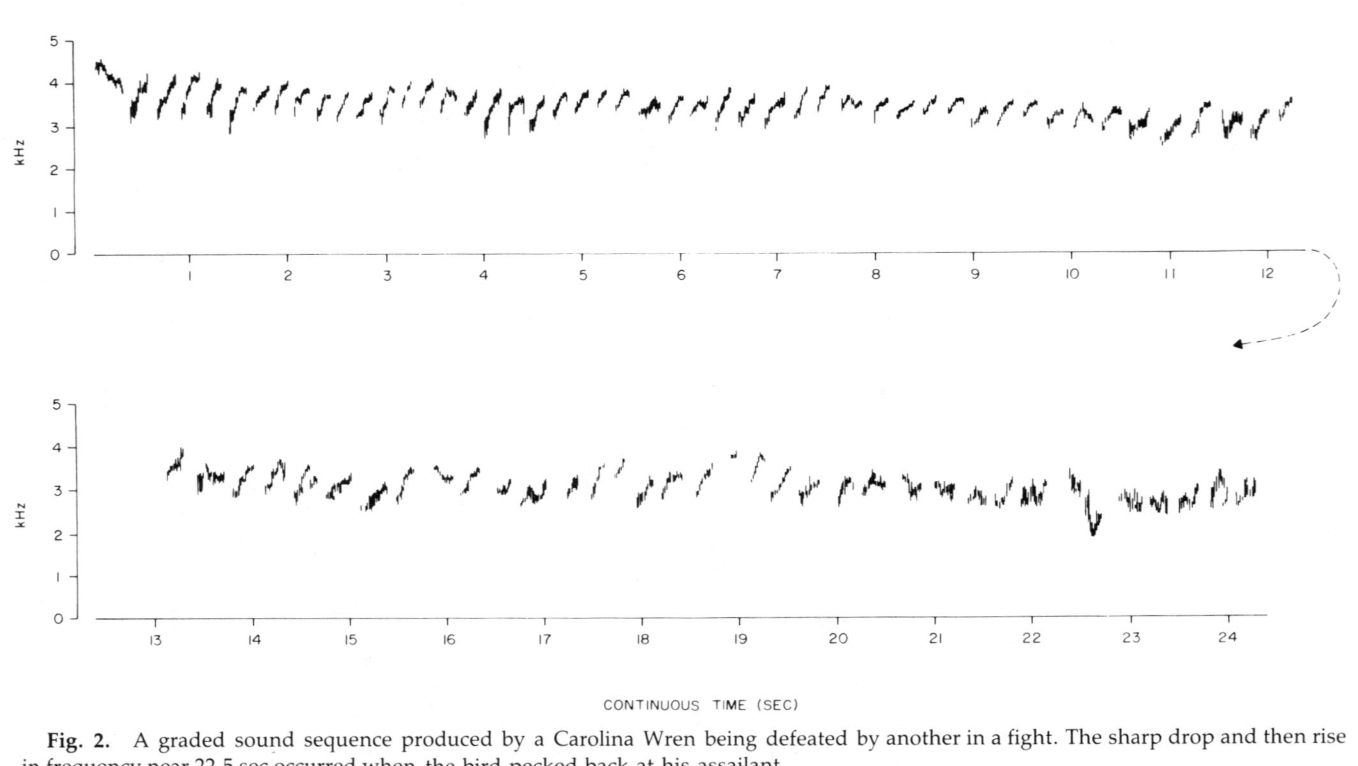

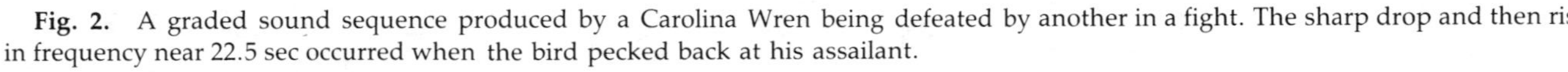

Fig. 2. A graded sound sequence produced by a Carolina Wren being defeated by another in a fight. The sharp drop and then rise in frequency near 22.5 sec occurred when the bird pecked back at his assailant.

In birds there may be, in addition, a relationship between the frequency and quality of vocalizations and discreteness. Low-frequency sounds may acquire a broadband spectrum and hence a harsh quality when tension on the vibrating membrane producing them falls low enough to produce nonharmonically related tones. If this observation were raised to the level of a prediction, a dilemma would arise, since in small birds their lowest frequency sounds are generally low in amplitude and used while senders are close to receivers. Sounds used at close distances tend to be graded apart from their frequency (Section III,A). Carolina Wren sounds show a tendency for harsh vocalizations to be used over short distances and exhibit the most grading (Table I). The question of why selection favors either graded or discrete vocalizations in particular contexts remains open. So far, the only general prediction is related to communication distance.

Discrete vocalizations may be combined to produce structurally complex vocal expressions. The *chickadee* call of the Black-capped Chickadee (*Parus atricapillus*) provides an example (illustrated in Ficken *et al.*, 1978). The call consists of two to four chevron-shaped notes (''chicka'') that decrease in frequency followed by still lower frequency and harsh ''dee'' sounds. These components are also used separately, which is why the call is termed compound (Ficken *et al.*, 1978). Using the motivation-structural code, I would describe the sound as a series of barks followed by low, harsh sounds. The barks indicate that the sender has perceived something of interest, with the decrease in their frequency indicating increasing aggressive tendencies. The last notes are indicative of an aggressive motivation but not at the aggressive endpoint, for this species' repertoire contains lower frequency and harsher sounds used in fighting (the *snarl* described in Ficken *et al.*, 1978). The motivation-structural code suggests that whatever stimulus elicited the call, the sender is interested in it, and it is not something that causes fear in the sender, or movement away from it, as indicated by the broad-band, aggressive, structure/mood relationship. The call is used most frequently by individuals in flocks during the nonbreeding season, apparently in situations of mild alarm (Ficken *et al.*, 1978). In the presence of stimuli that elicit a compound sound with these motivation-structual components it may benefit the sender to attract other chickadees, which, indeed, is the function of the *chickadee* call (Ficken *et al.*, 1978).

Figure 3 illustrates nine discrete calls of three different structures uttered rapidly in 2 sec by a captive 3-day-old Black Crake (*Limnocorax flavirostra*). The use of such differing sound structures in such a short time period would seem difficult to explain. In the wild, or when reared by parents in captivity, 3-day-old crakes are fed by their parents and the chicks compete for parental attention and the food they peck from the parent's beak (E. S. Morton, unpublished data). When this chick was not being fed by us, it uttered only the chevron-shaped barks recorded in Fig. 3; they sound like the *peeps* common to Domestic Fowl (*Gallus domesticus*). The crake uttered the two other sounds described in Fig. 3

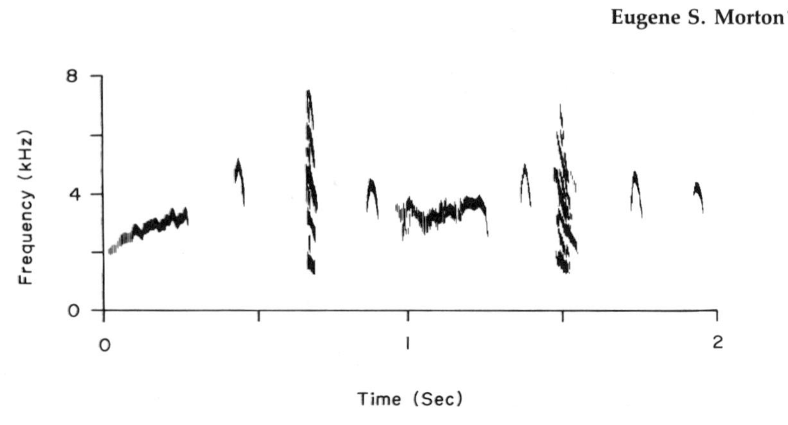

Fig. 3. Three structurally divergent sound types of a captive 3-day-old Black Crake (*Limnocorax flavirostra*) approaching a forceps holding food (see text).

only when we approached it with forceps holding food. From time 0 to 1 sec, the chick used a rising tonal sound when seeing the forceps, then a bark, then a broadband call dropping in frequency. This sequence is repeated again from 1 to 2 sec as indicated in Fig. 3. I believe it is the natural context of parental feeding plus sibling competition that explains this sequence of widely varying components from the motivation-structural code. The rising tone might be due to increasing fear or appeasement directed toward the parent, followed by a bark when the food (item of interest) is approached, followed by an aggressive sound structure just following food taking at both 0.7 and 1.5 sec (Fig. 3). The aggressive sounds might function to ward off siblings competing for the food.

C. Redundancy in Passerine Songs

1. Songs and the Motivation-Structural Code

The mechanisms by which songbirds produce their high-amplitude songs are not well known and are under debate. The debate is over the contribution of syringeal membrane vibration versus an aerodynamic whistle mechanism in song production (A. Gaunt, unpublished data). The whistle mechanism, using aeolian forces to produce sound, is capable of explaining the high source amplitudes songbirds produce in their far-carrying songs. But high source amplitude may be more important to our understanding of the functions and evolution of song than solely as an accouterment to increase sound propagation.

If songbirds use aeolian forces to produce song, then sound amplitude takes on greater importance in symbolizing large size than does sound frequency. For a whistle of a given size, the greater the airflow rate through it, the higher the sound amplitude produced. Also, the frequency of sound produced increases directly with airflow rate, as anyone who has blown a whistle knows. Therefore, a larger animal might symbolize its size advantage or aggressive motivation by illustrating its ability to produce a call with higher frequency and amplitude than those of another animal if the sound is produced by aeolian forces. Rats (*Rattus norvegicus*), which produce their ultrasonic calls using aeolian whistle mechanisms, use an 80-kHz call when aggressive and about a 20-kHz call when losing a fight (Roberts, 1975). Since the motivation-structural code stresses the relationship between aggressive motivation and low sound frequency, when the sounds are produced by vibrating vocal membranes, I would like to think that the use of high sound frequency by aggressive rats, using an aeolian sound production mechanism, is the exception that "proves the rule."

The importance of sound amplitude to our understanding of song function and evolution will remain little known until sound production mechanisms are better understood. However, Kroodsma (1979) argues convincingly that amplitude was related to dominance and to "leader/follower roles" during song matching in Long-billed Marsh Wrens (*Cistothorus palustris*).

2. *Redundancy and Song Function in the Carolina Wren*

Oscine song behavior is remarkable for its diversity of singing styles and repertoire sizes (Hartshorne, 1973). I want to focus on the singing behavior of the Carolina Wren, a species with a male song repertoire averaging 32 songs within a range of 27–41 for a sample of 15 males (Chu, 1979). What is the selective value and function of this *redundancy:* having more than one song type with seemingly identical functions? Only males sing in this species, the only member of the genus to reach the temperate zone latitudes, but in the 17 tropical species females also sing, often duetting with their mates (E. S. Morton, unpubl. data). Most tropical passerines are permanently pairbonded and their songs function in territorial defense (Morton, 1980). In species with female and male song, songs seem to be directed toward like-sexed territorial intruders (see Chapter 4 by Farabaugh, Volume 2).

Perhaps there has been a temperate zone bias in the determination of song function, when mate attraction is suggested as one main function of song? In the Carolina Wren, males may form pair bonds even when they are too young to sing fully developed songs and then continue singing throughout the year even while they are constantly paired (Morton and Shalter, 1977). Territorial defense is also a permanent feature of this wren's social system. The Carolina Wren ranges from Maine to Florida, west to central Oklahoma, south to southern Tamaulipas, Mexico, with separated populations in Yucatan, Mexico, and portions of Guatemala and Honduras.

The structural features of the song used in species recognition are not well known but, through playbacks, I determined that songs from widely separated populations are responded to by any Carolina Wren as conspecific. Yucatan wrens responded to songs from Maryland and Florida wrens and vice versa. But at the local population level, 66–75% of the 32 song types in a male's repertoire are held in common with one or more neighboring males. Repertoire sharing results from young males settling near established males and copying their songs (Helgeson, 1980), just as Kroodsma (1974) described for the closely related Bewick's Wren (*Thryomanes bewickii*). The percentage of songs held in common among males decreases directly with distance. Chu (1979) determined that males 100 km apart have only 20–25% of their song types held in common, if the sampled males are connected by continuous wren populations. If the wren distribution is not continuous, for example, if an island population is compared to a mainland population 3 km distant, only 12% of the songs are held in common. Some songs are unique to a single male wren, most songs in an individual's repertoire are held in common with neighbors, as mentioned, but about 3% of the songs have a much wider geographic range for reasons that are not clear. These few song types were found throughout the 140-km linear transect studied by Chu (1979).

With the above information in mind, I now describe how individual males use their songs before offering an hypothesis to explain the function and adaptive significance of redundancy in Carolina Wren singing.

Males repeat the same song type from a few to many times before they switch to another song type. The highest number of repetitions I have observed is the singing of one song type 206 times for a duration of 21 minutes without a break. A repetition of the same song type is called a bout (AAA . . . BBB . . . CCC . . .). Kroodsma (1977) has compared singing styles in North American wrens; the Carolina Wren has a song of low complexity along with the Cactus Wren (*Campylorhynchus brunneicapillus*) and the Rock Wren (*Salpinctes obsoletus*), since each song consists of the repetition of one syllable type. Carolina Wren song bouts are lengthier from awakening to about 1000 hr after which the bouts are shorter with more nonsinging time between each bout (Table II).

It takes an observer about 2 full days to record all of the song types in a male's repertoire, or about 75 bouts of singing (Chu, 1979; E. S. Morton, unpublished data). The important point is that the singing behavior does not seem to maximize the potential diversity of songs a male could sing. Thus, the question of why Carolina Wrens have so many song types does not seem to be answered by the antihabituation hypothesis (e.g., Krebs, 1976), which states that large song repertoires reduce the chance that listening conspecifics will ignore a male's songs through classical habituation. Certainly, that hypothesis does not explain the large size of the wren's repertoire: most commonly studied temperate zone passerines with redundant repertoires have fewer songs, usually about 5–20.

Could some songs differ in information from others, thus suggesting a "need" for a large repertoire? This has been suggested for the small song repertoires in other species, wherein unmated males sing a song type different from that of

TABLE II

Relationship of the Duration of Song Bouts to Time of Day in One Carolina Wren (*Thryothorus ludovicianus*)[a]

	Time of day (hr)		
	0612[b]–1012	1012–1412	1412–1812
No. of bouts	18	10	16
Mean no. of songs/bout	99* ± 50.9[c]	41 ± 19.8	25 ± 16.6
Mean duration of bout (sec)	521 ± 277.8	201 ± 109.3	158 ± 87.0
No. of different song types used	18	10	10

[a] Data for 11 March 1976.
[b] Time of first song.
[c] Mean ± standard deviation.
*Mean significantly higher than other time periods ($p < 0.005$) by t test.

mated males or males sing some song types while in the center of their territories and others when interacting with another male at a territorial boundary (Morse, 1970; Lein, 1978; Smith *et al.*, 1978). There is no indication of this for Carolina Wrens. Males do not sing one song type preferentially over others in their repertoire, position within the territory is unimportant in predicting which song type they will sing, and time of year is unimportant (E. S. Morton, unpublished data).

Neighbor recognition via song, wherein a male might not expend much energy reacting to a song it recognized as a neighbor's, does not explain the large repertoire size, although it could explain why neighboring males hold many songs in common (Falls, 1979). However, a male's neighbors are the greatest threat to his territory for 8 months of the year; territorial establishment by the young persists for only about 4 months in my Maryland study site (August through November; E. S. Morton, unpubl. data). In years of high wren density, about four to ten border clashes per day take place between an individual male and his neighbors. If a male is experimentally removed, neighbors take over the territory within 1 day (Morton and Shalter, 1977). Thus, there is no reason to think that neighbors are responded to differently than strangers such as has been reported in some species that are territorial for only the breeding season.

There is one situation where males sing song types chosen nonrandomly from their repertoire. When one male sings other males are apparently stimulated to sing. If the first singer happens to be near the territorial boundary shared with a neighbor, this neighboring male may begin singing the identical song type. This song matching occurs only about 10% of the time one male's singing stimulated others to sing. When song matching occurs, the neighbor male faces the first singer and often moves in his direction if the first singer keeps on singing (E. S.

Morton, unpubl. data). There is little doubt that song matching indicates that singing is being directed toward a specific individual.

The large repertoire and singing style of the Carolina Wren are not adequately explained by current hypotheses offered to explain redundancy in singing (e.g., Krebs, 1977; Payne, 1979). Mate attraction does not appear to be a function of song in this wren, for males who lose mates do not increase singing rates, young males incapable of singing fully developed songs may pair, and most of the time singing occurs with the male paired (E. S. Morton, unpublished data). The Carolina Wren singing behavior suggests to me a new hypothesis that, although not contradicting the others mentioned, explains or incorporates many heretofore isolated facts about passerine bird song. The hypothesis is termed the ''ranging hypothesis.'' I present it to explain singing behavior in the Carolina Wren, then suggest predictions from this hypothesis for other species.

3. *The Ranging Hypothesis*

The hypothesis attempts to explain: (1) the functions of song; (2) the sources of selection favoring large repertoires in the Carolina Wren; and (3) the preferential learning of neighbors' songs by young males.

A major assumption of the hypothesis is that natural selection operates differently on what I term singers and listeners, although obviously individual males are both. Another way of stating this is that the goals of singers and listeners differ; there is no selection favoring sharing of information, and singers and listeners are usually at odds with one another. The ranging hypothesis suggests that the two goals of singing are to disturb neighboring males, in particular to disrupt their foraging behavior, and/or to threaten individual males.

Listeners are under selection pressure not to be disturbed or threatened by the songs of singers. Selection on listeners favors their ability to determine, as accurately as possible, if singers are truly encroaching upon the listener's territory. Singers are selected to produce songs whose physical structures are adapted to propagate with the least possible degradation of source characteristics (source characteristics are the frequency and amplitude characteristics of the song as it emanates from the singer's mouth). By producing songs with this attribute, singers more often accomplish their goal to sound to the listener that they are sufficiently close to constitute a threat to the listener's territory, thereby disturbing him. The ranging hypothesis suggests that *if the listener has learned to sing his neighbors' songs, he can match this undegraded song stored in his brain to the song he hears.* Therefore, if the listener has learned neighbors' songs, he is able to determine more accurately his distance from the singer and therefore challenge the singer if it is near and ignore the singer if it is far away, i.e., respond to the song in a manner of benefit to the listener.

Since selection is suggested to fashion listener responses in the way just described, selection on singers would be expected to favor mechanisms to decrease listener ability to accurately determine singer distance. The ranging hypothesis suggests that over evolutionary time individual singers have done this by *increasing the number of different song types in their repertoire.* If a singer has some song types not also learned by a listener, the listener will not be able to determine accurately the singer's distance and the song will disturb the listener. Since male Carolina Wrens are apparently able to learn about 32 song types, a listener will not be able to learn all of the song types of all of its immediate neighbors. I envision the large repertoire size of the Carolina Wren to be the result of an "evolutionary arms race" between the adaptations of listeners to thwart the disturbing effects of songs and the adaptations of singers to continue disturbing listeners (Dawkins and Krebs, 1978, p. 309).

I have used terms such as "disturb" and "degraded song" in the formulation of the ranging hypothesis; I will now define these and provide supporting evidence for the hypothesis. How do Carolina Wrens, and other songsters, judge the distance from themselves to a singing conspecific? Songs decrease in amplitude as they propagate from their source due to spherical divergence (Morton, 1975). But amplitude is not a particularly precise method for judging distance for several reasons:

1. The singer may increase or decrease loudness by facing away from or toward the listener.
2. Amplitude can be distorted by wind and temperature gradients and hence not be a reliable indicator of distance.
3. With a high source amplitude (100–110 dB measured at 1 m, E. S. Morton, unpubl. data.) Carolina Wren songs may be expected to arrive at the listener through direct and indirect sound propagation pathways (e.g., reflections off the ground, vegetation), making loudness fluctuate.

An additional and reliable indication of the singer's distance is signal degradation due to differential attenuation of frequencies, reverberation, deflection, and scattering of the sound (Wiley and Richards, 1978). Richards (1978) found that the amount of degradation, but not song amplitude, was an important predictor of the response given to artificial song playbacks by male Carolina Wrens. He determined that undegraded songs evoked search behavior while a degraded version of the same song evoked only a mild singing response. The difference in energy expenditure responding to song playbacks must be substantial. Wrens responding to song playback by searching spend from 2 to 15 min actively flying and hopping from one bush or tree to another, after which they remain stationary and sing for a relatively short time (E. S. Morton, unpublished data). Wrens responding to a playback of song by singing do not move about, and soon return to foraging. Carolina Wren males respond differently to degraded and undegraded songs and these responses are in the direction predicted by the ranging hypothesis: more energy-consuming behaviors are elicited by undegraded songs.

Since song degradation increases directly with distance from the singer, we may assume that the response differences reflect the listener's adaptive response to the threat of territorial intrusion.

How significant is the disturbing effect of song on listeners? During the northern winters, unusually deep snow lasting for several days will lead to 80–90% mortality in Carolina Wrens (Wetmore, 1923; E. S. Morton, unpublished data). This mortality occurs perhaps once or twice per decade in eastern Maryland; between these die-offs, wren densities are high and territories are compressed to about two acres per pair. Therefore, survivors of crashes are essentially founders, hence rapid changes in gene frequencies are possible. A territory which includes fallen limbs, stream edges, or other structures that keep snow from covering the ground leaf litter where wrens forage is best for surviving such crashes. Carolina Wrens that are able to sing when they are satiated thereby may be able to interrupt their neighbors' foraging. The ranging hypothesis suggests that singing could increase neighbor mortality by reducing their foraging time, particularly in winter stress periods. Thus loud and persistent singing, large song repertoires, and the learning of neighbors' songs by newcomers become integrated with relative territorial quality and, in turn, to relative genetic fitness.

This suggests two predictions: (1) the amount of singing should decline with decreasing latitude since winter-induced food stress periods should be increasingly rare to nonexistent in tropical Carolina Wren populations; and (2) males on food-rich territories should sing more than males occupying relatively food-poor territories during winter stress conditions. Table III presents data on the first prediction. The number of songs sung by Maryland wrens during the early morning was much greater than the song output by Florida wrens. Wrens from the tropical Yucatan region did not sing at all. Males in the warmer climates responded to song playbacks with fewer songs than Maryland males (Table III). Thus two measures of the incidence of song use show decreasing song use with decreasing latitude.

The second prediction was tested in Maryland from January 4 to 15, 1981. During this period the ground was snow covered and temperatures were far below normal, ranging from 0°–30°F for night-time lows (mean 10°F) to 18°–41°F for daytime highs (mean 29°F). Normal mean high and low temperatures for this area are 43° and 28°F (National Weather Service records). Four pairs of Carolina Wrens occupied territories at the study site in Severna Park, Maryland. The territories were separated by unoccupied water or salt marsh but all four males could hear each other and earlier it was noticed that singing by one often stimulated the other males to respond. One pair was provisioned with mealworms (larvae of *Tenebrio* sp.), thus artificially increasing food availability, on alternate days during the January period. On days he was provisioned, the male sang persistently whereas no song was heard from the other three males

TABLE III

Latitudinal Variation in the Use of Song Outside the Breeding Season in Carolina Wrens (*Thryothorus ludovicianus*)

	Mean no. spontaneous songs per male, daybreak to 1 hr past	No. males	No. songs following playback ± SD[a]	No. males
Maryland (Nov.) (Arnold)	357[b]	11	32* ± 25.2882	15
Florida (Dec.) (Myakka St. Pk.)	53	6	13 ± 21.4738	8
Yucatan, Mexico (Sept. 17–21) (near Chichen Itza)	0	4[c]	8 ± 11.5866	4

[a] Six song playbacks made within male's territory were used as a stimulus; counts of songs refer only to that male, and do not include neighboring males. Counting continued until singing ceased.

[b] No. calculated from tape recordings of song output from several males simultaneously, therefore no standard deviations are calculated.

[c] Number of males determined through song playbacks after data on spontaneous songs were collected.

*$p < 0.05$ that Maryland birds sing fewer songs following playback than Florida birds (t test).

while I listened from 0730 to 0830 hr. On days he was not provisioned, no songs were heard from any of the males. The songs of the provisioned males elicited *cheer* calls from one male on one occasion (Table I). The results of this qualitative experiment show that wrens will sing when food is provisioned even though temperatures are abnormally low and that singing by neighbors is not necessary to stimulate singing by a provisioned, food-satiated male. It is likely, therefore, that under natural conditions males on territories with relatively high food availability will sing often and potentially increase their relative fitness by increasing mortality in neighbors on poorer territories.

If wrens do die during the winter, the remaining birds expand their territories. Wrens continually attempt to expand their territories, I suggest, because there is no optimum territory size. A territory of a size capable of supporting a pair for the winter may suddenly become unsuitable when the unpredictable abnormally deep and long lasting snows occur. Selection pressures arise from these rare snows that favor territorial expansion: the winners in the game wrens play are those that continue to push for larger winter territories for they attain a higher probability of surviving than wrens less genetically prone to do so. Verner's superterritory concept is important to mention here (Verner, 1977). Verner suggested that territories are larger than needed to provide sufficient resources because by inhibiting reproduction by others, the holder of a superterritory increases its relative fitness. Rothstein (1979) defined *inhibition* as traits that "reduce the absolute performance of other individuals and may or may not improve the absolute performance of the inhibitor." Rothstein pointed out that it is unlikely inhibitory traits that only reduce another individual's fitness will evolve. One function of song in Carolina Wrens, to disturb other males, should be viewed as a trait that indeed is selected to inhibit others, but that also aids the trait holder. As Rothstein further points out, "once a feature that both aids the trait holder and inhibits others becomes common, the inhibition may contribute enough to fitness to itself become a significant evolutionary factor in maintaining the trait" (Rothstein, 1979, p. 330).

For selection pressures to be adequately predicted by the ranging hypothesis, Carolina Wren singers should use songs that disturb listeners. Thus songs should be selected for that listeners might take as emanating from a disturbingly close singer. This implies more than that the song simply propagates well or is detectable as conspecific to a listener; it implies that songs should maintain their source characteristics, or remain undegraded, for as long as distance as possible. An alternative hypothesis suggests that a singer should use songs that degrade predictably with distance so that listeners can judge the singer's distance and avoid the singer "without risking an interaction" (Wiley and Richards, 1978, p. 91).

Sheri Lynn Gish and I recently performed a study to differentiate the efficacy of the ranging hypothesis and the "degrade predictably to inform" hypothesis (Gish and Morton, 1981). A test tape containing 50 Carolina Wren songs from two locations was rerecorded 50 m from a loudspeaker playback at sites where the songs were foreign, native, or neutral. The songs did not differ in their frequency ranges or average frequency. We calculated changes in each song's source characteristics by using changes in the proportional distribution of sound amplitude (in dB) within a song between its source and 50 m to arrive at a "change index." A change index of 0 equaled no change in source characteristics over the 50 m. By using proportional (fractional) representation of sound amplitude, we eliminated attenuation due to spherical divergence from the calculations and obtained a measure of the summed degradation effects due to reverberation, differential attenuation of frequencies, reflection, and refraction. We found that Carolina Wren songs native to a test site had lower change indices than songs foreign to the site; sites where none of the playback songs were native showed no trend favoring either song sample. Thus, wren songs are adapted for maximum transmission of source characteristics, as predicted by the ranging hypothesis but not the "degrade predictably" hypothesis. Singers, therefore, preferentially use songs that are difficult for listeners to ignore because these songs mask information from degradation about the singer's distance as much as possible.

The discerning reader might note that the ranging hypothesis now has produced two potential reasons why young male wrens preferentially learn neighbors' songs. As listeners, young males might learn neighbor's songs so that an undegraded version is available to facilitate ranging or estimating the distance of singers, as previously mentioned. It is also possible that young males learn only those neighbors' songs that propagate source characteristics well in the particular site in which they are establishing territories. In other words, young males are being choosy in selecting songs in order to become more effective as singers. In effect, selection should favor the learning of particularly "good" songs both because these will be the most effective when used by neighbors to disturb him (he will be a fit listener) and because these songs are well adapted to propagate in the specific site (he will be a fit singer). Hansen (1979) discusses one of these alternative hypotheses, suggesting that young males learn neighbors' songs to attain songs best adapted for sound propagation in their particular territorial sites.

The ranging hypothesis does predict an important function of song matching during countersinging between two males. In addition to the function of disturbing neighbors, the second major function of song repertories suggested was to threaten a singing neighbor (male 1) that was determined by the listener (male 2) to be close to their common territorial boundary. When male 1 sings close to the boundary and male 2 is near that border, male 2 responds either by singing the same song type used by male 1 or by using the *cheer* call, a call note held in common with all males (E. S. Morton, unpublished data). The *cheer* (Table I) is used by males in many intra- and interspecific alarm contexts. *Cheer* is replaced by *ti-dink* in male Carolina Wrens living in peninsular Florida north to southeast Georgia (E. S. Morton, unpublished data). The ranging hypothesis suggests the reason male 2 uses either the song type held in common with male 1 or the call note is to *provide precise ranging information to male 1*. In this single context, males near one another, male 2 is selected to provide male 1 with accurate information concerning his distance from male 1. If male 2 responded to male 1 with a random song choice, the ranging hypothesis predicts that, if male 1 did not "know" that song type, he would not receive accurate cues to male 2's distance. Male 2, by providing male 1 with an accurate cue to his proximity, is using song matching as a threat. The song-matching male 2 will often escalate this threat by moving closer to the territorial boundary (E. S. Morton, unpublished data). Boundary defense is the single context where it is in male 2's selfish interest to provide honest information in his song.

4. *Predictions of the Ranging Hypothesis for Other Species*

Many of the known facts and generally held impressions about passerine song are viewed in isolation. Some that I speculate on include song learning and loss of learning, singing outside of the breeding season in permanent residents and migratory species, repertoire size and individual song complexity, the stability of neighbor–neighbor relations as a selective force on song function, and the widespread use of song in relatively small birds with high metabolic rates. The ranging hypothesis provides an opportunity to bring such diverse aspects of song into a single conceptual framework. Hopefully, the ideas will stimulate others to think about birdsong in new ways. As with the motivation-structural code model, the ranging hypothesis is offered to suggest testable hypotheses about communication based on evolutionary theory.

a. Song as a Mechanism to Increase Relative Fitness. Many species begin singing regularly several months before breeding begins, when the cold temperatures and short days of winter still produce food stress conditions. For example, in North America the Song Sparrow (*Zonotrichia melodia*), Cardinal (*Cardinalis cardinalis*), Tufted Titmouse (*Parus bicolor*), and White-breasted Nuthatch (*Sitta carolinesis*) begin singing in mid-January in Washington, D.C., 3 months before breeding (E. S. Morton, unpublished data). Higgins (1979) studied the duration of morning song in the Song Thrush (*Turdus philomelos*), another winter songster. He was surprised that singing duration was longest during the four coldest mornings, but with warmer temperatures (5°–13°C) singing duration was positively correlated with temperature. The ranging hypothesis provides an explanation: song functions to disturb neighbors, thus song should be used when disturbance will provide the greatest fitness benefits to the singer. Climatically induced stress conditions provide the singer an opportunity to increase its relative fitness.

Singing during migration is also frequent for many species of New World warblers (Parulidae) and other tropical migrants returning north to breed. I speculate that this singing, as well, functions to disrupt the foraging of other males. If other males can thus be weakened, singing males might complete the long flight to breeding habitat ahead of others. This suggestion presumes that a high proportion of migrating conspecifics within hearing distance of the songs are heading toward the same breeding neighborhood.

b. Species Specificity in Song, Song Complexity, and the Ranging Hypothesis. The ranging hypothesis suggests that selection should favor songs that are difficult to learn. If a singer's songs are not copied correctly by learning males, this singer will not be accurately ranged by listeners and his songs will be relatively more disturbing. However, singers are constrained from continually producing new, different, or more complex songs because songs that are not easily recognized as conspecific will have little disturbing or threatening effects on listeners. I suggest that this constraint has resulted in species-distinctive properties in birdsongs. The ranging hypothesis predicts that species distinctiveness is a "bind" singers face, not a result of evolution to avoid dysgenic hybridization or time-wasting errors (Marler and Hamilton, 1966, p. 444). The ranging hypothesis places selection favoring species distinctiveness in song at the level of the individual male interacting with conspecifics, or conspecifics plus ecological competitors (Cody, 1973), rather than as a reproductive isolating mechanism operating at the species level (Marler, 1977, p. 57).

The ranging hypothesis predicts that song complexity should arise for the same reasons that favor increased repertoire size. Singers' evolutionary responses to song learning by listeners may be manifested either by increasing the number of differing song elements (i.e., increase complexity) or by increasing the number of song types. The path a species follows depends on the innate species-specific properties in its song(s). For example, the Carolina, Rock, and Marsh wrens have simple songs composed primarily of one repeated element (Kroodsma, 1977). They have evolved along the increased song repertoire route, with 27 to 100+ songs. The Bewick's Wren, in contrast, has fewer but more internally complex songs. Bewick's Wrens in Colorado average 10 songs per male while males from Arizona average 17.5 songs. However, the number of elements sung is the same in both populations: Colorado birds have longer songs than Arizona birds and the former have more elements per song (Kroodsma, 1973). The Bewick's Wren has taken the path toward increasing song complexity. In both cases, the ranging hypothesis suggests that the evolutionary reason is to increase singer effectiveness against a selective background of listener counteradaptations.

In addition, singers may improve their songs' disturbing effects by changing the amplitude of song during delivery while still maintaining song structure within species-specific constraints. The Ovenbird (*Seiurus aurocapillus*) changes the amplitude of its song during delivery (Falls, 1963). I suggest that singers thereby make it more difficult for listeners to judge their range.

c. To Learn or Not to Learn Songs? Learning songs through an interaction with experience and an innate template is not the rule among all passerine birds. Song learning does not seem to occur among the more primitive passerines such as the New World Dendrocolaptidae, Furnariidae, Formicariidae, and Tyrannidae, but to my knowledge no controlled learning experiments have been performed on these (see Chapter 1 on ontogeny, by Kroodsma, Volume 2). The suggestion that songs are innate comes from the observation that songs in species from the above families do not vary geograaphically and where variation is found it is suspected that speciation has occurred (E. Eisenmann, unpublished data). One such instance of suspected speciation was confirmed for an *Empidonax* flycatcher (Stein, 1963). The ranging hypothesis suggests a reason why selection has not favored song learning: song functions only to threaten other males, not to disturb them; song has only the song matching, honest distance-cue-providing function. Song learning is selected for when the disturbing function of song may lead to relative fitness advantages. Most of the passerine species purported not to learn songs are endemic to tropical climates where, as discussed above, disturing neighbors may not function to increase relative fitness.

In species that do learn songs, it is often mentioned that this learning ability disappears after songs have crystallized (Konishi and Nottebohm, 1969). Kroodsma (1980) suggests that social interactions may play a crucial role in determining "exactly where, when, and from whom songs are learned." The ranging hypothesis suggests that the end to song learning occurs when it is advantageous for a singer to become a song model for newcomers. An older male has the advantage of prior territorial occupancy that may be maintained against newcomers that are not effective in disturbing him through their songs if he is copied: he "knows" the song. If newcomers do not copy his songs exactly they may thwart this advantage. "Copying errors" or "drift" (Lemon, 1976) may thus be an adaptive trait in newcomers rather than inevitable developmental errors. Selection favoring when and from whom songs are learned or no longer learned may also be derived from demographic characteristics of the local population that I term neighborhood stability.

d. Song Redundancy and Neighborhood Stability. Small, often one-song repertoires, and high among-individual song variability are predicted to be adaptive in populations where neighbor changeover is high between breeding seasons or there are many newcomers seeking territories. The ranging hypothesis suggests that this is due to a lack of selection pressure favoring newcomers that learn the songs of neighbors with whom they interact for a short time. Instead, selection favors the development of individually distinctive songs to afford listeners only inaccurate distance cues. Examples are found in the Field Sparrow (*Spizella pusilla*) (Goldman, 1973) and White-throated Sparrow (*Zonotrichia albicollis*) (Falls, 1969). Neighbor recognition, wherein the playback of a neighbor's song elicits a reaction milder than that of the playback of a stranger's song, has been reported in species with high neighbor changeover (e.g., Falls, 1969). These studies are describing the adaptive response of the birds only as listeners; there is no information to suggest individually distinctive songs function to *promote* neighbor recognition. Perhaps the critical time for singers is during territorial establishment before neighbors have learned to respond adaptively and are more easily disturbed by songs new to them that they cannot effectively range.

Where neighborhood stability is high, selection may favor either increased repertoire size, as in the Carolina Wren, or single-song repertoires, depending upon the climate during the singing season. If the climate is harsh and unpredictable or territories vary greatly in quality, large repertoires should be favored because the disturbing function of song will favor them. If the climate is mild and stable, selection is predicted to favor low repertoire size. Now threat may be the important function of song. Song matching overbalances selection favoring disturbing in stable situations where the territorial boundaries are established between the same neighbors for long time periods. Dialects arise in this situation: young birds learn older males' songs, selection is against making copying errors, and songs are used to threaten neighbors over established territory borders.

Dialect species should respond more vigorously to native relative to foreign songs in contrast to species having distinctive individual songs. In the former, foreign songs are predicted to produce milder responses because no accurate distance cues are perceived and the bird responds as though the sound is degraded and of little threat. Milligan and Verner (1971) believe that White-crowned Sparrows (*Zonotrichia leucophrys*) respond less to foreign dialect songs.

IV. SUMMARY REMARKS

The evolutionary context in which I have placed singing behavior has yielded the prediction that songs function to disrupt conspecific males by providing poor distance assessments and/or to threaten them by providing accurate distance assessments. The information in song is predicted to be simple; species distinctiveness is important since this property is essential to the effectiveness of the behavior. Species distinctiveness is part of a feedback system constantly balancing, in some species, increased repertoire size and/or internal song complexity against its effects on conspecific recognition; it cannot function without recognition. The ranging hypothesis, and thus the discussion in this section, is concerned only with intrasexual selection. Selection of males through female mate choice is undoubtedly an important source of selection on male song, but reproductive isolation per se is probably of lesser importance than previously thought.

An overall review of the kinds of birds that use song supports the functions attributed to song by the ranging hypothesis. Attempts to increase relative fitness by disrupting neighbors, and described above, are largely found in small-sized species with relatively high metabolism and low energy storage capacity relative to the daily energy budget (Faaborg, 1977). A single hypothesis that encompasses such diverse aspects of what is known about passerine songs should, at least, challenge our assumptions.

ACKNOWLEDGMENTS

Many people have provided valuable criticism. I specially thank Luis Baptista, Sheri L. Gish, Judith L. Hand, John R. Krebs, Donald E. Kroodsma, Edward H. Miller, Robert B. Payne, Michael Robinson, and Neal G. Smith. Virginia Garber graciously and meticulously typed several versions of the manuscript. Grants from the Smithsonian Institution from the Research Awards Program, the Secretary's Fluid Research Fund, and a Trust Fund grant for communication research at the National Zoological Park supported this report.

REFERENCES

Barklow, W. E. (1979). Graded frequency variations of the tremolo call of the Common Loon (*Gavia immer*). *Condor* **81,** 53–64.

Barlow, G. W. (1977). Modal action patterns. *In* "How Animals Communicate" (T. A. Sebeok, ed.), pp. 98–134. Indiana Univ. Press, Bloomington.

Chu, P. R. (1979). Geographic variation in song of island and mainland populations of the Carolina Wren. M.S. Thesis, Univ. of Maryland, College Park.

Cody, M. L. (1973). Character convergence. *Annu. Rev. Ecol. Syst.* **4,** 189–212.

Darwin, C. (1872). "The Expression of the Emotions in Man and Animals." Appleton, London.

Davies, N. B., and Halliday, T. R. (1978). Deep croaks and fighting assessment in toads *Bufo bufo. Nature (London)* **274,** 683–685.

Dawkins, R., and Krebs, J. R. (1978). Animal signals: information or manipulation. *In* "Behavioural Ecology" (J. R. Krebs and N. B. Davies, eds.), pp. 282–309. Blackwell, Oxford.

Faaborg, J. (1977). Metabolic rates, resources, and the occurrence of non-passerines in terrestrial avian communities. *Am. Nat.* **111,** 903–916.

Falls, J. B. (1963). Properties of bird song eliciting responses from territorial males. *Proc. Int. Ornithol. Congr.* **13,** 259–271.

Falls, J. B. (1969). Functions of territorial song in the White-throated Sparrow. *In* "Bird Vocalizations" (R. A. Hinde, ed.), pp. 207–232. Cambridge Univ. Press, London and New York.

Fellers, G. M. (1979). Aggression, territoriality, and mating behavior in North American treefrogs. *Anim. Behav.* **27,** 107–119.

Ficken, M. S., Ficken, R. W., and Witkin, S. R. (1978). Vocal repertoire of the Black-capped Chickadee. *Auk* **95,** 34–48.

Frings, H., Frings, M., Cox, B., and Peissner, L. (1955). Auditory and visual mechanisms of the Herring Gull. *Wilson Bull.* **67,** 155–170.

Gish, S. L., and Morton, E. S. (1981). Structural adaptations to local habitat acoustics in Carolina wren songs. *Z. Tierpsychol.* **56,** 74–84.

Goldman, P. (1973). Song recognition by Field Sparrows. *Auk* **90,** 106–113.

Greenewalt, C. H. (1968). "Bird Song: Acoustics and Physiology." Smithsonian Inst. Press, Washington, D.C.

Hansen, P. (1979). Vocal learning: Its role in adapting sound structures to long-distance propagation, and a hypothesis on its evolution. *Anim. Behav.* **27,** 1270–1271.

Hartshorne, C. (1973). "Born to Sing: An Interpretation and World Survey of Bird Song." Indiana Univ. Press, Bloomington.

Helgeson, N. (1980). Development of song in the Carolina Wren. M.S. Thesis, Univ. of Maryland, College Park.

Higgins, R. M. (1979). Temperature-related variation in the duration of morning song of the Song Thrush, *Turdus ericetorum. Ibis* **121,** 333–335.

Jenni, D. A., Gambs, R. D., and Betts, J. (1974). Acoustic behavior of the northern Jacana. *Living Bird* **13,** 193–210.

Johnston, R. F. (1966). The adaptive basis of geographic variation in color of the Purple Martin. *Condor* **68,** 219–228.

Konishi, M., and Nottebohm, F. (1969). Experimental studies in avian vocalizations. *In* "Bird Vocalizations" (R. A. Hinde, ed.), pp. 29–48. Cambridge Univ. Press, London and New York.

Krebs, J. R. (1976). Habituation and song repertoires in the Great Tit. *Behav. Ecol. Sociobiol.* **1,** 215–227.

Krebs, J. R. (1977). The significance of song repertoires: The Beau Geste hypothesis. *Anim. Behav.* **25,** 475–478.

Kroodsma, D. E. (1973). Singing behavior of the Bewick's Wren: Development, dialects, population structure, and geographic variation. Ph.D. Thesis Oregon State Univ., Corvallis.

Kroodsma, D. E. (1974). Song learning, dialects, and dispersal in the Bewick's Wren. *Z. Tierpsychol.* **35,** 352–380.

Kroodsma, D. E. (1977). Correlates of song organization among North American wrens. *Am. Nat.* **111,** 995–1008.

Kroodsma, D. E. (1979). Vocal dueling among male Marsh Wrens: Evidence for ritualized expressions of dominance/subordinace. *Auk* **96,** 506–515.

Kroodsma, D. E. (1980). Aspects of learning in the ontogeny of bird song: where, from whom, when, how many, which, and how accurately? *In* "Ontogeny of Behavior" (G. Burghardt and M. Bekoff, eds.), pp. 215–230. Garland, New York.

Lein, M. R. (1978). Song variation in a population of Chestnut-sided Warblers (*Dendroica pensylvanica*): Its nature and suggested significance. *Can. J. Zool.* **56,** 1266–1283.

Lemon, R. E. (1976). How birds develop song dialects. *Condor* **77,** 385–406.

MacKay, D. M. (1972). Formal analysis of communicative processes. *In* "Nonverbal Communication" (R. A. Hinde, ed.), pp. 3–25. Cambridge Univ. Press, London and New York.

Marler, P. (1955). Characteristics of some animal calls. *Nature (London)* **176,** 6–8.

Marler, P. (1961). The logical analysis of animal communication. *J. Theor. Biol.* **1,** 295–317.

Marler, P. (1967). Animal communication signals. *Science* **157,** 769–774.

Marler, P. (1977). The evolution of communication. *In* "How Animals Communcate" (T. A. Sebeok, ed.), pp. 45–70. Indiana Univ. Press, Bloomington.

Marler, P., and Hamilton, W. J. (1966). "Mechanisms of Animal Behavior." Wiley, New York.

Marten, K., Quine, D., and Marler, P. (1977). Sound transmission and its significance for animal vocalization. II. Tropical forest habitats. *Behav. Ecol. Sociobiol.* **2,** 291–302.

Martin, W. F. (1972). Evolution of vocalization in the genus *Bufo*. *In* "Evolution in the Genus *Bufo*" (W. F. Blair, ed.), pp. 279–309. Univ. of Texas Press, Austin.

Miller, E. H. (1979). An approach to the analysis of graded vocalizations of birds. *Behav. Neural Biol.* **27,** 25–38.

Miller, E. H., and Baker, A. J. (1980). Displays of the Magellanic Oystercatcher (*Haematopus leudopodus*). *Wilson Bull.* **92,** 149–168.

Milligan, M., and Verner, V. (1971). Inter-populational song dialect discrimination in the White-crowned Sparrow. *Condor* **73,** 208–213.

Morris, D. (1956). The feather postures of birds, and the problem of the origin of social signals. *Behaviour* **9,** 75–113.

Morse, D. H. (1970). Territorial and courtship songs of birds. *Nature (London)* **226,** 659–661.

Morse, D. H. (1973). Interactions between tit flocks and sparrowhawks, *Accipiter nisus*. *Ibis* **115,** 591–593.

Morton, E. S. (1975). Ecological sources of selection on avian sounds. *Am. Nat.* **109,** 17–34.

Morton, E. S. (1977). On the occurrence and significance of motivation-structural rules in some bird and mammal sounds. *Am. Nat.* **111,** 855–869.

Morton, E. S. (1980). The ecological background of the evolution of close-range vocal sound structure. *Proc. Int. Ornithol. Congr.* **17,** 737–741.

Morton, E. S., and Shalter, M. D. (1977). Vocal response to predators in pairbonded Carolina Wrens. *Condor* **79,** 222–227.

Moynihan, M. (1956). Notes on the behaviour of some North American gulls. I. Aerial hostile behavior. *Behaviour* **10,** 126–178.

Moynihan, M. (1970). Control, suppression, decay, disappearance and replacement of displays. *J. Theor. Biol.* **29,** 85–112.

Payne, R. B. (1979). Song structure, behaviour, and sequence of song types in a population of Village Indigobirds, *Vidua chalybeata*. *Anim. Behav.* **27,** 997–1013.

Richards, D. G. (1978). Environmental acoustics and song communication in passerine birds. Ph.D. Thesis, Univ. of North Carolina, Chapel Hill.

Roberts, L. H. (1975). The rodent ultrasound production mechanism. *Ultrasonics* **13,** 83–88.

Rohwer, S., and Niles, D. M. (1979). The subadult plumage of male Purple Martins: Variability, female mimicry and recent evolution. *Z. Tierpsychol.* **51,** 282–300.

Rothstein, S. I. (1979). Gene frequencies and selection for inhibitory traits, with special emphasis on the adaptiveness of territoriality. *Am. Nat.* **113,** 317–331.

Ryan, M. J. (1980). Female mate choice in a neotropical frog. *Science* **209,** 523–525.

Sebeok, T. A. (1977). Zoosemiotic components of human communication. *In* "How Animals Communicate" (T. A. Sebeok, ed.), pp. 1055–1077. Indiana Univ. Press, Bloomington.

Smith, W. J. (1974). Communication, animal. *Encycl. Britannica* **4,** 1010–1019.

Smith, W. J. (1977). "The Behavior of Communicating." Harvard Univ. Press, Cambridge, Massachusetts.

Smith, W. J., Pawlukiewicz, J., and Smith, S. T. (1978). Kinds of activities correlated with singing patterns of the Yellow-throated Vireo. *Anim. Behav.* **26,** 862–884.

Stein, R. C. (1963). Isolating mechanisms between populations of Traill's Flycatchers. *Proc. Am. Philos. Soc.* **107,** 21–50.

Thielcke, G. A. (1976). "Bird Sounds." Univ. of Michigan Press, Ann Arbor.

Thomas, J. C. (1978). A design-interpretation analysis of natural English with applications to man–computer interaction. *Int. J. Man–Mach. Stud.* **10,** 651–668.

Trivers, R. L. (1972). Parental investment and sexual selection. *In* "Sexual Selection and the Descent of Man 1871–1971" (B. Campbell, ed.), pp. 136–179. Aldine, Chicago, Illinois.

Verner, J. (1977). On the adaptive significance of territoriality. *Am. Nat.* **111,** 769–775.

Wallace, B. (1973). Misinformation, fitness, and selection. *Am. Nat.* **107,** 1–7.

Waser, P. M., and Waser, M. S. (1977). Experimental studies of primate vocalization: Specializations for long distance propagation. *Z. Tierpsychol.* **43,** 239–263.

Wetmore, A. (1923). Present status of the Carolina Wren near Washington, D.C. *Auk* **40,** 134–135.

Wiley, R. H., and Richards, D. G. (1978). Physical constraints on acoustic communication in the atmosphere: Implications for the evolution of animal vocalizations. *Behav. Ecol. Sociobiol.* **3,** 69–94.

Wilson, E. O. (1972). Animal communication. *Sci. Am.* **227,** 52–60.

Zweifel, R. G. (1968). Effects of temperature, body size and hybridization on matings calls of toads, *Bufo a. americanus* and *Bufo woodhousii fowleri*. *Copeia 1968,* 269–284.

7

The Coding of Species-Specific Characteristics in Bird Sounds

PETER H. BECKER

ACOUSTIC COMMUNICATION IN BIRDS
VOLUME 1

ISBN 0-12-426801-3

I. INTRODUCTION

Acoustic communication is pronounced and well developed in birds and has important intra- and interspecific functions. One of the most consistent and striking characteristics of bird vocalizations is species specificity. This is related to the fact that most sounds are directed toward conspecifics whose reactions should be appropriate to the content of the message. Hence, the coding of species specificity is typically the basis for effective acoustical communication. Only when a sound signal is unequivocally that of the appropriate species can efficient transfer of information occur, and only then can it be subject to further refinement by selective forces.

However, selective pressures favoring species specificity are much stronger on some vocalizations than others. For example, species identification can be achieved at close range through optical signals in species with strong interindividual social bonds, and vocalizations may then be released from those functions; indeed, species identification in such circumstances may even be unnecessary (e.g., Estrildidae: Immelmann, 1959, 1968; Smith, 1965). Conversely, interspecific communication may become an important function, as in the case of certain alarm and flight calls (Marler, 1956, 1957) and in songs used in interspecific territoriality (Sections IV,D,1 and 2).

This chapter deals primarily with those vocalizations whose most important functional characteristic is species specificity. These are, above all, those sounds which are broadcast over long distances to conspecifics, and those which are implicated in sexual isolation. These include mainly territorial songs, which usually have pronounced species-distinguishing characteristics (e.g., Fig. 1) and may function as premating isolating mechanisms (see Section IV,A). Because of their conspicuousness and other practical advantages, such signals are well studied from the viewpoint of species-specific coding, and are the main subject of this chapter.

As species-specific releasers, signals must be stable over time, conspicuous, easily perceived, and not easily confused with other signals (Lorenz, 1935). The analysis of acoustical characteristics which encode species-specific information in various species, and of similarities and differences in the releasing mechanisms between species and populations, should reveal the main features of the principles mentioned (Sections II and III). Furthermore, acoustical characteristics by which syntopic species distinguish each other's songs will be described (Section IV).

Bird vocalizations have many functions (reviewed in Thielcke, 1970b). For songs, these include territorial advertisement and defense, formation and maintenance of the pair bond, coordination of reproductive cycles within pairs and populations, individual recognition, and advertisement of motivation. The coding of information serving these functions requires a certain level of complexity,

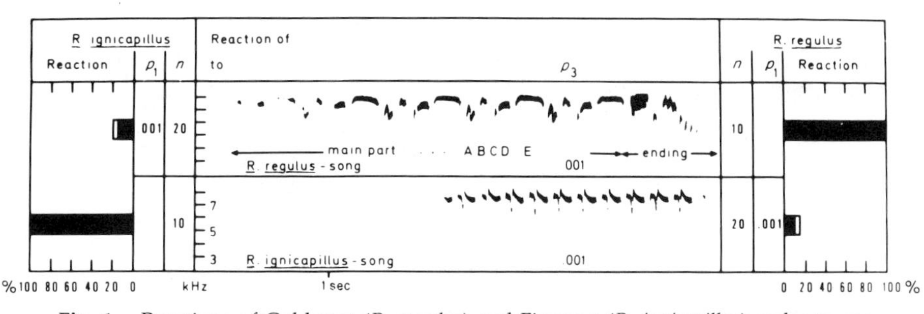

Fig. 1. Reactions of Goldcrest (*R. regulus*) and Firecrest (*R. ignicapillus*) males to conspecific and heterospecific songs in Bodanrück, southwestern Germany, where the species coexist syntopically. Goldcrest song consists of a sterotyped main part in which a group of elements (ABCDE) is repeated, and a variable ending. The song of the Firecrest contains one repeated element (in this song: element F) or phrases of different elements (see Fig. 5). The numbers of males tested in the field (n) are given. First the test song and subsequently the population-typical control song were played to each male. The test criterion was the male's approach to at least 20 m from the speaker. The number of males responding to the test song is expressed as a percentage of the number reacting to the control. The proportion of males approaching the test song, but less closely (> 10–20 m) than the control (≤ 10 m) is shown as the white part of the column (in Goldcrest and Firecrest only). p values are given where $p \leq 0.05$ between reactions to test and control (p_1, by one-tailed sign-test), between reactions of one species to different test songs (p_2, see other figure legends), and between reactions of two species to the same test song (p_3; p_2 and p_3, by two-tailed χ^2 or Fisher's exact test). (From Becker, 1976; for more details, see reference.)

which probably is promoted by female's choice in some species (see p. 241). Complexity, however, may conflict with the need for precise transmission of species identity (Marler, 1959, 1960, 1961). How are these conflicts resolved (Section V)?

In Section VI, I consider the programming of species-specific releasers and releasing mechanisms, whose functions are the maintenance of releaser stability and of the effectiveness and continuity of acoustic communication.

II. SPECIES-SPECIFIC PARAMETERS

A. Problems

Bird vocalizations in all their varied forms are the result of diverse influences and selection pressures which will be dealt with in other chapters. Such influences include body size, phylogeny, sound function, habitat, the effects of sympatric species' sounds, and the continuity of tradition in learned sounds.

The manifold sounds of birds are characterized by a large number of different combinations of acoustical characteristics. But only some characteristics of a species' song act as releasers for a distinct function, and these may differ from species to species (Section II,D). This concentration of intraspecific releasers in certain parameters guarantees the species specificity of songs and permits the coding of further information in other characteristics (Section V).

Which characteristics function as species-typical cues, how do they differ interspecifically, and in what combinations do they occur? Are there preferences across species for certain acoustical cues? These are the principal subjects of this section. The term "parameter" hereafter denotes a characteristic which functions as a releaser.

B. Remarks on Methodology

Most studies of species-typical sound parameters have concentrated on songs in oscines. The species referred to only by name in the text of this section (II) are listed alphabetically below. I will not repeat the literature citations, except where necessary.

Bonelli's Warbler (*Phylloscopus bonelli:* Brémond, 1972a, 1976b), Brown Thrasher (*Toxostoma rufum:* Boughey and Thompson, 1976), Chaffinch (*Fringilla coelebs:* Brémond, 1972b), Chiffchaff (*Phylloscopus collybita:* G. Schubert, 1971; Krammer, 1971, 1982a; Becker *et al.*, 1980b), Collared Dove (*Streptopelia decaocto:* Gürtler, 1973), Common Yellowthroat (*Geothlypis trichas:* Wunderle, 1979), Nightjar (*Caprimulgus europaeus:* Abs, 1963), Robin (*Erithacus rubecula:* Brémond, 1967, 1968a, 1976a), Winter Wren (*Troglodytes troglodytes:* Brémond, 1968b, 1976c, 1978), Field Sparrow (*Spizella pusilla:* Goldman, 1973), Firecrest (*Regulus ignicapillus:* Becker, 1976), Goldcrest (*Regulus regulus:* Becker, 1976), Golden-winged Warbler (*Vermivora chrysoptera:* Ficken and Ficken, 1973), Gray Catbird (*Dumetella carolinensis:* Boughey and Thompson, 1976), Indigo Bunting (*Passerina cyanea:* Emlen, 1972; Shiovitz, 1975; Shiovitz and Lemon, 1980), Marsh Tit (*Parus palustris:* Romanowski, 1979), Ovenbird (*Seiurus aurocapillus:* Falls, 1963), Red-winged Blackbird (*Agelaius phoeniceus:* Beletsky *et al.*, 1980), Rufous-sided Towhee (*Pipilo erythrophthalmus:* Ewert, 1980), Song Sparrow (*Zonotrichia melodia:* Peters *et al.*, 1980), Spotted Sandpiper (*Actitis macularia:* Heidemann and Oring, 1976), Swamp Sparrow (*Zonotrichia georgiana:* Peters *et al.*, 1980), White-throated Sparrow (*Zonotrichia albicollis:* Falls, 1963, 1969), Willow Tit (*Parus montanus:* Romanowski, 1979), Willow Warbler (*Phylloscopus trochilus:* M. Schubert, 1971; Helb, 1973), Wood Duck (*Aix sponsa:* Gottlieb, 1974; Miller and Gottlieb, 1976), Wood Lark (*Lullula arborea:* Tretzel, 1965), Wood Warbler (*Phylloscopus sibilatrix:* Brémond, 1972a), Yellowhammer (*Emberiza citrinella:* Thielcke, 1970a).

Studies of song have been restricted to only one of its functions, namely that of repelling conspecific rivals from the territory. Territorial behavior of males is easily released through playback of recorded songs, so species-specific parameters can be determined by manipulating natural or synthetic songs. The attractive effect of maternal calls on ducklings was used in the study of parameters in the Wood Duck. Species-specific parameters important in intersexual selection have not been evaluated, largely because of practical difficulties, for in most species it is mainly the males which defend the territory; females either react not at all to playback songs or together with their males, so that the evaluation of the female's response is not possible in the field.

C. Importance of Song Characteristics as Species-Specific Parameters

1. *Song Length*

Many species show an increase in song length and a shortening of the inter-song interval under conditions of excitement (Section V,D). In Bonelli's Warbler artificially long test songs do not affect the responsiveness of territorial males. Shortened songs do not influence responsiveness either, as long as the remaining song parts contain species-specific information. Thus 80% of Chiffchaff males will react to a single song element containing all critical parameters (see Section II,C,6,a). Bonelli's Warbler, Firecrest, Goldcrest, Indigo Bunting, Red-winged Blackbird, and Yellowhammer also respond well to partial songs. In contrast, only weak reactions are shown by Marsh Tits to a single element, by Wood Larks to a syllable, and by Willow Warblers to a phrase; Robins do not react to playback of a single motif. This lowered responsiveness results from the absence of two essential parameters, syntax and cadence (for the Marsh Tit only) from these song fragments (Sections II,C,3 and 5).

Song length is not in itself a species-typical parameter. But the generally observed decrease in reaction to shortened songs demonstrates the importance of repetition and redundancy in assuring transmission of the information in order that the message can be received and act as a releaser.

2. *Amplitude Pattern of Song*

In none of the species studied is the pattern of amplitude over song important in affecting responsiveness of males (Robin, Goldcrest, Ovenbird, White-throated Sparrow, Yellowhammer). The possible releasing value of amplitude changes in the Goldcrest (element D is the loudest in the song) is unconfirmed, due to the strong reaction of all tested males to a song without element D and of 80% of males to artificial songs without amplitude changes (Fig. 9).

Intensity of a song is susceptible to gross disturbances in the channel (e.g., through habitat structure, weather, and movements of both senders and receivers), and is probably not suitable as a species-specific parameter. For a long-distance signal, as in territorial song, amplitude assures, however, that song information can reach territorial neighbors. In the Reed Warbler (*Acrocephalus scirpaceus*), Heuwinkel (1978) shows the relation between amplitude and average territory size.

3. *Intervals between Elements*

The duration of intervals is an important parameter in the songs of many species (as in Common Yellowthroat, Nightjar, Field Sparrow, Rufous-sided Towhee, Song Sparrow, and species indicated in Fig. 5). In contrast to the lengthening of intervals, male reaction is diminished only slightly by irregular pauses in the Marsh Tit and the Ovenbird, and by missing intervals in the Willow

Warbler. A possible explanation for these observations lies in the process of "short-term retention" by the receiver: if another species-typical element does not closely follow and reinforce the first, no reaction occurs (M. Schubert, 1971); the actual length of the interval is not critical. Only 40% of Firecrest males respond to a song without pauses, presumably because they cannot resolve the essential structure of the elements (Section II,C,6a).

Other species show no diminution of reaction when interval length is altered (Collared Dove, Spotted Sandpiper, Swamp Sparrow, Wood Warbler, and species indicated in Fig. 5). Frequency changes between the motifs of Robins and the elements of Goldcrests convey the necessary information (see Section II,C,5); in the other species individual elements do so.

Another characteristic which sometimes determines rhythm within the song is alternation between long and short elements, and in the Goldcrest this is presumably a species-specific parameter (Figs. 1 and 8). Similarly, Tretzel (1965) assumes that the variable length of the syllable-building elements itself serves as a releaser in the Wood Lark.

4. *Frequency Range and Tonality*

The frequency range of song is an important characteristic in some species (Fig. 5; also Collared Dove, Common Yellowthroat, Nightjar, Spotted Sandpiper). However, frequency can be varied somewhat without diminishing reaction to playback (Bonelli's Warbler, Chiffchaff, Robin, Willow Warbler).

Species often react more weakly to songs of decreased frequency than to songs in which frequency has been increased by the same amount (Chiffchaff, Robin, Firecrest, White-throated Sparrow, Willow Warbler, Wood Lark, Yellowhammer; in contrast see Bonelli's Warbler). This suggests that sound perception by birds follows harmonic steps: a frequency increase of a certain amount (in kHz) is perceived as less of a change than an equivalent decrease.

The filtration of songs (removal of higher or lower frequencies, or both) has only a slight influence on response (Chiffchaff, Firecrest, Goldcrest, Willow Warbler, Yellowhammer). Harmonics and tonality are thus fairly unimportant in these songs (but see Section II,C,6,d). In the Robin, however, the filtration of the lower and especially the higher frequencies reduces response strength, since the parameter "frequency change" (Section II,C,5) has been weakened. In the White-rumped Shama (*Copsychus malabaricus*: Kneutgen, 1969) and the Hill Myna (*Gracula religiosa:* Bertram, 1970) it is assumed that tonality is the most important species-specific parameter.

5. *Syntax*

The songs of most species have a characteristic syntax which often has releasing functions, as in Brown Thrasher, Chaffinch, Winter Wren, Ovenbird, Rufous-sided Towhee, Willow Warbler, and Wood Lark (Fig. 5).

A particular syntactical order also allows the formation of a parameter important for many species, that of frequency changes between elements or motifs. Thus, only a few male Goldcrests react to a sequence of elements of the same type (Fig. 8E) or to combinations of different elements all of about the same frequency (Fig. 8B,E). Scrambled rearrangement of notes while maintaining frequency change results in good responses, but when frequency variation is reduced through arrangement of these same notes in phrases, the stimulus value is diminished (Becker, 1976). Strong, normal reactions are released in this species only when the neighboring elements within the song differ clearly in frequency. Similarly, of two test songs made with artificial sounds, and differing in the presence or absence of such frequency changes, Goldcrest males react only to the former (Fig. 9). Frequency changes are also an important parameter in the Robin, in which the variety of motifs is also necessary, and in the Winter Wren, the White-throated Sparrow, and presumably in the Golden-winged Warbler.

In several species elements and phrases can be rearranged into new combinations without reducing the releasing effect of the song (Fig. 5, also Common Yellowthroat and Red-winged Blackbird). Chiffchaff males are attracted nearly as well to test songs containing only a single element or repetitions of that element as to natural songs. This repetition of one element is the basic plan for songs of Bonelli's Warbler, Firecrest, Marsh Tit, and Willow Tit (Fig. 5), and the species-specific message is conveyed at least partly by the structure of the elements themselves.

6. *Structure of Elements*

In the majority of species the structure of single elements plays a decisive role as a species-specific releaser.

a. Frequency Modulation. Frequency modulations characterize the songs of most species, and this often takes the form of sudden frequency shifts within elements. In many species, e.g., in the Firecrest and Chiffchaff, elements are divided into more or less strongly modulated parts (Figs. 1 and 5).

While unmodulated, artificial elements elicit no reaction in Firecrest males, most react when modulations are introduced into the song (Fig. 2). In many Firecrest elements frequency modulation is coupled with a marked modulation of amplitude (see Becker, 1976). But as the frequency-modulated artificial elements of Figs. 2b and c are not subdivided by amplitude modulation, the enhancement of the reaction value is probably due to the frequency modulation. In some Firecrest elements, however, the reaction-releasing effect of frequency modulation is perhaps strengthened through amplitude modulation.

In the Chiffchaff, the rapid frequency modulation in the beginning of elements (see Fig. 5), in contrast to the second part of notes, is critical in species recogni-

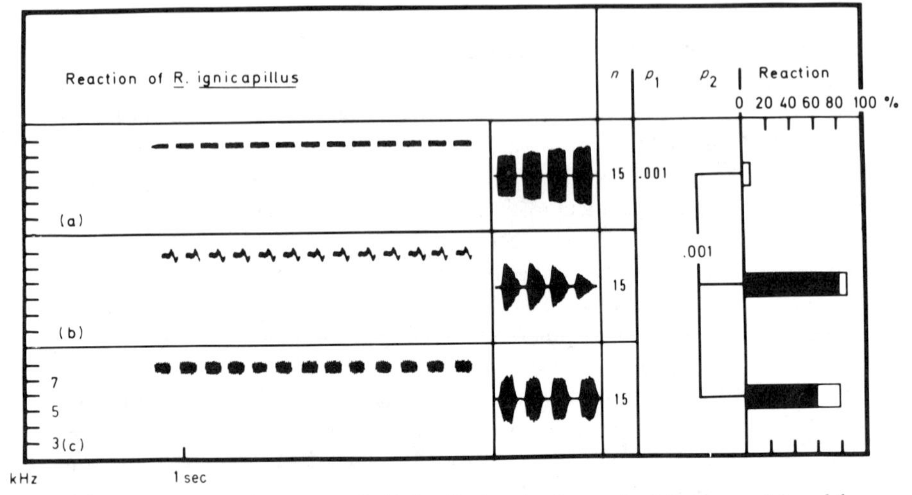

Fig. 2. Reactions of Firecrest (*R. ignicapillus*) males to constant-frequency (a) and frequency-modulated artificial test songs (b, c). For methods see Fig. 1. The last four elements are also shown as oscillograms. For more details, see Fig. 1 and Becker (1976).

tion. The beginning has to be characterized by a distinct slope of frequency change to elicit responses of males (Fig. 3). Furthermore, the elements responded to must begin at around 7 kHz (Fig. 4). If they start at a lower frequency, they must be accentuated with a change in the frequency slope to elicit a strong reaction. When normal songs are played backward, the slope of elements is reversed, and Chiffchaff males do not react at all (similarly in the Spotted Sandpiper and the Marsh Tit). Descending frequency modulation in maternal calls is also critical in evoking responses from young Wood Ducks.

These examples show the importance of fine structure of sounds for species identification. Frequency modulation is also an important cue in the Chaffinch, Indigo Bunting, several *Phylloscopus* species, and Golden-winged Warbler, but subtle frequency modulations applied to the frequency pattern of Indigo Bunting elements have little effect on response (Shiovitz and Lemon, 1980).

In the Goldcrest frequency modulations in the main part of the song are not used as a parameter, as shown by the good reaction of males to unmodulated, artificial songs (Fig. 9). There are also examples of unmodulated sounds that serve as parameters: the constant-frequency ending of elements is the releasing component in songs of the Willow Tit, not the modulated beginning, and the White-throated Sparrow scarcely reacts to artificially introduced modulations in its elements.

b. Amplitude Modulation. The modulation of sound pressure, which is often coupled with frequency modulation, is not known to serve as a species characteristic, though Krammer (1971, 1982a) assumes so for elements of the Chiffchaff. This seems unlikely, however, based on experiments that show similar responsiveness to artificial elements with fixed frequency modulation but varied amplitude modulation (Becker *et al.*, 1980b). Amplitude modulation is also unimportant as a species-specific parameter in the Robin, Winter Wren, Goldcrest, Indigo Bunting, Marsh Tit, and Willow Warbler (Firecrest: see Section II,C,6,a), presumably because amplitude patterns are subject to various distortions in the channel of transmission, and therefore are not suitable as species-specific parameters in long-distance signals.

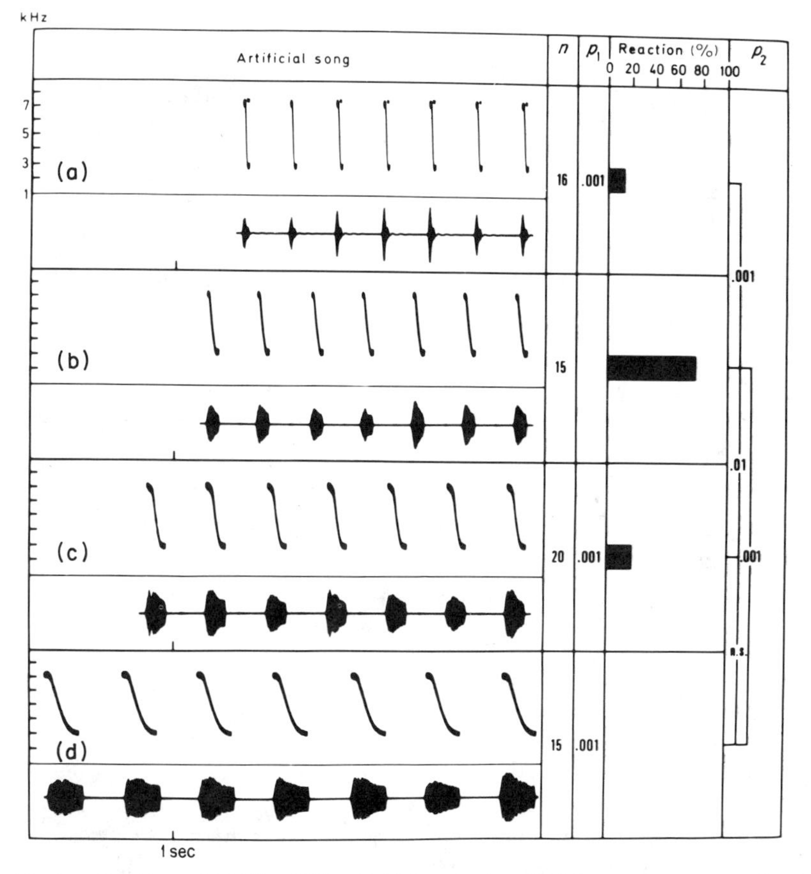

Fig. 3. Reactions of Chiffchaff (*P. collybita*) males to artificial songs consisting of beginning parts of elements with different slopes of frequency decrease. The lengths of elements b and c lie within the range for natural elements. The test songs are shown as sonagrams and oscillograms. The control song contained one natural element repeated seven times. For more details, see Fig. 1 (approach ≤ 10 m). (From Becker *et al.*, 1980b).

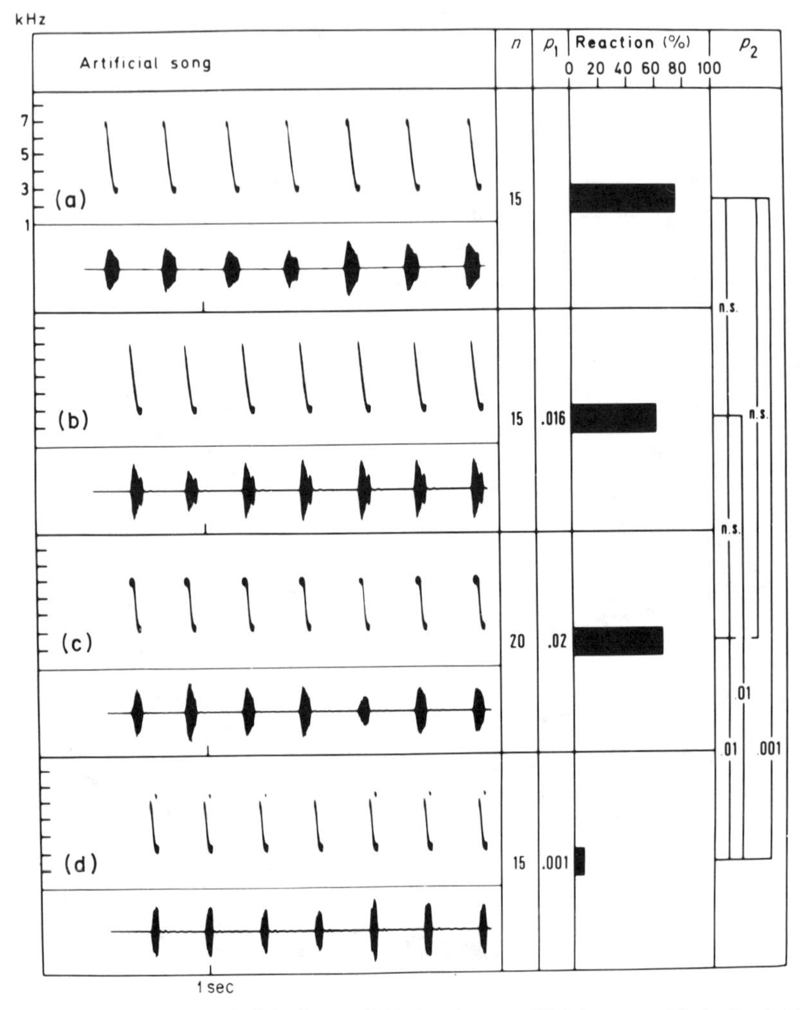

Fig. 4. Reactions of Chiffchaff (*P. collybita*) males to artificial songs with the beginning part of elements altered. The alterations are: (a) with or (b) without an accentuated beginning part at about 7 kHz; (c) with and (d) without an accentuated beginning part at about 6 kHz. For further details, see Figs. 1 and 3. (From Becker *et al.*, 1980b.)

c. Transients. Transients in the sounds of musical instruments play an important part in their precise identification by the human ear (Trendelenburg, 1961). In the Robin and the Winter Wren song elements must have a clearly defined beginning, unmasked by other frequencies (Brémond, 1976a,c). Transients may be important in the Wood Lark and the Chiffchaff (G. Schubert, 1971; but see Becker *et al.*, 1980b). In general, however, the structure or presence of transients is not an important parameter, as shown by the good reaction of males to test songs which: (1) consist of elements cut off at the beginning or of artificial elements which do not contain the species-specific transients (Chiffchaff, Collared Dove, Common Yellowthroat, Firecrest, Goldcrest, Indigo Bunting, Marsh Tit, White-throated Sparrow, Willow Tit, Willow Warbler); or (2) are played backward [Robin, Firecrest, Goldcrest, Indigo Bunting (Thompson, 1969), Yellowhammer].

d. Tonality. The releasing value of song elements in the Bonelli's Warbler and the White-throated Sparrow is diminished if other frequencies, not necessarily harmonic, are added. It is presumably also true of many other species that elements must consist of pure tones (Section II,C,4). On the other hand the presence of a second, higher and unrelated frequency band in the territorial call of the Collared Dove is required for species recognition.

D. Synopsis

Species-typical sounds have many unique characteristics, but only a few of these serve as species-specific releasers for a distinct function. Such parameters occur in all songs of the species and vary little (Falls, 1969; Emlen, 1972; Shiovitz, 1975; Brémond, 1976b). These trends support the suggestions of Schleidt (1976), according to whom optimal distinctiveness is achieved through few characteristics with small variability. It does not necessarily follow, however, that all such characteristics are species-specific parameters, as the studies of Emlen (1972) and Brémond (1976b) point out. These and other characteristics which are not species-specific parameters can encode other information.

Characteristics which function as parameters have enough contrast with the noises of the channel to ensure that they can reach the receiver with minimal distortion and loss of information. In most species the parameters are derived from frequency range, syntax, element structure, and the length of intervals between elements (Fig. 5). Amplitude patterns and some aspects of fine structure of the elements (e.g., transients) are subject to a variety of disturbances in the channel, and appear not to be used as parameters.

Species use a multitude of parameter combinations to specify their songs. For example, considering just syntax and element structure, some species use mainly syntax as a parameter (Fig. 5, also Brown Thrashers and Winter Wrens), and fine structure of elements is unimportant; in other species the opposite is true (lower part of Fig. 5). Between these two extremes many intergradations can be realized. The detailed structure of a parameter generally differs from species to species; for example, the parameter element structure occurs as unmodulated tones in the White-throated Sparrow, and as modulated sounds in Firecrest and Indigo Bunting.

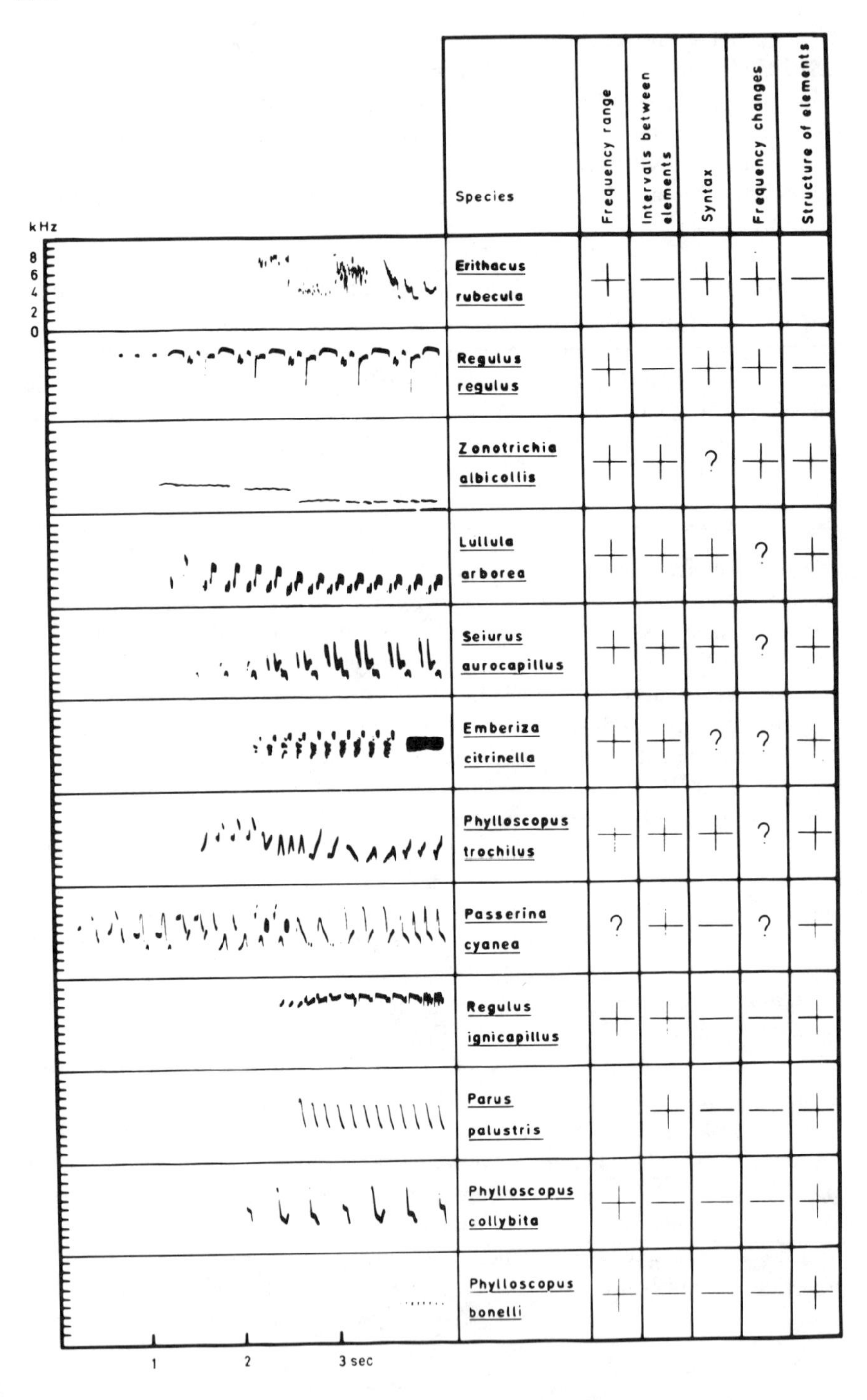

Species	Frequency range	Intervals between elements	Syntax	Frequency changes	Structure of elements
Erithacus rubecula	+	—	+	+	—
Regulus regulus	+	—	+	+	—
Zonotrichia albicollis	+	+	?	+	+
Lullula arborea	+	+	+	?	+
Seiurus aurocapillus	+	+	+	?	+
Emberiza citrinella	+	+	?	?	+
Phylloscopus trochilus	+	+	+	?	+
Passerina cyanea	?	+	—	?	+
Regulus ignicapillus	+	+	—	—	+
Parus palustris		+	—	—	+
Phylloscopus collybita	+	—	—	—	+
Phylloscopus bonelli	+	—	—	—	+

Although one parameter may assume greater significance than the others, most species use a complex of parameters for species recognition (e.g., Brown Thrasher, Robin, Firecrest, Goldcrest, Indigo Bunting, Willow Warbler, Wood Lark). The parameters strengthen each other in their releasing effectiveness (Brémond, 1968a; M. Schubert, 1971; Becker, 1976) or have additive effects (Shiovitz and Lemon, 1980: Additive-redundant model of song recognition). If a parameter is weakened below its threshold value or lacking, the other parameters may still ensure the releasing effect of the sound. Response strength may depend on "how many cues are present, and how closely each cue fits its feature detector" (Shiovitz and Lemon, 1980).

In many species' songs, contrasts and structuring are built in, including alternation between modulated and unmodulated parts of the same element (Section II,C,6) or between long and short elements (Section II,C,3), or altering of elements, phrases, motifs, and frequency changes (Section II,C,5). Such contrasts often not only serve as parameters but also lead to maximal distinctiveness from channel noise, and call the receiver's attention to the message (Krammer, 1982b; Tembrock, 1975; see also below).

Redundancy is achieved through the combination of parameters and their repetition in the form of elements, phrases, element groups or motifs, and songs. Redundancy is therefore both correlational and sequential. The importance of correlational redundancy may lie in the minimization of habituation in receivers (Hartshorne, 1956; Schleidt, 1973, 1977); this function is also met through structuring and contrasts in song. Sequential redundancy may be most important in maintaining or enhancing the responsiveness of the receiver, and in by-passing competing noise (Schleidt, 1973, 1977).

Intrinsic, unvarying properties characterize all the songs of a species, but does this principle also apply to species with very complex songs? Marler (1960) claims it does, and this is supported by studies of the Robin in which the various motifs (several hundred per individual) are constructed according to the same syntactical principles (see Section II,C,5). In this way even complex songs are species-specific. Boughey and Thompson (1976, p. 88) assume that mimids recognize the highly variable songs of their own species by the average value of several characteristics, and that "the birds like human observers in the field may have to hear several units of song before they can make a definite species identification." The recognition process is presumably facilitated through the rapid and constant singing of these species.

The great interspecific differences in song complexity suggest that species

---

Fig. 5. The significance of five song characteristics as species-specific parameters in some passerine species. (+) Important; (−) not important as species-specific parameter. A question mark indicates that the significance of the characteristic is still unclear. The parameter "frequency change" refers to changes in frequency between neighboring elements or motifs. For literature citations, see Section II,B.

may differ in the selectivity of their releasing mechanisms. The extent of the "signal variation tolerance" (Shiovitz and Lemon, 1980) presumably determines the limits of variability of a species' song.

Nothing is known about the similarities or differences in species-specific parameters of the various vocalizations of a species. Even in the song itself different combinations of characteristics can occur, for example, in the Goldcrest (Fig. 1): the stereotyped main part of the song is followed by a variable ending which also acts as a releaser. But this ending has other traits, and perhaps parameters different from those of the main part, and it presumably also fulfills other functions (see also Sections V,B and C).

Our present knowledge of species-specific sound parameters is derived almost entirely from the reactions of males to the song as a territorial advertisement. The second important function of song is sexual, and serves in the context of mate attraction, and formation and maintenance of the pair bond. Remarkably little is known about the reactions of females to song (e.g., Payne, 1973a,b; Kroodsma, 1976; Kling and Stevenson-Hinde, 1977). Do females react to the same species-specific parameters as males? Along with the study of species-specific parameters in nonsong vocalizations, such questions await future research.

III. POPULATION-SPECIFIC CHARACTERISTICS

The structure of songs and calls varies geographically in many species (Chapter 6, Volume 2). Do parameters show corresponding changes?

A. Song Variations and Responsiveness

The reactions of Goldcrest males in central Europe to the species' song recorded elsewhere are summarized in Fig. 6. The responsiveness of males declines with decreasing structural similarity of the test song to that of the local population. Frequency constrasts and the alternation of long and short elements are the essential parameters near Bodanrück (Section II,C,3 and 5). The songs of London and Eaux-Bonnes birds have sufficient frequency changes to attract most males. But significantly fewer central European males respond to the songs from Rouen and Soria, which show much weaker frequency changes (see also Fig. 8). The tested songs are reacted to fully by males of the corresponding populations, so it follows that releasers and releasing mechanisms are different from those in central Europe. Goldcrest males also respond more weakly to endings of foreign songs (Becker, 1977b). Clearly, individuals from different populations could experience problems in communication, though these could be overcome through learning ("mixed dialect singers": Becker, 1977b; Baptista, 1974, 1975, 1977; Conrads and Conrads, 1971).

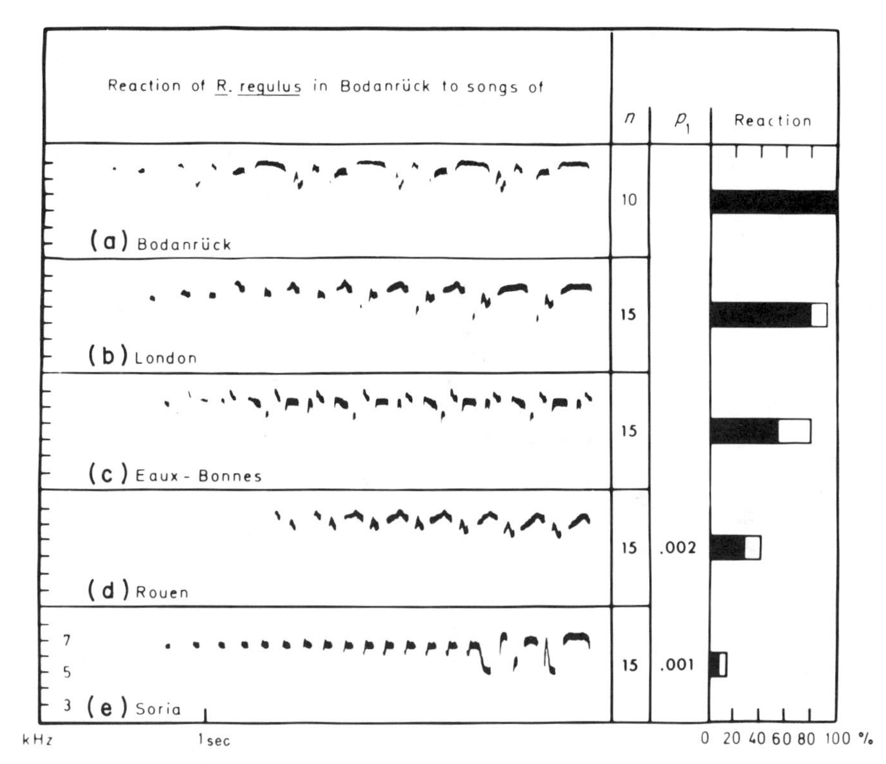

Fig. 6. Reactions of Goldcrest (*R. regulus*) males in Bodanrück, southwestern Germany, to the main parts of songs from other populations [Eaux-Bonnes, Pyrenees; Rouen, western France; Soria, northern Spain; for exact locations of the populations, see Becker (1977b)]. For further details, see Fig. 1. The song from Bodanrück was used as the control.

In a similar fashion, Chiffchaff males from central Europe respond very weakly to songs from conspecifics in Spain and the Canary Islands; indeed, they react about as strongly as to songs of other species. The important parameters have obviously been altered in these song forms (Thielcke *et al.*, 1978; Becker *et al.*, 1980b).

B. Differences in Mutual Reponse to Interpopulational Song Variations

Chiffchaff males in central Europe respond weakly to conspecific songs from Spain, but males in Spain react well to songs from central Europe (Fig. 7) (Thielcke *et al.*, 1978). This phenomenon also occurs in the Blue Tit (*Parus caeruleus*) in central Europe and the Canary Islands (Becker *et al.*, 1980a) and in the Goldcrest in central Europe and Soria, Spain (Becker, 1977b); songs from

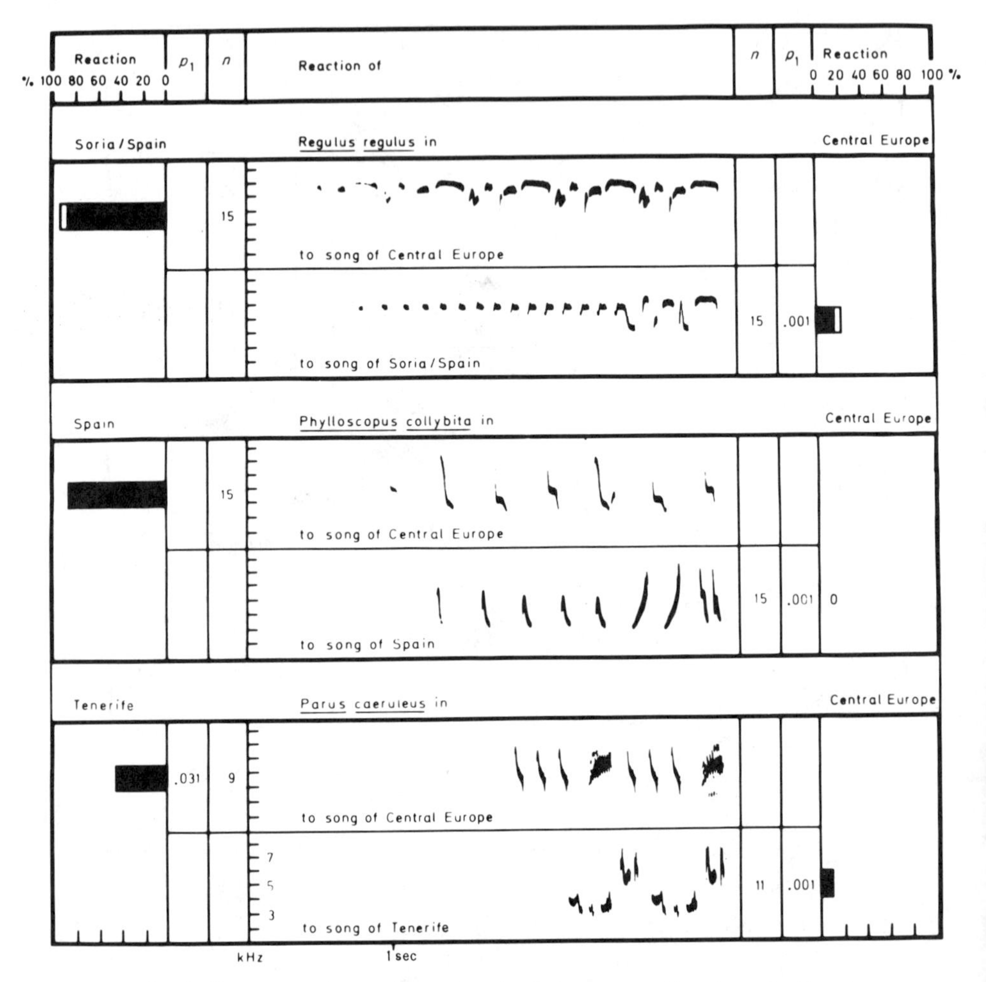

Fig. 7. Differences in the reactions of males in two populations to each others' songs in Goldcrest (*R. regulus*), Chiffchaff (*P. collybita*), and Blue Tit (*P. caeruleus*). One tested population lives in central Europe, the other in the periphery of the species' range. In Tenerife, of 6 Blue Tit males responding with song to the control, 4 also started to sing during playback of the central European song ($p_1 > 0.05$), but only 1 among 12 central European males did so during playback of the song of Tenerife ($p_1 < 0.001$). The population-typical song was a control in each case. For Goldcrest and Blue Tit: approach $\leq$ 20 m; for Chiffchaff: approach $\leq$ 10 m; for further details, see Fig. 1 [After Becker (1977b), Thielcke *et al.* (1978), and Becker *et al.* (1980a).]

central Europe include effective releasers for a great part of the males tested in these geographically peripheral populations, but the reverse is not true. These findings suggest that releasing mechanisms in the peripheral populations respond to characteristics in central European songs. Hypothetically this situation could evolve as follows: if the peripheral populations descended from central European stock, as seems likely, changes in peripheral songs could lead to a corresponding broadening of the releasing mechanism which now reacts to "new" as well as to "old" parameters. This broadened releasing mechanism should not evolve in the ancestral population. Thus, in the Goldcrest, the Soria song has an apparent vestigial element group with frequency change at the end of the song's main part (Fig. 7). Soria males therefore react to central European songs because these contain frequency changes, but the vestigial element group in Soria song is insufficient to release a reaction from central European males.

At the periphery of a species' range interpopulation song differences can be especially pronounced (through the effects of disruption of tradition, isolation, small population size, etc.). Thus, different song variations can affect the song of a peripheral population in the case of exchange of individuals between populations. These and other factors, such as small population size and density, could lead to a greater song variability within peripheral, isolated populations (e.g., Thielcke, 1965; Ward, 1966; Lemon, 1975; Becker, 1977b; Becker *et al.*, 1980a), which itself demands responsiveness to a greater variety of sounds than in the center of a species' range.

C. Concluding Remarks

From the preceding examples and from studies of other species in which males respond weakly to songs of other populations (e.g., Lemon, 1967, 1974; Thielcke, 1969b, 1973b; Milligan and Verner, 1971; Harris and Lemon, 1974; Kreutzer, 1974; Romanowski, 1978), it follows that song parameters can differ among populations, and that population-specific songs are not necessarily effective as species-specific signals in all populations of the species. In such cases we must specify that we are dealing with "population-specific" releasers.

Species identification may be transmitted in other ways when individuals are exchanged between populations, for example, through widely distributed song types, as in Cardinals (*Cardinalis cardinalis:* Lemon, 1966, 1967) and Firecrests (Becker, 1977b), or through song endings as in the Goldcrest (Becker, 1977b). Similarly, innate calls, optical releasers, or behavioral sequencing may permit species identification in such cases. The occurrence of "mixed dialect singers" suggests one way in which communication problems may be overcome (Section III,A).

Species-specific parameters are consistent throughout all studied populations in some species, in spite of geographic song variation, such that the capacity to react to song variations is assured (e.g., Firecrest: Becker, 1977b).

IV. INTERSPECIFIC DISCRIMINATION

A. Introduction

The more acoustically active species living together in a habitat, the greater become the problems of acoustic confusion and the demands for precision in the perception apparatus and for the characterization of vocalizations. And if sounds differ between species, there should be less confusion in communication. Therefore we expect that effective sound divergence and discrimination among species should evolve in a complex sound environment.

Interspecific discrimination of sounds minimizes unnecessary expenditure of energy, interspecific conflicts, and hybridization (see below) and in general makes intraspecific acoustical communication more efficient. In view of the territorial function of most bird songs, species discrimination also permits territorial overlap among syntopic species and may therefore be important in regulating their coexistence.

Closely related, sympatric species often differ markedly in their vocalizations. This seems to support the views that selection can produce strong divergence (see Section IV,C,1) and that the vocalizations may serve in species isolation (see Nicolai, 1968). Many studies have shown that related species can discriminate one another's vocalizations, so such calls may be isolating mechanisms (e.g., *Catharus* and *Hylocichla:* Dilger, 1956; Stein, 1956; *Catharus:* Raitt and Hardy, 1970; *Certhia:* Thielcke, 1962; *Empidonax:* Johnson, 1963; Stein, 1963; *Luscinia:* Göransson *et al.*, 1974; *Zonotrichia:* Peters *et al.*, 1980; *Myiarchus:* Lanyon, 1963; *Parus:* Ludescher, 1973; Romanowski, 1979; Martens, 1975; *Passerina:* Thompson, 1969; Emlen, 1972; Emlen *et al.*, 1975; *Phylloscopus:* Thielcke and Linsenmair, 1963; Thielcke *et al.*, 1978; *Cardinalis*: Lemon and Herzog, 1969; *Regulus:* Becker, 1976, 1977a; *Sturnella:* Lanyon, 1957; Szijj, 1966; Rohwer, 1973; *Toxostoma* and *Dumetella:* Boughey and Thompson, 1976; *Vermivora:* Gill and Lanyon, 1964; Gill and Murray, 1972; Murray and Gill, 1976). But the effectiveness of song as an isolating mechanism remains hypothetical because we know so little about the ultimately decisive reactions of females. Only for the Paradise Whydah (*Vidua paradisaea*) is there good evidence that the mimetic songs of males and the responses of females are behavioral isolating mechanisms (Payne, 1973a,b).

B. Discrimination Cues

1. *Song Characteristics Used in Discrimination*

In their sympatric areas in central Europe and in coniferous forests of southern Europe, Firecrests and Goldcrests generally live in the same habitat and have overlapping territories (Becker, 1977a). Because of their close relationship their vocalizations occur in the same high-frequency range, and consist partly of rather similar elements. Nevertheless, these sibling species can discriminate one another's songs (Fig. 1) and calls (Becker, 1977a).

Responses of Goldcrest males to test songs from Goldcrest elements become less frequent as frequency differences between adjacent song elements are reduced (Fig. 8; Section II,C,5). Firecrest males respond in just the reverse way; indeed, test song E is a releaser for significantly more heterospecific than conspecific males. Two artificial songs differing in the presence or absence of frequency change give the following results: Goldcrest males react only to songs with frequency changes, and Firecrest males only to songs without them (Fig. 9). These experiments thus show the importance of frequency changes as a key characteristic by which the two species discriminate one another's songs.

In the species pair Willow Tit–Marsh Tit, the frequency constant parts of elements in the Willow Tit song serve as discrimination cues (Romanowski, 1979). The removal of these extinguishes the Willow Tit response and induces strong reactions from the sibling species, whose own elements correspond to the frequency-descending part of the elements of the Willow Tit. In their sympatric range, Bonelli's Warbler and Wood Warbler seem to use differences in frequency and element structure as discrimination cues (Brémond, 1972a, 1976b). Mainly syntactical differences prevent the Winter Wren from reacting to the

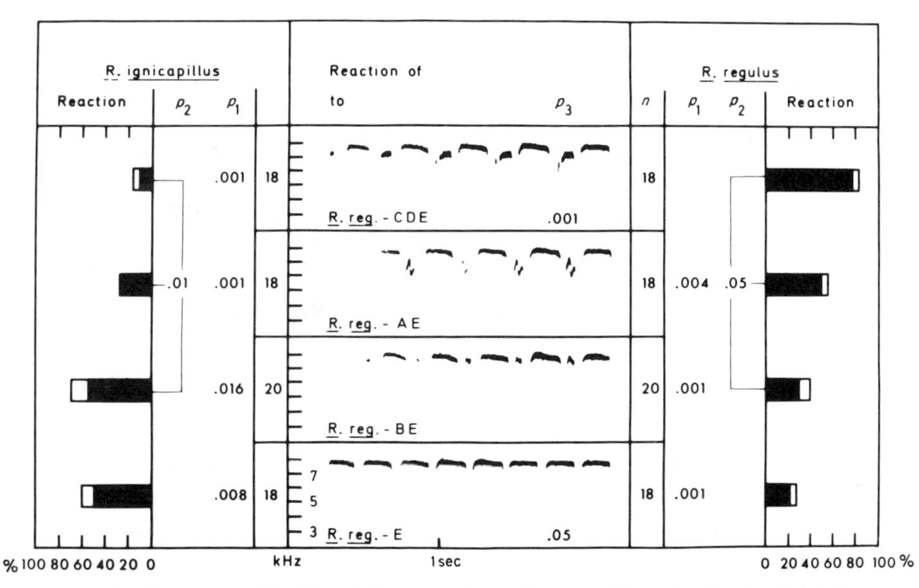

Fig. 8. Reactions of Goldcrest (*R. regulus*) and Firecrest (*R. ignicapillus*) males to four test songs with decreasing frequency differences between elements. All elements are from Goldcrest song (see Fig. 1). (From Becker, 1976.)

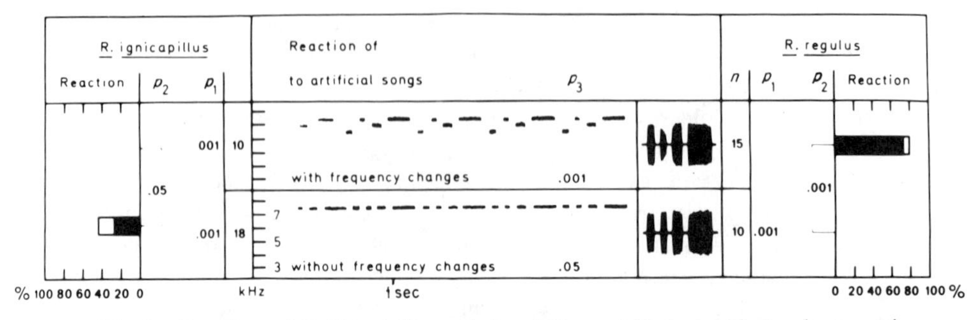

Fig. 9. Reactions of Goldcrest (*R. regulus*) and Firecrest (*R. ignicapillus*) males to artificial test songs with and without frequency changes between elements. The last four elements are also shown as oscillograms. For further details, see Fig. 1. (From Becker, 1976.)

rather similar elements of the Nightingale (*Erithacus megarhynchos*), which sings in the same habitat (Brémond, 1968b). The Brown Thrasher clearly discriminates its songs from those of the Gray Catbird on the basis of the number of repetitions of each sound (Boughey and Thompson, 1976). Song and Swamp sparrows both use differences between syllable types of their songs in discrimination, but temporal differences serve for song distinction in the first species only (Peters *et al.*, 1980).

2. *Artificial Songs with Elements Taken from Other Species*

Tests using artificially mixed songs show clearly that heterospecific sounds or their parts can reduce the reactions to species-specific sounds. Foreign elements introduced into the songs of Wood Lark (Tretzel, 1965), Willow Warbler (M. Schubert, 1971), and Firecrest and Goldcrest (Becker, 1976) cause a decrease of male responsiveness if syntax is disrupted or if these elements differ from conspecific elements, or both. Thus the response rate in the *Regulus* species is not altered by the exchange of their song elements E (Fig. 10b,c), but it is altered by the exchange of Goldcrest E with Firecrest F (Fig. 10b,d) or Goldcrest CD with Firecrest G (Fig. 10b,e). Slight differences in frequency (Fig. 10e) or species-atypical sharply modulated elements (Fig. 10d) reduce the reactions of Goldcrests. Similarly, frequency changes in the test songs shown in Figs. 10c and d reduce the reactions of Firecrests.

Heterospecific song portions placed at the beginning or end of species-typical song lower responsiveness of Willow Warbler males, because the critical parameter syntax is disrupted (M. Schubert, 1971; Helb, 1973). This is not true for males of the *Regulus* species, for the parameters are not masked by the heterospecific song parts (Fig. 10a). Elements of Willow Warbler songs introduced into the song of Chiffchaffs do not reduce responses in males tested, because each note of the Chiffchaff encodes sufficient information (G. Schubert, 1971) (Section II,C,6,a).

3. *Geographic Differences in Discrimination Cues*

Falls (1963) suggests that characteristics which are not important as discrimination cues in one part of a species' range may be useful elsewhere, where a species with similar songs occurs.

A possible example of this is the two *Regulus* species in Soria, Spain. They can distinguish each other's songs, although frequency changes within elements, the main factor in discrimination in central Europe, occur only as a vestige in the songs of Soria Goldcrests (Fig. 6) (Section III,B) (Becker, 1977a). Other characteristics of the species' songs are presumably used as discrimination cues.

Possibly the sounds of sympatric species and the whole sound environment have influenced which sound characteristics are used as parameters and discrimi-

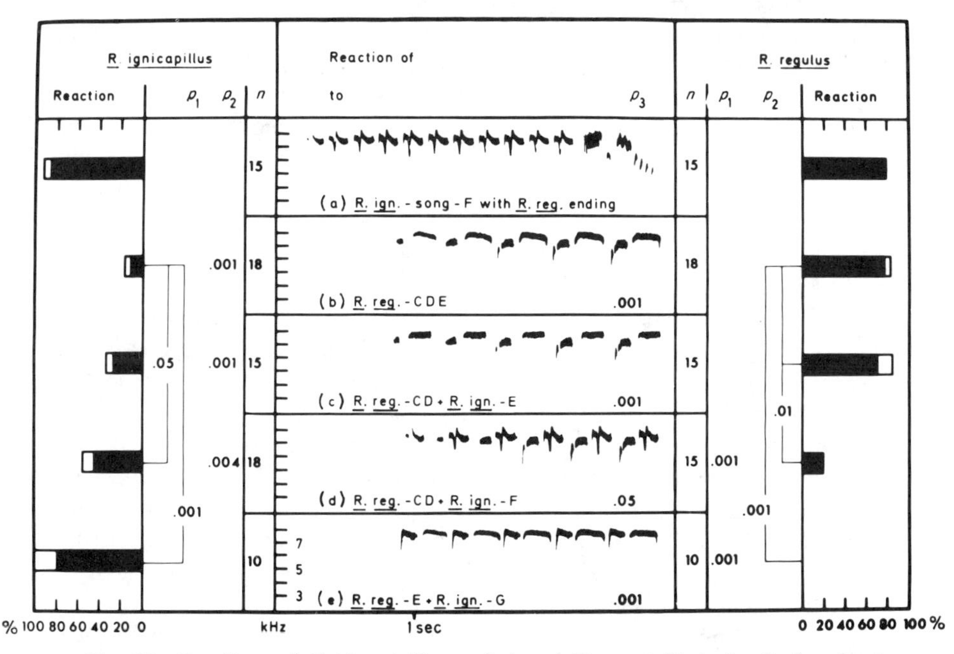

Fig. 10. Reactions of Goldcrest (*R. regulus*) and Firecrest (*R. ignicapillus*) males to mixed songs containing elements of both species: (a) Firecrest song with element F and Goldcrest ending (see Fig. 1), (b) Goldcrest element combination CDE and (c–e) combinations of Goldcrest and Firecrest elements, which originate from other Firecrest songs (see Becker, 1974, 1977b). For further details, see Fig. 1; for interpretations, see text. (From Becker, 1976.)

nation cues by a species in a distinct area (e.g., Marler, 1960; Emlen, 1972; Peters *et al.*, 1980) (Sections III and IV,C,3).

4. *Rejection Markers and Discrimination in Versatile Singers*

Examples discussed above show that many syntopic species are able to discriminate each other's sounds. Brémond (1976b) has termed those characteristics of sounds which are used to identify them from heterospecifics as "rejection markers." For example, the rejection markers for Goldcrests are "no frequency changes" and for Firecrests, "frequency changes" in respect to one another (Sections IV,B,1 and 2). These allow the recipient to discriminate at once and to choose an appropriate behavior. On account of the complexity of their songs, it is likely that in versatile singers many more difficulties exist in achieving such direct discrimination, because rejection markers have to be more complex, too (but see the Robin, p. 225). It seems, rather, that they must receive a substantial portion of sounds before they can determine species identity. This process surely takes time, and is hindered by disturbances in the acoustical channel. According to Boughey and Thompson (1976) this could be the reason why the mimids sing so rapidly and constantly. In this way they compensate for the inefficiency of having to hear several units of their highly variable songs before making a definite species identification. This could be one explanation for the positive correlation between versatility and continuity of singing pointed out by Hartshorne (1956).

5. *Interspecific Contrast: A Prerequisite for Discrimination?*

The mimids are one example of species which can discriminate one another despite similarity in song structure. The Cardinal and Pyrrhuloxia (*Cardinalis sinuatus*) can also distinguish their very similar songs easily (Lemon and Herzog, 1969). Coal Tit (*Parus ater*) males from southwestern Germany discriminate their own songs from those of the Great Tit (*Parus major decolorans*) from eastern Afghanistan which sound nearly identical (Thielcke, 1969b). Likewise, central European Chiffchaffs do not react to aggressive songs (Erregungsstrophe) of Spanish conspecifics, although these appear similar (Thielcke *et al.*, 1978). Also, Firecrests and Goldcrests in Soria, Spain, easily discriminate one another's songs though they seem to be more similar there than in central Europe (Figs. 1 and 6). The alarm calls of these sibling species differ only in the rate of frequency modulation, but are still discriminated by most males (less well in the Firecrest; Becker, 1977a).

These observations show that similarities determined by the human ear or by technical analysis need not agree with perception as achieved by birds. It is clearly necessary to support our estimates of structural similarity of vocalizations (as in studies of character convergence or loss of contrast) with tests using the reactions of the birds themselves.

C. Strategies for More Effective Interspecific Sound Discrimination

Interspecific sound discrimination can be improved through (1) increased contrast between sounds of the species, with corresponding changes in the releasing mechanisms, (2) temporal separation of vocal activities of syntopic species, and (3) sharpened discrimination ability of the birds (increasing selectivity of the releasing mechanism).

1. Character Displacement

Support for the first possibility may come from sympatric close relatives with strong vocal differences. It has often been assumed that selection pressure leads to divergence in vocal characteristics in such cases (see also Chapter 8, this volume). To test this assumption comparisons of the corresponding species' songs in sympatric and allopatric areas are required.

Support for the hypothesis of character displacement comes from just a few examples (e.g., Marler and Boatman, 1951; Lack and Southern, 1949; Marler, 1959, 1960; Marler and Hamilton, 1972), but these have not been based on detailed song analyses or studies of reactions of the species in sympatry and allopatry (as for example in Chiffchaff and Willow Warbler: Thielcke and Linsenmair, 1963; Thielcke *et al.*, 1978; Blue Tit, Canary Islands: Becker *et al.*, 1980a; Goldcrest of the Azores: Knecht and Scheer, 1971; Becker, 1978a). These and other studies provide no support for the hypothesis (e.g., *Certhia:* Thielcke, 1965, 1969a, 1973b; *Vermivora:* Gill and Murray, 1972; *Regulus:* Becker, 1977a,b). It is not known whether song differences are more pronounced or frequent between species in sympatry than between allopatric species (Thielcke, 1969b). It thus remains to be shown that sympatry with one (or more) species with similar songs causes changes such that interspecific contrasts in acoustical releasers develop or are reinforced, or alternately disappear in allopatry. In any case, we know that great contrasts are not prerequisites for discrimination by sound (Section IV,B,5).

2. Temporal Shifts in Song Activity

That some syntopic species differ in their daily singing rhythms is well known; but the reasons for these differences are still unclear. Ficken *et al.* (1974), however, describe a means of facultatively minimizing acoustical interference. Syntopic breeding populations of Least Flycatcher (*Empidonax minimus*) and Red-eyed Vireo (*Vireo olivaceus*) stagger their daily singing schedules so that they overlap little. Seasonal differences in the peaks of singing occur to some extent among species, but in seasonal breeders there are probably strict limits on how much shift is tolerable.

3. *Reinforcement of Discriminatory Behavior*

Another means by which interspecific sound discrimination is improved without displacement of the releaser is shown in studies by Gill and Murray (1972) on Blue-winged Warbler (*Vermivora pinus*) and Golden-winged Warbler and by Becker (1977a) on Firecrest and Goldcrest. In allopatric zones only 2 of 9 Blue-winged Warbler males distinguish the species-specific song from that of the congener, as compared with 14 of 18 in areas of sympatry. In addition, 53 and 64%, respectively, of males in two allopatric populations of the Firecrest react to the Goldcrest song, significantly more than in three areas of sympatry [10, 13, and 20% (Fig. 1)]. An intermediate reaction value of Firecrests (33%) was found in another area of sympatry, where the Goldcrest was the species with the much lower population density.

Less song contrast in allopatric areas has not been detected in these cases, so the cause of the stronger reaction in allopatry may be a diminished selectivity of the releasing mechanisms. This means that Blue-winged Warblers and Firecrests in their respective sympatric areas learn to distinguish and not react to heterospecific song through coexistence and encounters with males of the related species. In this way experience with heterospecific sounds probably sharpens the response to species-specific song, makes the releasing mechanism more selective (''contrast reinforcement'' in the releasing mechanism), and improves discriminatory behavior.

The two other species of these two species pairs, however, Golden-winged Warbler and Goldcrest, show no greater responsiveness to the heterospecific song in the allopatric range; obviously in these species a reinforcement of discrimination ability in sympatry is unnecessary. This is also absent in Icterine and Melodious Warblers (*Hippolais icterina, H. polyglotta:* Ferry and Deschaîntre, 1974), which are, however, interspecifically territorial, and respond to each other's songs in both allopatric and sympatric areas.

D. Reduction in Species Specificity

1. *Mixed Singers*

The ability to incorporate heterospecific sounds into the song repertoire occurs in an array of birds, mostly among closely related species (e.g., *Ficedula:* Löhrl, 1955; *Sturnella:* Lanyon, 1957; *Pipilo:* Marshall, 1964; *Phylloscopus:* Gwinner and Dorka, 1965, Schubert, 1969; *Certhia:* Thielcke, 1972; *Vermivora:* Gill and Murray, 1972; *Sylvia:* Bergmann, 1973; *Thryomanes* and *Troglodytes:* Kroodsma, 1973; *Parus:* Ward and Ward, 1974; *Passerina:* Emlen *et al.*, 1975; *Acrocephalus:* Lemaire, 1977; *Regulus:* Becker, 1977a; *Emberiza:* Kreutzer, 1979). In addition to species-typical songs, these mixed singers are able to sing heterospecific songs or mixed songs composed of parts of both species' songs. Where the *Regulus* species are sympatric the development of mixed singing is favored in the species with the lower population density, and can be explained through the greater probability of hearing the songs of the sibling species in lieu of conspecific song during the song learning process (Becker, 1977a). In the laboratory, interspecific song learning occurs where juvenile males are insufficiently exposed to songs of conspecifics (Lanyon, 1960; Thorpe, 1961; Thiecke, 1972).

Mixed singing is often tied to loss of species specificity, since it acts as a releasing signal on conspecifics and individuals of the species whose sounds have been appropriated (e.g., Thielcke, 1972; Kroodsma, 1973; Murray and Gill, 1976; Becker, 1977a). Mixed singers usually also react to learned heterospecific song components, for as a rule they react to the song that they have learned (e.g., Hinde, 1958; Stevenson, 1967; Thielcke, 1972; Lemaire, 1977; Becker, 1977a). As a consequence mixed singers are often interspecifically territorial.

Thorough investigations of mixed singers and interspecifically territorial individuals have shown, however, that singing behavior is not necessarily coupled with the reaction to sounds. Some males, in fact, react to sounds of another species, though they themselves do not reproduce these sounds (Thielcke, 1972; Emlen *et al.*, 1975; Murray and Gill, 1976; Becker, 1977a); other males do not respond to a song, though they sing it themselves (Murray and Gill, 1976). This suggests that the releasing mechanism for the reaction to sounds (and the individual's territorial response) is derived through a learning process independent of song learning.

2. *Character Convergence in Sounds and Interspecific Territoriality*

Species may exclude each other if they occupy a homogeneous ecologically marginal habitat (Mayr, 1967). In such cases resemblance by appearance or voice could be selectively advantageous because convergence facilitates interspecific territoriality (Marler, 1960; Cody, 1969; but see Murray, 1971, 1976). Cody cites several examples of apparent convergence in song (Cody, 1969, 1974; Cody and Brown, 1970; Cody and Walter, 1976; Grant, 1966; Lanyon, 1957), but these have not been substantiated or were incorrectly interpreted in the first place (Murray, 1976; Becker, 1977a). Only the reactions of the birds allows us to decide whether sounds are really more similar for the populations in question than in other regions and are involved in the maintenance of interspecific territoriality (Section IV,B,5). Similarly, other studies have failed to provide evidence in support of this hypothesis (Rohwer, 1973; Ferry and

Deschaîntre, 1974; Murray, 1976; Ordal, 1976; Catchpole, 1977; Becker, 1977a). Only Emlen *et al.* (1975) show the adaptive value of vocal convergence for interspecific territoriality in Indigo and Lazuli buntings (*Passerina amoena*), achieved through reciprocal exchange of song figures.

These authors indicate, however, that interspecific territoriality can be achieved through increased responsiveness to heterospecific song in areas of sympatry, through individual encounters and experience; song structure need not be similar. This mechanism for interspecific song reactions is often mentioned and should be given more attention (see Section IV,D,1) (Catchpole, 1977). It is a counterpart to the reinforcement of discriminatory behavior (Section IV,C,3), and emphasizes in a similar way the importance of individual experience.

3. *Imitations*

Many songbird species integrate imitations of heterospecifics into their songs (see Chapter 3, Volume 2). Imitations in the songs of the Crested Lark (*Galerida cristata:* Tretzel, 1965) and the Marsh Warbler (*Acrocephalus palustris:* Lemaire, 1975) are so exact that they act as releasers for the imitated species if taken from the song and tested in the field. On the other hand, the song of the mimic does not release a territorial response in the imitated species, presumably because imitations of any one species occur too infrequently and are presented in abbreviated form [Lemaire (1974, 1975); Mockingbird (*Mimus polyglottos*): Howard (1974)]. Harcus (1977) presumes, however, that imitation in the Chorister Robin-chat (*Cossypha dichroa*) is important for interspecific territoriality as well as for advertising species identity and territory.

V. CONFLICTS WITH THE CODING OF OTHER INFORMATION

A. Preliminary Remarks

Species specificity requires clarity and stability in vocalizations. However, the transmission of further information in the same vocalizations (e.g., individual identity or motivational state) demands clearly distinctive individual or motivational markers. This informational load can be encoded only through increased complexity and individual and contextual variability. By these means conflicts with the need for species specificity may arise (Marler, 1960).

However, these problems do not affect sounds which do not require species specificity [see Section I; another example is the "luring" contact call of the Common Murre adults (*Uria aalge*), which has to encode only individual characteristics of the parents for their identification by their young (Tschanz, 1968)].

According to Marler (1961, p. 304) "some bird songs appear to convey both species-specific and individual information by relegating the stereotyped and variable properties to different parameters of the song." (This use of the term "parameter" differs from that of this chapter.) Subsequent studies have contributed data on this matter.

B. Different Parts of the Song for Encoding Different Messages

The beginning of the song of some species is markedly stereotyped, while the ending is more plastic and shows individual characteristics [e.g., Chaffinch, Reed Bunting (*Emberiza schoeniclus*), Willow Warbler, Yellowhammer: Schubert (1969); Tree Pipit (*Anthus trivialis*): Bjerke (1971); Whitethroat (*Sylvia communis*): Bergmann (1973); Hazel Grouse (*Bonasa bonasia*); Bergmann *et al.* (1975); Indigo Bunting: Emlen (1972); Shiovitz (1975); Goldcrest: Becker (1974, 1976)]. On the contrary, in Red-winged Blackbird song only the ending trill serves for species recognition (Beletsky *et al.*, 1980). Therefore, one may presume that the coding of species and individual characteristics devolves upon different parts of the song, thereby minimizing conflicts between messages. In Golden-winged Warbler, Indigo Bunting, and Goldcrest, the beginning of the song functions in gaining the attention of conspecifics (Ficken and Ficken, 1973; Shiovitz, 1975; Becker, 1976; Wiley and Richards, Chapter 5, this volume). In the Goldcrest the song ending is important in communication within the pair (Becker, 1976; Thaler, 1979) (Fig. 1). In the Indigo Bunting motivation is encoded in the number of terminal elements in the song. These observations point to a need for more precise experimental analysis.

C. Individuality

Individual recognition by sound occurs in many species (Chapter 8, Volume 2), and the existence of some individually constant and interindividually varying characteristics is presupposed. Yet seldom have species-specific parameters and individual-specific cues been studied in the same species. Some exceptions are White-throated Sparrow (Falls, 1963, 1969; Brooks and Falls, 1975a,b), Indigo Bunting (Emlen, 1971, 1972), Ovenbird (Weeden and Falls, 1959; Falls, 1963), and Golden-winged Warbler (Ficken and Ficken, 1973). In all these studies characteristics that encode individual identity are also parameters in species specificity. Thus, for example, in the White-throated Sparrow the frequency range is critical for species identity and, within that range, individual-typical frequency differences are used in encoding individual identity. The possible conflicts in coding seem to be resolved in a way other than assumed by Marler (1960, 1961): within the range of variation in a species-specific parameter there remains adequate room for the coding of individual characteristics.

D. Motivation and Fitness

Encoding of motivational information requires intraindividual variability, as, for example, in the following characteristics.

1. Vocalizing Rate

In many species excitation can be expressed in the number of vocalizations per unit time (song: e.g., Falls, 1963, 1969; Tretzel, 1965; M. Schubert, 1971; Helb, 1973; Emlen, 1972; Ficken and Ficken, 1973; Becker, 1974; Heidemann and Oring, 1976; Kroodsma and Verner, 1978; calls: e.g., van den Elzen, 1977; Becker, 1977a); thus males utter more songs after territorial encounters or playback experiments than in other circumstances. Moreover, the singing rate is an indication of a male's pairing status; paired males sing less than unpaired males (e.g., Curio, 1959; Marshall, 1960; Blase, 1960; Diesselhorst, 1968; Thompson, 1972; Catchpole, 1973).

2. Song Length

In conflict situations males may increase song length (e.g., Falls, 1963; M. Schubert, 1971; Helb, 1973; G. Schubert, 1971; Emlen, 1972; Becker, 1974), or shorten it (e.g., Tretzel, 1965; Falls, 1969; G. Schubert, 1971; M. Schubert, 1971; Helb, 1973; Becker, 1974, 1976).

3. Amplitude

In behavioral conflicts amplitude may be reduced (G. Schubert, 1971; M. Schubert, 1971; Helb, 1973; Emlen, 1972; Ficken and Ficken, 1973; Becker, 1976).

4. Frequency

Excitation increases the frequency of the song (e.g., Brémond, 1968a; Thompson, 1972; van den Elzen, 1977) and broadens the frequency range in the Golden-winged Warbler (Ficken and Ficken, 1973).

5. Complexity

Songs of Robin males include more complex motifs when they are likely to attack (Brémond, 1968a). The most variable song of the Chestnut-sided Warbler (*Dendroica pensylvanica*), the "jumbled song," is significant for territorial encounters (Lein, 1978).

In these examples changes in one sound characteristic are often accompanied by changes in another. For example, in the alarm call of the Common Loon (*Gavia immer*), variations in duration, frequency, amplitude, and intervals between calls are coupled (Barklow, 1979).

The assumption that graded vocalizations actually encode different motivations of the sender and release appropriate reactions in the receiver may be accurate, but has seldom been demonstrated (e.g., von Haartman, 1953).

Comparison of these characteristics with species-specific parameters (Section II) shows that characteristics reserved for species identity rarely encode motivational information. The physical complexity and redundancy of most sounds leaves enough traits free from the function of species identification to serve in coding for other information (Section II,D). As in individual-specific coding, however, the range of the species-specific parameter frequency can be utilized to encode motivation.

Marler's assumption about individual characterization thus seems to apply to the coding of motivation: segregation of information among different components of sounds allows the simultaneous transmission of information about species, individual, and motivational state.

The repertoire size of the Mockingbird males appears to play a significant role in determining territory rank, and possibly influences how early a male acquires a mate (Howard, 1974). In the Sedge Warbler (*Acrocephalus schoenobaenus:* Catchpole, 1980), too, pairing date is inversely correlated with syllable repertoire size. Female Canaries (*Serinus canaria*) come into breeding condition more in response to complex than to simple songs (Kroodsma, 1976). Therefore, complexity in song may encode information about a male's "quality" (age, learning abilities, attentiveness, etc.), and seems to be promoted through selection by a female's choice of a mate (e.g., Sedge Warbler: Catchpole, 1973, 1975, 1980).

E. Conclusion

Our limited knowledge about encoding different kinds of information in the same vocalization indicates that species specificity is not disturbed by other requirements. Several species have developed different songs for different social situations (e.g., Estrildidae: Immelmann, 1968; Sossinka and Böhner, 1980; *Vidua chalybeata:* Payne, 1979). While one song serves in territorial advertisement, another occurs in sexual contexts (Peking Robin, *Leiothrix lutea:* Thielcke and Thielcke, 1970; Thrush Nightingale, *Erithacus luscinia:* Sorjonen, 1977; Melodious Grassquit, *Tiaris canora:* Baptista, 1978). Males of *Dendroica* species and the American Redstart (*Setophaga ruticilla*) react differently to their different song types, which probably vary with motivation and function (Ficken and Ficken, 1962, 1970; Morse, 1970; Lein, 1978). It would be interesting to test whether species specificity has equal importance and is encoded similarly in the different songs.

Finally, it must be stressed that all these kinds of information potentially available in a song are lost or become irrelevant if the song does not convey the songster's species (Falls, 1969; Emlen, 1972).

VI. THE PROGRAMMING OF SPECIES SPECIFICITY

To what extent are parameters and releasing mechanisms passed on through heredity and tradition? The ontogeny of releasers is discussed elsewhere (e.g., Chapters 1 and 2, Volume 2), so I will discuss here mainly the programming of releasing mechanisms.

A. Species-Specific Releasers

The motor patterns which generate bird sounds are innate according to present knowledge of non-passerine sounds and most calls in passerines, or partly learned as in song and some calls in passerines (see Chapters 1 and 2, Volume 2; e.g., Marler and Hamilton, 1972; Thielcke, 1977). Correspondingly, species-specific parameters are encoded in the genome or transmitted through learning from conspecifics, or both.

For example, the innate aggressive call of Kaspar-Hauser Willow Warblers evokes reactions from wild birds as well as do calls from the wild birds themselves (M. Schubert, 1971, 1976). Males of the oscine species Brown-headed Cowbird (*Molothrus ater*) reared in isolation develop abnormal songs which still contain the necessary parameters to release copulatory responses in females—an important adaptation for a parasitic species (King and West, 1977).

In contrast, experiments in which wild birds are exposed to Kaspar-Hauser songs show the importance of tradition for the development of passerine song and its parameters: the majority of songs developed in auditory isolation release little or no reaction compared with the natural songs [Short-toed Tree Creeper (*Certhia brachydactyla*) and Brown (Common Tree) Creeper (*C. familiaris*): Thielcke (1970c); Willow Warbler: M. Schubert (1971); Coal Tit: Thielcke (1973a); Indigo Bunting: Shiovitz (1975); Marsh Tit: Becker (1978b); Chiffchaff: G. Thielcke (unpublished data); Grasshopper Warbler (*Locustella naevia*): P. H. Becker (unpublished data)]. Some Kaspar-Hauser songs have a releasing effect, however: in the Coal Tit they seem to be developed from alarm calls which are probably innate (Thielcke, 1973a).

B. Species-Specific Releasing Mechanisms

Few studies address the subject of the relative importance of heredity and experience in determining the species-specific releasing mechanism. Some answers to this question can be obtained from experiments on vocal learning.

1. Non-Passerines

Most sounds in non-passerines appear to be innate, but this does not necessarily indicate that the releasing mechanisms for species identification are also hereditary: young birds can be imprinted upon the sounds of other species, as shown by the possible exchange of chicks in the Arctic Tern (*Sterna paradisaea*) and the Common Tern (*S. hirundo;* Busse, 1977).

In some species a preference of the young for species-specific vocalizations is apparent, even in the absence of prior experience with them [e.g., Wood Duck, Mallard (*Anas platyrhynchos*), and Domestic Fowl (*Gallus domesticus*): Gottlieb (1965, 1966); Herring Gull (*Larus argentatus*) and Ring-billed Gull (*L. delawarensis*): Evans (1973, 1975); Common Tern: Busse (1977); Neubauer (1978); Arctic Tern: Busse (1977)]. But as shown by Gottlieb (1971, 1977), prenatal experience with self-generated sounds in the egg can strengthen these preferences. Through prenatal or early posthatching experience with the vocalizations of parents and through visual stimuli, the preference for species-specific sounds is increased (Evans, 1972; Impekoven, 1976; Busse and Busse, 1977). Alternatively, this preference can be masked through imprinting on another species (Busse, 1977).

2. *Passerines*

An example of an inborn releasing mechanism is in female Brown-headed Cowbirds, which react innately only to the songs of male conspecifics (King and West, 1977).

With Thielcke (1970c, 1977), I recognize three kinds of tradition in the songs of birds. The White-crowned Sparrow (*Zonotrichia leucophrys*) is in the first group. They learn their song and will accept a tape as a song model (Marler and Tamura, 1962, 1964; Marler, 1970). The males do not learn heterospecific songs, however, whether they are offered alone or in choice experiments. Accordingly, young White-crowned Sparrows may be able to innately recognize certain parameters in the species-typical song. Swamp Sparrows learn by the same method (Peters *et al.*, 1980). Also, other species that can learn song from a tape may have a similar system (Chaffinch: Thorpe, 1958; Konishi and Nottebohm, 1969; Cardinal: Dittus and Lemon, 1969; Blackbird (*Turdus merula*): Messmer and Messmer, 1956; Thielcke and Thielcke, 1960; Song Sparrow: Peters *et al.*, 1980). The Song Sparrow, Chaffinch, and Blackbird learn other species' songs, too; yet the two last-mentioned species seem to prefer conspecific song.

As studies by Hinde (1958) and Brémond (1972b) suggest and Shiovitz and Lemon (1980) suppose, a largely innate releasing mechanism can be refined through learning, making experienced birds more selective in song identification than young. Nevertheless in the Swamp Sparrow, Peters *et al.* (1980) did not see any indication of an improvement of the song recognition process by learning during ontogeny.

In the second kind of tradition, individuals learn songs only from known birds which raised them [Bullfinch (*Pyrrhula pyrrhula*): Nicolai (1959); Zebra Finch (*Poephila guttata*): Immelmann (1967, 1969); host-species songs of the Fischer's

Whydah (*Vidua fischeri*): Nicolai (1973)]. The appropriate releasing mechanism and the song are developed together. This is also the case in the Brown Creeper and the Short-toed Tree Creeper (third kind of tradition, Thielcke, 1970c, 1972, 1977): The males do not learn from a tape, but only from conspecifics (or birds of a related species), with which visual contact is unnecessary. Evidently the young learn the song only if it is heard from an adult which is accepted as model, presumably by communication through innate calls.

Other studies suggest that in many oscine species the releasing mechanisms are primarily determined through learning. Some examples are: the greater reaction of Chaffinch males to their own songs (Hinde, 1958; Stevenson, 1967); the greater reaction to population-typical song than to foreign song variants (Section III); mixed singers, which react to learned conspecific and heterospecific songs (Section IV,D,1), and individuals for which not only conspecific songs but songs of other species are releasers, although the latter are not sung (Section IV,D,1 and 2). The releasing mechanism for species-typical song can probably become more selective through learning processes (Section IV,C,3).

In spite of fragmentary evidence of the subject it is obvious that different possibilities for reciprocal tuning of releasers and releasing mechanisms are realized in evolution.

Reliance upon learning in the entrainment of the releasing mechanism could have some negative effects (e.g., in misimprinting). But learning of the releasing mechanism has great advantages and is necessary when sounds are learned: through learning, releasers and releasing mechanisms are optimally adjusted to each other, and this opens new possibilities for the species to adapt quickly in its acoustical behavior to altered ecological conditions (e.g., mixed songs, interspecific responses). Furthermore, learning of acoustical signals and entrainment of their respective releasing mechanisms are prerequisites for rapid changes in sounds, which are possible through deprivation of tutoring (withdrawal of learning, Thielcke, 1970c, 1973b) and which may have promoted the rapid adaptive radiation of the passerines.

ACKNOWLEDGMENTS

I am very grateful to C. S. Adkisson for his friendly assistance in translation of the manuscript and for helpful comments, and to my wife Gabriele for her patience in drawing the figures and typing the manuscript. G. Thielcke's ideas form the basis and stimulation for my studies. I thank him for his constant cooperation and for his many suggestions and comments with regard to this chapter.

REFERENCES

Abs, M. (1963). Field tests on the essential components of the European Nightjar's song. *Proc. Int. Ornithol. Congr.* **13,** 202–205.

Baptista, L. F. (1974). The effects of songs of wintering White-crowned Sparrows on song development in sedentary populations of the species. *Z. Tierpsychol.* **34,** 147–171.

Baptista, L. F. (1975). Song dialects and demes in sedentary populations of the White-crowned Sparrow (*Zonotrichia leucophrys nuttalli*). *Univ. Calif. (Berkeley), Publ. Zool.* **105,** 1–52.

Baptista, L. F. (1977). Geographic variation in song and dialects of the Puget Sound White-crowned Sparrow. *Condor* **79,** 356–370.

Baptista, L. F. (1978). Territorial, courtship and duet songs of the Cuban Grassquit (*Tiaris canora*). *J. Ornithol.* **119,** 91–101.

Barklow, W. E. (1979). Graded frequency variations of the tremolo call of the Common Loon (*Gavia immer*). *Condor* **81,** 53–64.

Becker, P. H. (1974). Der Gesang von Winter- und Sommergoldhähnchen (*Regulus regulus, R. ignicapillus*) am westlichen Bodensee. *Vogelwarte* **27,** 233–243.

Becker, P. H. (1976). Artkennzeichnende Gesangsmerkmale bei Winter- und Sommergoldhähnchen (*Regulus regulus, R. ignicapillus*). *Z. Tierpsychol.* **42,** 411–437.

Becker, P. H. (1977a). Verhalten auf Lautäusserungen der Zwillingsart, interspezifische Territorialität und Habitatansprüche von Winter- und Sommergoldhähnchen (*Regulus regulus, R. ignicapillus*). *J. Ornithol.* **118,** 233–260.

Becker, P. H. (1977b). Geographische Variation des Gesanges von Winter- und Sommergoldhähnchen (*Regulus regulus, R. ignicapillus*). *Vogelwarte* **29,** 1–37.

Becker, P. H. (1978a). Vergleich von Lautäusserungen der Gattung *Regulus* (Goldhähnchen) als Beitrag zur Systematik. *Bonn. Zool. Beitr.* **29,** 101–121.

Becker, P. H. (1978b). Der Einfluss des Lernens auf einfache und komplexe Gesangsstrophen der Sumpfmeise (*Parus palustris*). *J. Ornithol.* **119,** 388–411.

Becker, P. H., Thielcke, G., and Wüstenberg, K. (1980a). Versuche zum angenommenen Kontrastverlust im Gesang der Blaumeise (*Parus caeruleus*) auf Teneriffa. *J. Ornithol.* **121,** 81–95.

Becker, P. H., Thielcke, G., and Wüstenberg, K. (1980b). Der Tonhöhenverlauf ist entscheidend für das Gesangserkennen des mitteleuropäischen Zilpzalps (*Phylloscopus collybita*). *J. Ornthol.* **121,** 229–244.

Beletsky, D., Chao, S., and Smith, D. G. (1980). An investigation on song-based species recognition in the Red-winged Blackbird (*Agelaius phoeniceus*). *Behaviour* **73,** 189–203.

Bergmann, H.-H. (1973). Die Imitationsleistung einer Mischsänger-Dorngrasmücke (*Sylvia communis*). *J. Ornithol.* **114,** 317–338.

Bergmann, H.-H., Klaus, S., Müller, F., and Wiesner, J. (1975). Individualität und Artspezifität in den Gesangsstrophen einer Population des Haselhuhns (*Bonasa bonasia bonasia* L., Tetraoninae, Phasianidae). *Behaviour* **55,** 94–114.

Bertram, B. (1970). The vocal behavior of the Indian Hill Mynah, *Gracula religiosa*. *Anim. Behav. Monogr.* **3,** Part 2, 81–189.

Bjerke, T. (1971). Song variation in the Tree-Pipit, *Anthus trivialis*. *Sterna* **10,** 97–116.

Blase, B. (1960). Die Lautäusserungen des Neuntöters (*Lanius c. collurio* L.), Freilandbeobachtungen und Kaspar-Hauser-Versuche. *Z. Tierpsychol.* **17,** 293–344.

Boughey, M. J., and Thompson, N. S. (1976). Species specificity and individual variation in the songs of the Brown Thrasher (*Toxostoma rufum*) and Catbird (*Dumetella carolinensis*). *Behaviour* **57,** 64–90.

Brémond, J. C. (1967). Reconnaissance de schémas réactogènes liés à l'information contenue dans le chant territorial du Rouge-Gorge (*Erithacus rubecula*). *Proc. Int. Ornithol. Congr.* **14,** 217–229.

Brémond, J. C. (1968a). Recherches sur la sémantique et les éléments vecteurs d'information dans les signaux acoustiques du Rouge-Gorge. *Terre Vie* **68,** 109–220.

Brémond, J. C. (1968b). Valeur spécifique de la syntaxe dans le signal de défense territoriale du Troglodyte (*Troglodytes troglodytes*). *Behavior* **30,** 66–75.

Brémond, J. C. (1972a). Recherche sur les paramètres acoustiques assurant la reconnaissance spécifique dans les chants de *Phylloscopus sibilatrix, Phylloscopus bonelli* et d'un hybride. *Gerfaut* **62,** 313–323.

Brémond, J. C. (1972b). Comparaison entre l'apprentissage du chant chez le jeune Pinson (*Fringilla coelebs*) et les éléments réactogènes du chant territorial de l'adulte. *Rev. Comp. Anim.* **6,** 191–195.

Brémond, J. C. (1976a). Rôle des phénomènes transitoires dans la reconnaissance spécifique du chant du Rouge-Gorge (*Erithacus rubecula*). *Experientia* **32,** 460–462.

Brémond, J. C. (1976b). Specific recognition in the song of Bonelli's Warbler (*Phylloscopus bonelli*). *Behaviour* **58,** 99–116.

Brémond, J. C. (1976c). Les phénomènes transitoires et la reconnaissance du chant chez le Troglodyte (*Troglodytes troglodytes*). *Behav. Processes* **1,** 145–152.

Brémond, J. C. (1978). Acoustic competition between the song of the Wren (*Troglodytes troglodytes*) and the songs of other species. *Behaviour* **65,** 89–98.

Brooks, R. J., and Falls, J. B. (1975a). Individual recognition by song in White-throated Sparrows. I. Discrimination of songs of neighbors and strangers. *Can. J. Zool.* **53,** 879–888.

Brooks, R. J., and Falls, J. B. (1975b). Individual recognition by song in White-throated Sparrows. III. Song features used in individual recognition. *Can. J. Zool.* **53,** 1749–1761.

Busse, K. (1977). Prägungsbedingte akustische Arterkennungsfähigkeit der Küken der Flussseeschwalben und Küstenseeschwalben *Sterna hirundo* L. und *S. paradisea* Pont. *Z. Tierpsychol.* **44,** 154–161.

Busse, K., and Busse, K. (1977). Prägungsbedingte Bindung von Küstenseeschwalbenküken (*Sterna paradisea* Pont.) an die Eltern und ihre Fähigkeit, sie an der Stimme zu erkennen. *Z. Tierpsychol.* **43,** 287–294.

Catchpole, C. K. (1973). The functions of advertising song in the Sedge Warbler (*Acrocephalus schoenobaenus*) and the Reed Warbler (*A. scirpaceus*). *Behaviour* **46,** 300–320.

Catchpole, C. K. (1975). Temporal and sequential organisation of song in the Sedge Warbler (*Acrocephalus schoenobaenus*). *Behavior* **59,** 226–246.

Catchpole, C. K. (1977). Aggressive responses of male Sedge Warblers (*Acrocephalus schoenobaenus*) to playback of species song and sympatric species song, before and after pairing. *Anim. Behav.* **25,** 489–496.

Catchpole, C. K. (1980). Sexual selection and the evolution of complex songs among European warblers of the genus *Acrocephalus*. *Behaviour* **74,** 149–166.

Cody, M. L. (1969). Convergent characteristics in sympatric species: A possible relation to interspecific competition and aggression. *Condor* **71,** 222–239.

Cody, M. L. (1974). "Competition and the Structure of Bird Communities." Princeton Univ. Press, Princeton, New Jersey.

Cody, M. L., and Brown, J. H. (1970). Character convergence in Mexican finches. *Evolution* **24,** 304–310.

Cody, M. L., and Walter, H. (1976). Habitat selection and interspecific interactions among Mediterranean sylviid warblers. *Oikos* **27,** 210–238.

Conrads, K., and Conrads, W. (1971). Regionaldialekte des Ortolans (*Emberiza hortulana*) in Deutschland. *Vogelwelt* **92,** 81–100.

Curio, E. (1959). Verhaltensstudien am Trauerschnäpper. *Z. Tierpsychol.*, Beiheft No. 3.

Diesselhorst, G. (1968). Struktur einer Brutpopulation von *Sylvia communis*. *Bonn. Zool. Beitr.* **19,** 307–321.

Dilger, W. C. (1956). Hostile behavior and reproductive isolating mechanisms in the avian genera *Catharus* and *Hylocichla*. *Auk* **73,** 313–353.

Dittus, W. P. J., and Lemon, R. E. (1969). Effects of song tutoring and acoustic isolation on the song repertoires of Cardinals. *Anim. Behav.* **17,** 523–533.

Emlen, S. T. (1971). The role of song in individual recognition in the Indigo Bunting. *Z. Tierpsychol.* **28,** 241–246.

Emlen, S. T. (1972). An experimental analysis of the parameters of bird song eliciting species recognition. *Behaviour* **41,** 130–171.

Emlen, S. T., Rising, J. D., and Thompson, W. L. (1975). A behavioral and morphological study of sympatry in the Indigo and Lazuli Buntings of the Great Plains. *Wilson Bull.* **87,** 145–179.

Evans, R. M. (1972). Development of an auditory discrimination in domestic chicks (*Gallus gallus*). *Anim Behav.* **20,** 77–87.

Evans, R. M. (1973). Differential responsiveness of young Ring-billed Gulls and Herring Gulls to adult vocalizations of their own and other species. *Can. J. Zool.* **51,** 759–770.

Evans, R. M. (1975). Responsiveness of young Herring Gulls to adult "mew" calls. *Auk* **92,** 140–143.

Ewert, D. N. (1980). Recognition of conspecific song by the Rufous-sided Towhee. *Anim. Behav.* **28,** 379–386.

Falls, J. B. (1963). Properties of bird song eliciting responses from territorial males. *Proc. Int. Ornithol. Congr.* **13,** 259–271.

Falls, J. B. (1969). Functions of territorial song in the White-throated Sparrow. *In* "Bird Vocalizations" (R. A. Hinde, ed.), pp. 207–232. Cambridge Univ. Press, London and New York.

Ferry, C., and Deschaîntre, A. (1974). Le chant, signal interspécifique chez *Hippolais icterina* et *polyglotta*. *Alauda* **42,** 289–311.

Ficken, M. S., and Ficken, R. W. (1962). The comparative ethology of the wood warblers: A review. *Living Bird* **1,** 103–122.

Ficken, M. S., and Ficken, R. W. (1970). Responses of four warbler species to playback of their two song types. *Auk* **87,** 296–304.

Ficken, M. S., and Ficken, R. W. (1973). Effect of number, kind and order of song elements on playback responses of the Golden-winged Warbler. *Behaviour* **46,** 114–128.

Ficken, R. W., Ficken, M. S., and Hailman, J. P. (1974). Temporal pattern shifts to avoid acoustic interference in singing birds. *Science* **183,** 762–763.

Gill, F. B., and Lanyon, W. E. (1964). Experiments on species discrimination in Blue-winged Warblers. *Auk* **81,** 53–64.

Gill, F. B., and Murray, B. G. (1972). Discrimination behavior and hybridization in sympatric Blue-winged and Golden-winged Warblers. *Evolution* **26,** 282–293.

Göransson, G., Högstedt, G., Karlsson, J., Köllander, H., and Mifstraud, S. (1974). Sångens roll för revirhållandet hos näktergal *Luscinia luscinia*—några experiment med playback teknik. *Var Fagelvärd* **33,** 201–209.

Goldman, P. (1973). Song recognition by Field Sparrows. *Auk* **90,** 106–113.

Gottlieb, G. (1965). Imprinting in relation to parental and species identification by avian neonates. *J. Comp. Physiol. Psychol.* **59,** 345–356.

Gottlieb, G. (1966). Species identification by avian neonates: Contributory effect of perinatal auditory stimulation. *Anim. Behav.* **14,** 282–290.

Gottlieb, G. (1971). "Development of Species Identification: An Inquiry into the Prenatal Determinants of Perception." Univ. of Chicago Press, Chicago, Illinois.

Gottlieb, G. (1974). On the acoustic basis of species identification in wood ducklings (*Aix sponsa*). *J. Comp. Physiol. Psychol.* **87,** 1038–1048.

Gottlieb, G. (1977). The call of the duck. *Nat. Hist.* **86,** 40–47.

Grant, P. R. (1966). The coexistence of two wren species of the genus *Thryothorus*. *Wilson Bull.* **78,** 266–278.

Gürtler, W. (1973). Artisolierende Parameter des Revierrufs der Türkentaube (*Streptopelia decaocto*). *J. Ornithol.* **114,** 305–316.

Gwinner, E., and Dorka, V. (1965). Beobachtungen an Zilpzalp-Fitis-Mischsängern. *Vogelwelt* **86,** 146–151.
Harcus, J. L. (1977). The functions of mimicry in the vocal behaviour of the Chorister Robin. *Z. Tierpsychol.* **44,** 178–193.
Harris, M. A., and Lemon, R. E. (1974). Songs of Song Sparrows: Reactions of males to songs of different localities. *Condor* **76,** 33–44.
Hartshorne, C. (1956). The monotony threshold in singing birds. *Auk* **73,** 176–192.
Heidemann, M. K., and Oring, L. W. (1976). Functional analysis of Spotted Sandpiper (*Actitis macularia*) song. *Behaviour* **56,** 181–193.
Helb, H.-W. (1973). Analyse der artisolierenden Parameter im Gesang des Fitis (*Phylloscopus trochilus*) mit Untersuchungen zur Objektivierung der analytischen Methode. *J. Ornithol.* **114,** 145–206.
Heuwinkel, H. (1978). Der Gesang des Teichrohrsängers (*Acrocephalus scirpaceus*) unter besonderer Berücksichtigung der Schalldruckpegel- (''Lautstärke''-) Verhältnisse. *J. Ornithol.* **119,** 450–461.
Hinde, R. A. (1958). Alternative motor patterns in Chaffinch song. *Anim. Behav.* **6,** 211–218.
Howard, R. D. (1974). The influence of sexual selection and interspecific competition on Mockingbird song (*Mimus polyglottos*). *Evolution* **28,** 428–438.
Immelmann, K. (1959). Experimentelle Untersuchungen über die biologische Bedeutung artspezifischer Merkmale beim Zebrafinken (*Taeniopygia castanotis*). *Zool. Jahrb., Abt. Syst. Oekol. Tiere* **86,** 437–592.
Immelmann, K. (1967). Zur ontogenetischen Gesangsentwicklung bei Prachtfinken. Verh. Dtsch. Zool. Ges. Göttingen 1966, 320–332.
Immelmann, K. (1968). Zur biologischen Bedeutung des Estrildidengesanges. *J. Ornithol.* **109,** 284–299.
Immelmann, K. (1969). Song development in the Zebra Finch and other Estrildid Finches. *In* ''Bird Vocalizations'' (R. A. Hinde, ed.), pp. 61–74. Cambridge Univ. Press, London and New York.
Impekoven, M. (1976). Responses of Laughing Gull chicks (*Larus atricilla*) to parental attraction and alarm calls, and effects of prenatal auditory experience on the responsiveness to such calls. *Behaviour* **56,** 250–278.
Johnson, N. K. (1963). Biosystematics of sibling species of Flycatchers in the *Empidonax hammondii-oberholseri-wrightii* complex. *Univ. Calif.* (*Berkeley*) *Publ. Zool.* **66,** 79–238.
King, A. P., and West, M. J. (1977). Species identification in the North American cowbird: Appropriate responses to abnormal song. *Science* **195,** 1002–1004.
Kling, J. W., and Stevenson-Hinde, J. (1977). Development of song and reinforcing effects of song in female Chaffinches. *Anim. Behav.* **25,** 215–220.
Knecht, S., and Scheer, M. (1971). Die Vögel der Azoren. *Bonn. Zool. Beitr.* **22,** 275–296.
Kneutgen, J. (1969). ''Musikalische'' Formen im Gesang der Schamadrossel (*Kittacincla macroura* Gm.) und ihre Funktion. *J. Ornithol.* **110,** 246–285.
Konishi, M., and Nottebohm, F. (1969). Experimental studies in the ontogeny of avian vocalizations. *In* ''Bird Vocalizations'' (R. A. Hinde, ed.), pp. 29–48. Cambridge Univ. Press, London and New York.
Krammer, K. (1971). Anpassung von Lautsignalen des Zilpzalps an die Kanalkapazität. *Naturwissenschaften* **58,** 417.
Krammer, K. (1982a). Die akustischen Parameter und ihre Relationen im Reviergesang des Zilpzalps (*Phyllosocopus collybita*). Unpubl. ms.
Krammer, K. (1982b). Zur Feinstruktur der Vogellaute. Unpubl. ms.
Kreutzer, M. (1974). Réponses comportementales de mâles Troglodytes (Passeriformes) à des chants spécifiques de dialectes différents. *Rev. Comp. Anim.* **8,** 287–295.

Kreutzer, M. (1979). Étude du chant chez le Bruant zizi (*Emberiza cirlus*). *Behaviour* **71,** 291–321.
Kroodsma, D. E. (1973). Coexistence of Bewick's Wrens and House Wrens in Oregon. *Auk* **90,** 341–352.
Kroodsma, D. E. (1976). Reproductive development in a female songbird: differential stimulation by quality of male song. *Science* **192,** 574–575.
Kroodsma, D. E., and Verner, J. (1978). Complex singing behaviours among *Cistothorus* wrens. *Auk* **95,** 703–716.
Lack, D., and Southern, H. N. (1949). Birds on Tenerife. *Ibis* **91,** 607–626.
Lanyon, W. E. (1957). The comparative biology of the Meadowlarks (*Sturnella*) in Wisconsin. *Publ. Nuttall Ornithol. Club* No. 1.
Lanyon, W. E. (1960). The ontogeny of vocalizations in birds. *In* "Animal Sounds and Communication" (W. E. Lanyon and W. N. Tavolga, eds.), Publ. No. 7, pp. 321–347. Am. Inst. Biol. Sci., Washington, D.C.
Lanyon, W. E. (1963). Experiments on species discrimination in *Myiarchus* Flycatchers. *Am. Mus. Novit.* No. 2126, 1–16.
Lein, R. (1978). Song variation in a population of Chestnut-sided Warblers (*Dendroica pensylvania*): its nature and suggested significance. *Can. J. Zool.* **56,** 1266–1283.
Lemaire, F. (1974). Le chant de la Rousserolle verderolle (*Acrocephalus palustris*): étendue du répertoire imitatif, construction, rhythmique et musicalité. *Gerfaut* **64,** 3–28.
Lemaire, F. (1975). Le chant de la Rousserolle verderolle (*Acrocephalus palustris*): fidélité des imitations et relations avec les espèces imitées et avec les congénères. *Gerfaut* **65,** 3–28.
Lemaire, F. (1977). Mixed song, interspecific competition and hybridization in the Reed and Marsh warblers (*Acrocephalus scirpaceus* and *palustris*). *Behaviour* **63,** 215–240.
Lemon, R. E. (1966). Geographic variation in the song of Cardinals. *Can. J. Zool.* **44,** 413–428.
Lemon, R. E. (1967). The response of Cardinals to songs of different dialects. *Anim. Behav.* **15,** 538–545.
Lemon, R. E. (1974). Song dialects, song matching and species recognition by Cardinals, *Richmondena cardinalis*. *Ibis* **116,** 545–548.
Lemon, R. E. (1975). How birds develop song dialects. *Condor* **77,** 325–406.
Lemon, R. E., and Herzog, A. (1969). The vocal behavior of Cardinals and Pyrrhuloxias in Texas. *Condor* **71,** 1–15.
Löhrl, H. (1955). Beziehungen zwischen Halsband- und Trauerfliegenschäpper (*Musicapa albicollis* und *M. hypoleuca*) in demselben Brutgebiet. *Acta Congr. Int. Ornithol., 11th, 1954,* pp. 334–336.
Lorenz, K. (1935). Der Kumpan in der Umwelt des Vogels. *J. Ornithol.* **83,** 137–213, 289–413.
Ludescher, F. B. (1973). Sumpfmeise (*Parus p. palustris* L.) und Weidenmeise (*P. montanus salicarius* Br.) als sympatrische Zwillingsarten. *J. Ornithol.* **114,** 3–56.
Marler, P. (1956). Über die Eigenschaften einiger tierlicher Rufe. *J. Ornithol.* **97,** 220–227.
Marler, P. (1957). Specific distinctiveness in the communication signals of birds. *Behavior* **11,** 13–39.
Marler, P. (1959). Developments in the study of animal communication. *In* "Darwin's Biological Work: Some Aspects Reconsidered" (P. R. Bell, ed.), pp. 150–206. Cambridge Univ. Press, London and New York.
Marler, P. (1960). Bird songs and mate selection. *In* "Animal Sounds and Communication" (W. E. Lanyon and W. N. Tavolga, eds.), Publ. No. 7, pp. 348–367. Am. Inst. Biol. Sci., Washington D.C.
Marler, P. (1961). The logical analysis of animal communication. *J. Theor. Biol.* **1,** 295–317.
Marler, P. (1970). A comparative approach to vocal learning: song development in White-crowned Sparrows. *J. Comp. Physiol. Psychol., Monogr. 71,* 1–25.
Marler, P., and Boatman, D. J. (1951). Observations on the birds of Pico, Azores. *Ibis* **93,** 90–99.

Marler, P., and Hamilton, W. J. (1972). "Tierisches Verhalten." BLV, Munich.

Marler, P., and Tamura, M. (1962). Song "dialects" in three populations of White-crowned Sparrows. *Condor* **64,** 368–377.

Marler, P., and Tamura, M. (1964). Culturally transmitted patterns of vocal behavior in sparrows. *Science* **146,** 1483–1486.

Marshall, J. T. (1960). Interrelations of Abert and Brown towhees. *Condor* **62,** 49–64.

Marshall, J. T. (1964). Voice in communication and relationship among Brown Towhees. *Condor* **66,** 345–356.

Martens, J. (1975). Akustische Differenzierung verwandtschaftlicher Beziehungen in der *Parus* (*Periparus*)—Gruppe nach Untersuchungen im Nepal–Himalaya. *J. Ornithol.* **116,** 369–433.

Mayr, E. (1967). "Artbegriff und Evolution." Parey, Berlin.

Messmer, E., and Messmer, I. (1956). Die Entwicklung der Lautäusserungen und einiger Verhaltensweisen der Amsel (*Turdus merula merula* L.) unter natürlichen Bedingungen und nach Einzelaufzucht in schalldichten Räumen. *Z. Tierpsychol.* **13,** 341–441.

Miller, D. B., and Gottlieb, G. (1976). Acoustic features of Wood Duck (*Aix sponsa*) maternal calls. *Behaviour* **57,** 260–280.

Milligan, M. M., and Verner, J. (1971). Interpopulational song dialect discrimination in the White-crowned Sparrow. *Condor* **73,** 208–213.

Morse, D. H. (1970). Territorial and courtship songs of birds. *Nature* (*London*) **226,** 659–661.

Murray, B. G. (1971). The ecological consequences of interspecific territorial behavior in birds. *Ecology* **52,** 414–423.

Murray, B. G. (1976). A critique of interspecific territoriality and character convergence. *Condor* **78,** 518–525.

Murray, B. G., and Gill, F. B. (1976). Behavioral interactions of Blue-winged and Golden-winged warblers. *Wilson Bull.* **88,** 231–254.

Neubauer, W. (1978). Experimentelle Untersuchungen zur akustischen und visuellen Kommunikation an der Flusseeschwalbe (*Sterna hirundo* L.) unter besonderer Berücksichtigung der Jungenaufzucht. *Beitr. Vogelkd.* **24,** 1–71.

Nicolai, J. (1959). Familientradition in der Gesangsentwicklung des Gimpels (*Pyrrhula pyrrhula* L.). *J. Ornithol.* **100,** 39–46.

Nicolai, J. (1968). Die Schnabelfärbung als potentieller Isolationsfaktor zwischen *Pytilia phoenicoptera* Swainson und *Pytilia lineata* Hinglin (Fam. Estrildidae). *J. Ornithol.* **109,** 450–461.

Nicolai, J. (1973). Das Lernprogramm in der Gesangsausbildung der Strohwitwe *Tetraenura fischeri* Reichenow. *Z. Tierpsychol.* **32,** 113–138.

Ordal, J. M. (1976). Effect of sympatry on Meadowlark vocalizations. *Condor* **78,** 100–101.

Payne, R. B. (1973a). Behavior, mimetic songs and song dialects, and relationships of the parasitic Indigo birds (*Vidua*) of Africa. *Ornithol. Monogr.* **11,** 1–333.

Payne, R. B. (1973b). Vocal mimicry of the Paradise Whydahs to the songs of their hosts (*Pytilia*) and their mimics. *Anim. Behav.* **21,** 762–771.

Payne, R. B. (1979). Song structure, behaviour, and sequence of song types in a population of Village Indigobirds, *Vidua chalybeata*. *Anim. Behav.* **27,** 997–1013.

Peters, S. S., Searcy, W. A., and Marler, P. (1980). Species song discrimination in choice experiments with territorial male Swamp and Song sparrows. *Anim. Behav.* **28,** 383–404.

Raitt, R. J., and Hardy, J. H. (1970). Relationships between two partly sympatric species of Thrushes (*Catharus*) in Mexico. *Auk* **87,** 20–57.

Rohwer, S. A. (1973). Significance of sympatry to behavior and evolution of Great Plains Meadowlarks. *Evolution* **27,** 44–57.

Romanowski, E. (1978). Der Gesang von Sumpf- und Weidenmeise (*Parus palustris* und *Parus montanus*)—Variation und Funktion. *Vogelwarte* **29,** 235–253.

Romanowski, E. (1979). Der Gesang von Sumpf- und Weidenmeise (*Parus palustris* und *Parus montanus*)—reaktionsauslösende Parameter. *Vogelwarte* **30,** 48–65.

Schleidt, W. M. (1973). Tonic communication: Continual effects of discrete signs in animal communication systems. *J. Theor. Biol.* **42,** 359–386.

Schleidt, W. M. (1976). On individuality: The constituents of distinctiveness. *Perspect. Ethol.* **2,** 299–310.

Schleidt, W. M. (1977). Tonic properties of animal communication systems. *Ann. N.Y. Acad. Sci.* **290,** 43–50.

Schubert, G. (1971). Experimentelle Untersuchungen über die artkennzeichnenden Parameter im Gesang des Zilpzalps. *Behaviour* **38,** 289–314.

Schubert, M. (1969). Untersuchungen über die akustischen Parameter von Zilpzalp-Fitis-Mischgesängen. *Beitr. Vogelkd.* **14,** 354–368.

Schubert, M. (1971). Untersuchungen über die reaktionsauslösenden Signalstrukturen des Fitis-Gesanges, *Phylloscopus trochilus t.* L., und das Verhalten gegenüber arteigenen Rufen. *Behaviour* **38,** 250–288.

Schubert, M. (1976). Das akustische Repertoire des Fitislaubsängers (*Phylloscopus t. trochilus*) und seine erblichen und durch Lernen erworbenen Bestandteile. *Beitr. Vogelkd.* **22,** 167–200.

Shiovitz, K. A. (1975). The process of species-specific song recognition by the Indigo Bunting, *Passerina cyanea,* and its relationship to the organization of avian acoustical behavior. *Behaviour* **55,** 128–179.

Shiovitz, K. A., and Lemon, R E. (1980). Species identification of song by Indigo Buntings as determined by response to computer generated sounds. *Behaviour* **74,** 167–199.

Smith, W. J. (1965). Message, meaning, and context in ethology. *Am. Nat.* **99,** 405–409.

Sorjonen, J. (1977). Seasonal and diel patterns in the song of the Thrush Nightingale, *Luscinia luscinia* in SE Finland. *Ornis Fenn.* **54,** 101–107.

Sossinka, R., and Böhner, J. (1980). Song types in the Zebra Finch, *Poephila guttata castanotis. Z. Tierpsychol.* **53,** 123–132.

Stein, R. C. (1956). A comparative study of "advertising song" in the *Hylocichla* thrushes. *Auk* **73,** 503–512.

Stein, R. C. (1963). Isolating mechanisms between populations of Traill's Flycatchers. *Proc. Am. Philos. Soc.* **107,** 21–50.

Stevenson, J. G. (1967). Reinforcing effects of Chaffinch song. *Anim. Behav.* **15,** 427–432.

Szijj, L. J. (1966). Hybridization and the nature of the isolating mechanisms in sympatric populations of meadowlarks (*Sturnella*) in Ontario. *Z. Tierpsychol.* **23,** 677–690.

Tembrock, G. (1975). "Biokommunikation." Vieweg, Braunschweig.

Thaler, E. (1979). Das Aktionssystem von Winter- und Sommergoldhähnchen (*Reguls regulus, R. ignicapillus*) und deren ethologische Differenzierung. *Bonn Zool. Mon.* **12,** 1–151.

Thielcke, G. (1962). Versuche mit Klangattrappen zur Klärung der Verwandtschaft der Baumläufer *Certhia familiaris* L., *C. brachydactyla* Brehm und *C. americana* Bonaparte. *J. Ornithol.* **103,** 266–271.

Thielcke, G. (1965). Gesangsgeographische Variation des Gartenbaumläufers (*Certhia brachydactyla*) im Hinblick auf das Artbildungsproblem. *Z. Tierpsychol.* **22,** 542–566.

Thielcke, G. (1969a). Geographic variation in bird vocalizations. *In* "Bird Vocalizations" (R. A. Hinde, ed.), pp. 311–339. Cambridge Univ. Press, London and New York.

Thielcke, G. (1969b). Die Reaktion von Tannen- und Kohlmeise (*Parus ater, P. major*) auf den Gesang nahverwandter Formen. *J. Ornithol.* **110,** 148–157.

Thielcke, G. (1970a). "Vogelstimmen." Springer-Verlag, Berlin and New York.

Thielcke, G. (1970b). Die sozialen Funktionen der Vogelstimmen. *Vogelwarte* **25,** 204–229.

Thielcke, G. (1970c). Lernen von Gesang als möglicher Schrittmacher der Evolution. *Z. Zool. Syst. Evolutionsforsch.* **8,** 309–320.

Thielcke, G. (1972). Waldbaumläufer (*Certhia familaris*) ahmen artfremdes Signal nach und reagieren darauf. *J. Ornithol.* **113,** 287–295.

Thielcke, G. (1973a). Uniformierung des Gesangs der Tannenmeise (*Parus ater*) durch Lernen. *J. Ornithol.* **114,** 443–454.

Thielcke, G. (1973b). On the origin of divergence of learned signals (songs) in isolated populations. *Ibis* **115,** 511–516.

Thielcke, G. (1977). Die Programmierung von Vogelgesängen. *Vogelwarte* **29** (Sonderheft), 153–159.

Thielcke, G., and Linsenmair, K. E. (1963). Zur geographischen Variation des Gesanges des Zilpzalps, *Phylloscopus collybita,* in Mittel- und Südwesteuropa mit einem Vergleich der Gesänge des Fitis, *Phylloscopus trochilus. J. Ornithol.* **104,** 372–402.

Thielcke, G., and Thielcke, H. (1970). Die soziale Funktion verschiedener Gesangsformen des Sonnenvogels (*Leiothrix lutea*). *Z. Tierpsychol.* **27,** 177–185.

Thielcke, G., Wüstenberg, K., and Becker, P. H. (1978). Reaktionen von Zilpzalp und Fitis (*Phylloscopus collybita, P. trochilus*) auf verschiedene Gesangsformen des Zilpzalps. *J. Ornithol.* **119,** 213–226.

Thielcke, H., and Thielcke, G. (1960). Akustisches Lernen verschieden alter schallisolierter Amseln (*Turdus merula* L.) und die Entwicklung erlernter Motive ohne und mit künstlichem Einfluss von Testosteron. *Z. Tierpsychol.* **17,** 211–244.

Thompson, W. L. (1969). Song recognition by territorial male Buntings (*Passerina*). *Anim. Behav.* **17,** 658–663.

Thompson, W. L. (1972). Singing behavior of the Indigo Bunting, *Passerina cyanea. Z. Tierpsychol.* **31,** 39–59.

Thorpe, W. H. (1958). The learning of song patterns by birds, with special reference to the song of the Chaffinch *Fringilla coelebs. Ibis* **100,** 535–570.

Thorpe, W. H. (1961). "Bird Song. The Biology of Vocal Communication and Expression in Birds," Monographs in Experimental Biology, No. 12. Cambridge Univ. Press, London and New York.

Trendelenburg, F. (1961). "Einführung in die Akustik." Springer-Verlag, Berlin and New York.

Tretzel, E. (1965). Artkennzeichnende und reaktionsauslösende Komponenten im Gesang der Heidelerche (*Lullula arborea*). *Verh. Dtsch. Zool. Ges. Jena 1965,* pp. 367–380.

Tschanz, B. (1968). Trottellummen. Die Entstehung der persönlichen Beziehung zwischen Jungvogel und Eltern. *Z. Tierpsychol., Beiheft* No. 4.

van den Elzen, R. (1977). Die Lautäusserungen der Bartmeise, *Panurus biarmicus,* als Informationssystem. *Bonn. Zool. Beitr.* **28,** 304–323.

von Haartmann, L. (1953). Was reizt den Trauerfliegenschnäpper zu füttern? *Vogelwarte* **16,** 157–164.

Ward, R. (1966). Regional variation in the song of the Carolina chickadee. *Living Bird* **5,** 127–50.

Ward, R., and Ward, D. A. (1974). Songs in contiguous populations of Black-capped and Carolina chickadees in Pennsylvania. *Wilson Bull.* **86,** 344–356.

Weeden, S. J., and Falls, J. B. (1959). Differential responses of male Ovenbirds to recorded songs of neighboring and more distant individuals. *Auk* **76,** 343–351.

Wunderle, J. M. (1979). Components of song used for species recognition in the common yellowthroat. *Anim. Behav.* **27,** 982–996.

8

Character and Variance Shift in Acoustic Signals of Birds

EDWARD H. MILLER

I. INTRODUCTION

Ecological and behavioral interactions among species commonly lead to coevolution. Indeed, it could be claimed that most phenotypic characteristics include adaptations for coexistence with other species. Some of these adaptations are complex and highly specific, as between termites and their digestive symbionts. Others are relatively simple general-purpose adaptations, such as cryptic appearance and behavior of vulnerable prey species. The complexity and specificity of such coadaptations depend largely upon the extent to which individual fitness in coexisting species is influenced by their interaction, although numerous other factors may be involved. In this chapter I am concerned with those characteristics of sound signals which have evolved through interaction (or its absence) among related and unrelated species.

Concepts central to this chapter have been used extensively in evolutionary and ecological studies. A species may exhibit displacement (either *contraction* or *divergent shift*, or both) in its use of a limiting resource when syntopic with a

ACOUSTIC COMMUNICATION IN BIRDS
VOLUME 1

ISBN 0-12-426801-3

superior or more abundant competitor.[1,2] Conversely, a species may show *expansion* (competitive release) or *convergent shift,* or both, when a competitor is rarer or absent. Such responses are probably facultative and nongenetic at first, and acquire a genetic basis over time (MacArthur and Wilson, 1967). The same processes are presumed to act upon morphological characters important in resource use. Thus, characters may exhibit divergent shift or reduced variation in competitive environments (Brown and Wilson, 1956; van Valen, 1965). Examples of divergent character shift and shift of character variance are shown in Fig. 1.

All of these ideas have been extended to animal signals [e.g., Marler, 1957, 1960; Smith, 1977; Thielcke (1973) and others use the term ''contrast reinforcement'' when referring to character displacement in bird sounds]. A good example of divergent shift in a visual signaling structure involves two damselfly species, *Calopteryx aequabilis* and *C. maculata*. Wing pigmentation is important in courtship in these species, and males of *C. maculata* use this cue to identify conspecific females, especially in areas of sympatry with *C. aequabilis* (Waage, 1975). Females of both species have dark and fairly similar wing pigmentation in allopatry, and diverge in sympatry (particularly *C. aequabilis:* see Fig. 2A; see also Waage, 1979).

Species often diverge unequally in sympatry. This is obvious for *C. aequabilis* and for chorus frogs: *Pseudacris feriarum* exhibits strong divergence in pulse rate of male advertisement calls in sympatry with *P. nigrita* (Fig. 2B) (Fouquette, 1975).

There are few clear examples of competitive release of signals or signaling structures. However, various authors have suggested that songs of birds in simple communities (e.g., on small or isolated islands) may differ in complexity from conspecific song in more diverse communities. For example, the Bluc Tit (*Parus caeruleus*) has a more complex song on Tenerife, Canary Islands, than on the mainland (Figs. 3 and 4; see Section II,A,2).

The concepts of displacement and expansion implicate important and widespread selection pressures and have a broad diversity of applications. Hereafter, I distinguish shifts in location and variance of acoustic characters as *character shift* and *variance shift,* respectively. Variance shift includes shifts in complexity and information content of sounds. Either or both kinds of shift can result from acoustic relations among syntopic species.

[1]Necessary criteria for invoking divergent shift have been reviewed by Grant (1972). These include knowledge of character states in allopatry and sympatry, and of why differences exist; accentuated differences in sympatry can result for many reasons unrelated to competition or interference between species.

[2]This is the conventional view. Thomson (1980) and others point out that a resource need not be limiting to population size for niche shifts to take place.

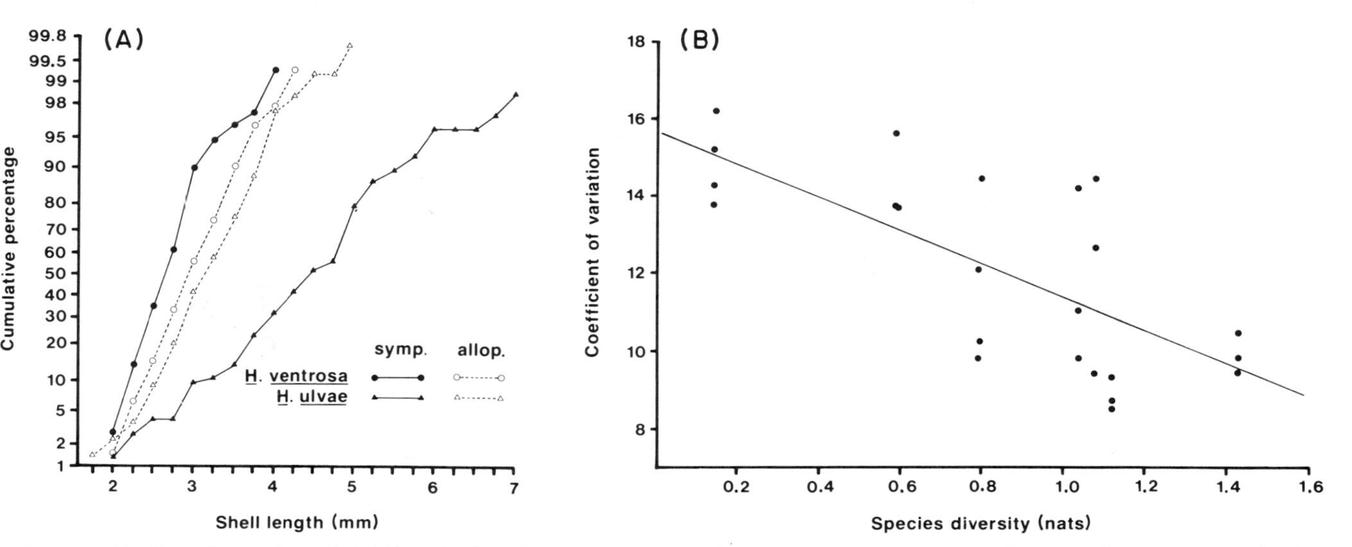

Fig. 1. (A) Cumulative plots of shell length of mud snail species (*Hydrobia*) in sympatry (symp.) and allopatry (allop.) in Denmark. The size of particles ingested by these deposit feeders varies directly with body size, as estimated by shell length. (After Fig. 1 of Fenchel, 1975; based on raw data of T. Fenchel.) (B) Relationship of the coefficient of variation in mandible size in workers of an ant species (*Veromessor pergandei*), to species diversity (in ants) of sympatric granivorous ant species. Size of seeds handled by workers of *V. pergandei* varies with mandible size. Thus, colonies with different-sized workers prey upon a great diversity of seeds and upon a large fraction of available seeds. (After Fig. 2 of Davidson, 1978; based on raw data of D. W. Davidson.)

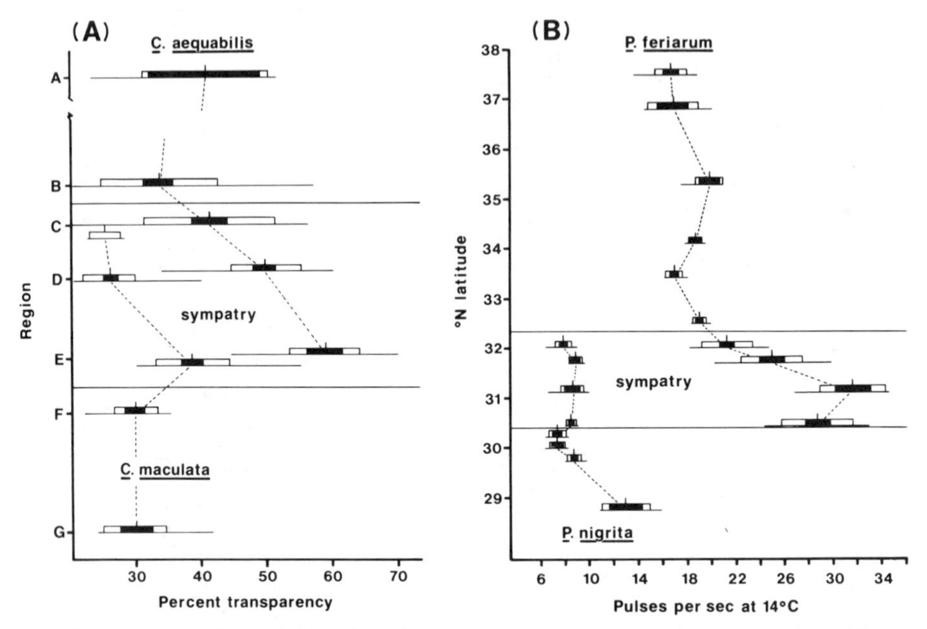

Fig. 2. (A) Hubbs–Hubbs plot of wing transparency in females of two damselfly species (*Calopteryx*) in sympatry and allopatry. The horizontal lines separate allopatric and sympatric locations. (After Fig. 3 of Waage, 1975.) (B) Hubbs–Hubbs plot of pulse rate in advertisement calls of male chorus frogs (*Pseudacris*) in sympatry and allopatry. The horizontal lines separate allopatric and sympatric locations. (After Fig. 2e of Fouquette, 1975.)

II. EVOLUTIONARY ORIGINS OF SPECIES SPECIFICITY

A. Selective Pressures and Constraints Affecting Species Specificity

1. *Hybridization*

The concept of character displacement as applied to sound signals developed from Peter Marler's writings in the 1950s, particularly on the issue of whether or not the structure of animal signals is arbitrary. Many songbird species emit calls when they perceive predators, and these calls have physical features which make callers difficult to locate. This suggests that the characteristics are not arbitrary, but have evolved under selective forces favoring nonlocatability (Marler, 1955).[3] It is a short step to consider that the structure of more complex sounds, like songs, is also adaptive, especially considering their importance to mate attraction and pair-bond formation in the presence of related species.

[3]But see Shalter (1978).

Reproduction between heterospecifics is generally assumed to lower their fitness, since no or few hybrids are typically produced, and these may be inviable, sterile, or have reduced fertility (Dobzhansky, 1970; Mayr, 1963). Such effects should lead to divergence in characters like song, which function as premating isolating mechanisms[4] in mate attraction and pair formation. But what fraction of species-specific acoustic characters has evolved in sympatry as a direct result of selective pressures surrounding hybridization, and what fraction has evolved in allopatry?

With Müller (1942), Mayr advocates the view that "the greatest part of the genetic basis of the isolating mechanisms is an incidental by-product of the genetic divergence" between species which took place during their allopatric origins (Mayr, 1976, p. 133).[5] Nevertheless, he acknowledges that behavioral (and other prezygotic) isolating mechanisms might be readily strengthened in sympatry, especially in taxa like birds whose mating is behaviorally complex (MacArthur and Wilson, 1967; Mayr, 1942, 1963, 1976; Mecham, 1960). Conversely, Fisher (1958), Dobzhansky (1937), and others view isolating mechanisms as adaptations serving to reduce hybridization: "Their evolutionary functions are . . . limitation or prevention of . . . gene exchange between species" (Dobzhansky *et al.*, 1977, p. 172). Furthermore, "sexual isolation in animals may very well be bolstered by making the species' traits easily recognizable to conspecific individuals" (Dobzhansky, 1951, p. 211). Thus, both schools agree that character differences at the level of reproductive behavior may be reinforced when species become sympatric, though they disagree about the extent of reinforcement.

The degree of elaboration of acoustic isolating mechanisms can only be estimated with knowledge about the incidence of hybridization and its costs to individual fitness, the nature of gene flow to and from areas of sympatry, the stage of speciation when isolating mechanisms are acquired, the heritability of relevant ethological (acoustic) characters, and the kinds of nonselective processes and selective pressures unrelated to hybridization which lead to species specificity (Bigelow, 1965; White, 1978).[6] Nevertheless, it is reasonable to

[4]The term "isolating mechanism" was introduced by Dobzhansky (1937). It has an unfortunate adaptive connotation, but is too widely used to be replaced. I use it here to simply indicate how reproductive isolation is maintained, regardless of the relative importance of adaptive and adventitious components.

[5]The mode of speciation (allopatric, sympatric, or parapatric) influences the importance and elaboration of isolating mechanisms, a matter I will not discuss here (Bush, 1975; Templeton, 1980).

[6]Other sorts of information can be used to infer selection for isolating mechanisms, however. Intergeneric hybrids in Parulidae, Pipridae, and Trochilidae are surprisingly common relative to intrageneric hybrids. It can be deduced that isolating mechanisms among closely related species (e.g., congeners) have been well developed due to selection against hybridization (Banks and Johnson, 1961; Parkes, 1961, 1978).

assume that signals used early in mate attraction and courtship are under the greatest selective pressure from related species, if any signals are, because they deter interbreeding with minimal loss of time and energy (Liley, 1966; Nuechterlein, 1981).[7] Because such sounds are typically complex, variable, and strongly differentiated among related species of birds, they must be the most effective as acoustic isolating mechanisms. For precisely the same reasons, they are the most difficult to judge adaptedness of.

Hybridization and introgression between Blue-winged and Golden-winged warblers (*Vermivora pinus* and *V. chrysoptera*) are well documented (Berger, 1958; Gill, 1980; Mayr, 1942; Parkes, 1951; Short, 1962, 1963). These have resulted from range modifications in both species since the late 1800s, particularly in *V. pinus,* whose range now overlaps extensively that of the more northern *V. chrysoptera*. The species show minor ecological differences in nesting habitat and foraging characteristics, but their nesting territories often overlap locally (Confer and Knapp, 1981; Ficken and Ficken, 1968a,b,c; Gill and Murray, 1972a; Murray and Gill, 1976; Parkes, 1951). In addition, male *V. pinus* arrive on the nesting grounds each spring before male *V. chrysoptera* (Ficken and Ficken, 1968a,b,c; Murray and Gill, 1976). Each species possesses two kinds of learned songs (I and II) which are generally considered to have epigamic and intrasexual functions, respectively, and each male sings only one version of each (Gill and Lanyon, 1964; Gill and Murray, 1972a,b; Kroodsma, 1981; Chapter 1, Volume 2). Altogether, songs of the two species include about seven types of components [termed A-G by Gill and Murray (1972a,b); other sonagrams are in Bondesen (1977), Ficken and Ficken (1967), Kroodsma (1981), Lanyon and Gill (1964), and Meyerriecks and Baird (1968)]. Song I should be the most sensitive indicator of shift resulting from maladaptive hybridization. In *V. pinus* this generally has two components (A, B), each occurring once, and in *V. chrysoptera* there are three to five repetitions of one component (C). Some component variants are unique to one or the other location, but no systematic trends emerge. Both A and B in song I of *V. pinus* are essentially identical in allopatry (Long Island, New York) and sympatry (Michigan) with *V. chrysoptera* (Gill and Murray, 1972b; Lanyon and Gill, 1964). The only suggestion of shift lies in the absence of component C from song of sympatric *V. pinus,* and the greater variation in song I patterns of *V. pinus* in allopatry than in sympatry (Table I). These data alone need not imply variance shift, and may just represent geographic variation (Gill and Murray, 1972a). Nevertheless, various lines of evidence suggest that isolating mechanisms are under reinforcing selection, for hybrids have low fitness, and positive assortative mating and back crossing prevail (Ficken and Ficken, 1968a; Gill and Murray, 1972a). Furthermore, *V. pinus* and

[7]Signals which occur later in mate attraction and courtship sequences should become increasingly divergent with time, however (Alexander, 1968).

TABLE I

Forms of Type I ("Epigamic") Songs of Blue-Winged and Golden-Winged Warblers (*Vermivora pinus* and *V. chrysoptera*) in Allopatry and Sympatry[a]

	Component sequence in song					
	AB/ABDB[b]	ABA	AAA	ACB	AC	CCCC
V. pinus allopatric	6	0	1	7	1	0
V. pinus sympatric	46	2	1	0	0	0
V. chrysoptera sympatric	1	0	0	0	0	20

[a] Data are from Gill and Murray (1972a,b), and include songs classified spectrographically and by ear. Identity of singers was therefore often judged by gross plumage phenotype. See remarks of Short (1969).

[b] These song patterns are not distinguishable by human ear.

V. chrysoptera have rapid courtship and a brief nesting season, which should favor species-specific displays in mate attraction (Ficken and Ficken, 1968a,b). The weak evidence for shift in epigamic song of these species suggests that song is sufficiently unimportant as an isolating mechanism, or is already sufficiently effective (different), that its further evolutionary change is unnecessary (or immeasurable?). Furthermore, song type I varies much less geographically than does II, at least in *V. pinus* (Kroodsma, 1981). In any case, song is only part of a suite of isolating mechanisms in *V. pinus* and *V. chrysoptera,* which are very different in plumage and which have prominent visual displays, especially at close range (Baird, 1967; Ficken and Ficken, 1962, 1968b; Meyerriecks and Baird, 1968).

An instructive contrast with the preceding example is provided by the Indigo and Lazuli buntings, *Passerina cyanea* and *P. amoena*. Like the *Vermivora* species, males of these species have distinctive species-specific plumage, and hold type A territories whose occupancy they advertise with learned song (Emlen, 1972; Emlen *et al.,* 1975; Rice and Thompson, 1968). *Passerina amoena* is broadly distributed in western North America, and *P. cyanea* in the east; ranges overlap in the Great Plains, where hybridization and interspecific territoriality occur (Emlen *et al.,* 1975; Sibley and Short, 1959). Song components in allopatric populations differ, and have been described in detail by Emlen (1972), Rice and Thompson (1968), and Thompson (1968, 1969, 1970, 1972, 1976). The species also differ in song in several other ways in allopatry: *P. cyanea* songs tend to be longer, slower in cadence, and more complex in temporal organization. In an area of hybridization near Chadron, Nebraska, songs of the two species contain various proportions of typical allopatric song components. Thus, components typical of allopatric *P. cyanea* comprised 14–100% of songs of four (phenotypic) *P. cyanea,* and 0–100% of songs of 14 (phenotypic) *P. amoena* (Emlen *et al.*, 1975, Table 1). Song cadence of the minority species there, *P. cyanea,* is faster than it is in allopatry. No other significant interpretable differences were found. Only indirect evidence of selection against hybridization could be sequestered: hybrids were uncommon; introgression was virtually undetectable; five birds collected in an ecologically suboptimal area were probably hybrids; and two "impure" pairings suffered delayed breeding as compared with four others. In brief, there is no evidence of contraction or divergent shift in song, despite intensive and detailed study, and even though various lines of evidence suggest that it should occur. This may be because hybridization is minimized through interspecific territoriality, which sets signaling needs met by convergent character shift (convergence in song cadence of *P. cyanea* and sharing of song syllables, in sympatry).

The preceding examples concern well-studied species that can hybridize and whose complex song is probably an effective isolating mechanism. Even so, no strong patterns suggesting contraction or divergent shift in song characteristics are apparent. These results are typical of intensive studies on various taxa and suggest that differences among related species are generally not strongly reinforced as isolating mechanisms (e.g., Grant, 1975; Heth and Nevo, 1981; Martens, 1975; Selander and Giller, 1961; Thielcke *et al.*, 1978; Watson and Littlejohn, 1978). In any case, interpretation is a major problem even with strong evidence of shift among related species. Such species are most likely to interfere with one another acoustically because of their relatedness, which imparts generally high similarity in habitats, activity rhythms, acoustic characteristics, sound functions, and so on. Consequently, shift is perhaps most likely to occur among related species, but for reasons unrelated to hybridization. [den Boer (1980) argues similarly, for high ecological similarity within groups of Carabidae (Coleoptera).]

2. *Other Considerations of Species Specificity*

The view that species-specific characteristics of song evolved to reduce hybridization fits nicely with the observation that songs differ greatly among species, even among closely related species with very similar external characteristics [e.g., sympatric species of *Empidonax* (Johnson, 1963, 1980; Stein, 1958, 1963) and *Phylloscopus* (Marler, 1957, 1960)]. It also makes sense of the observation that song is much more highly species-specific than simple calls are (Güttinger and Nicolai, 1973; Marler, 1957; Thielcke, 1970). The broad correlation among relatedness, similarity in appearance and in simple calls, and differences in complex songs, suggests that character shift is important and pervasive. However, there are general reasons for expecting this pattern to prevail even without pressure for divergence.

It is a truism that species differ phenotypically, and that differences vary directly with the complexity (dimensionality) of the characteristics under consideration. Songs must therefore vary more among and within species, and correspondingly must have higher species specificity than simple calls, on this basis alone. Furthermore, long-distance acoustic advertisement by birds is often under the dominant influence of sexual selection, which can cause rapid evolutionary change and marked divergence among species. Such divergence should be particularly marked if song is learned, since learning increases interindividual variation upon which sexual selection can work. Thus, strong differences among long-distance complex sounds of related species are to be expected, even in the absence of ecological, behavioral, or reproductive interactions. Simple calls should be more similar across species than should songs, but this by itself does not imply that selection has promoted greater species specificity in the latter than in the former. Also, evidence for evolution or adaptive alteration of signals to reduce interaction among species must be much stronger than simple listings of species differences, although such are often interpreted as adaptations in that sense (see also Section III).

The need for species specificity has also been used to explain examples of variance shift. Thus, in depauperate avifaunas where species recognition is not a problem, song may be more complex than in richer bird communities. The most widely cited examples of this trend are species in the Canary Islands (Lack and Southern, 1949; Marler, 1960; Marler and Boatman, 1951; Marler and Hamilton, 1966). For example, song of central European *Parus caeruleus* generally consists of just two simple, sequentially graded series, each of a single note type (Fig. 3). Songs on Tenerife, Canary Islands, are composed of at least two note types; these are more structurally complex than in central Europe, have a much more complex sequential organization, and vary more among individuals (Fig. 4) (Becker *et al.*, 1980). Similarly, nonsong calls of the Goldcrest (*Regulus regulus*) "are of bewildering variety" compared with those in British populations (Marler and Boatman, 1951, p. 95).

The suggestion that complex songs should characterize impoverished avifaunas can be readily assessed with other island–mainland comparisons. Consider the Savannah Sparrow (*Ammodramus sandwichensis*), which nests commonly throughout much of North America and experiences a broad spectrum of avian communities and sound environments. Most males have only one song type (rarely two: J. B. Gollop, unpublished data), and are often polygynous in the well-differentiated population on Sable Island, Nova Scotia, where they are the only native breeding passerine (McLaren, 1972; Stobo and McLaren, 1975; limited polygyny also occurs elsewhere: Weatherhead, 1977, 1979; Welsh,

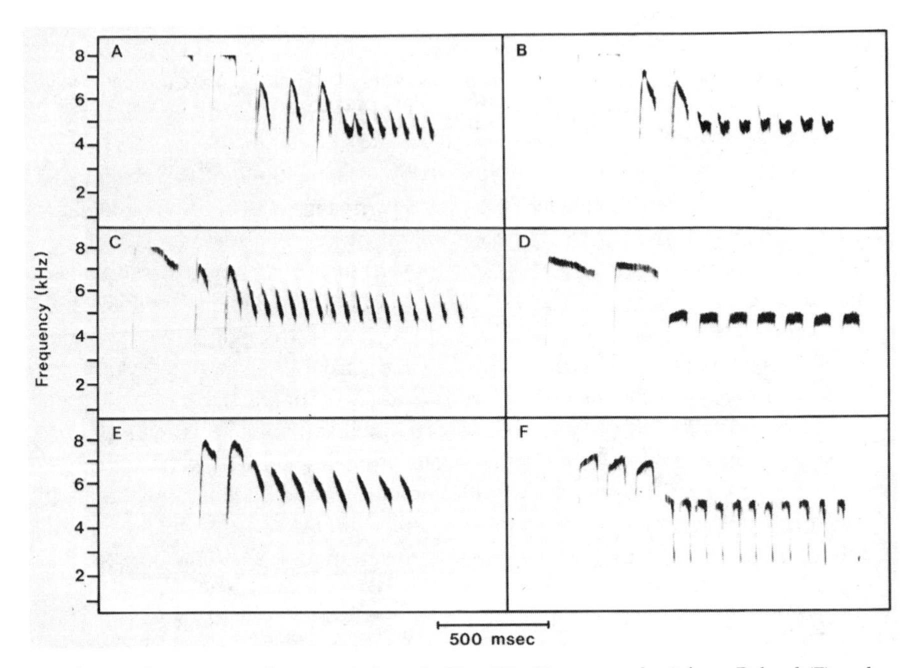

Fig. 3. Sonagrams of songs of six male Blue Tits (*Parus caeruleus*) from Poland (F) and West Germany (remainder). (Originals provided by H.-H. Bergmann; analyzing filter bandwidth, 300 Hz.)

1975). Thus, sexual selection and the near absence of other species could rapidly generate song complexity there. Individual song elements on Sable Island are not more complex than on the mainland (Fig. 5) (Chew, 1979, 1981; J. B. Gollop, unpublished data). However, five of six males on Sable Island had nine note types per song, and the sixth had ten ($\bar{Y} = 9.2$). Birds in the much more complex breeding avifauna near Churchill, Manitoba, had fewer note types per song: nine had seven, eighteen had eight, and one had nine ($\bar{Y} = 7.7$). This trend supports a correlation between avifaunal and song complexity, but is countered by data for communities of intermediate complexity: two males from the Ogilvie Mountains, Yukon Territory, had nine, one had ten, and five had eleven note types per song ($\bar{Y} = 10.4$; Fig. 5); and songs from birds in southern Ontario, mainland Nova Scotia, and Saskatchewan, show no systematic relationship of number of note types to avifaunal complexity (Chew, 1979, 1981; J. B. Gollop, unpublished data; see also Bradley, 1977).

Why should such an apparently simple test case yield such equivocal results? One reason is that the hypothesis being tested was inadequate: divergent predictions can pertain to variance shift. Thus, it can be advanced that complex songs should evolve in simple communities because of the release from normalizing pressure for species specificity; alternatively, complex species-specific sounds should evolve in complex communities to facilitate rapid, unequivocal species identification [on the Canary Islands, "Songs of several species are shorter and simpler . . . than those of their British counterparts" (Lack and Southern, 1949, p. 615)]. Another reason is that evolutionary changes in complexity can result from diverse selective and nonselective processes unrelated to community structure, including learning and sexual selection. One of these is simply sampling error and subsequent drift, resulting from low numbers of colonists, small population size, etc. Another has been suggested by Thielcke (1973), as the "withdrawal-of-learning" hypothesis. Consider the colonization of an island by a songbird species. Colonists are likely to be few in number and to include young birds, males of which have uncrystallized song. If these mature with insufficient exposure to species-typical song, their uncrystallized song could become characteristic of the population, and song of island adults could come to resemble that of mainland juveniles. This is the likeliest explanation for the complex song form

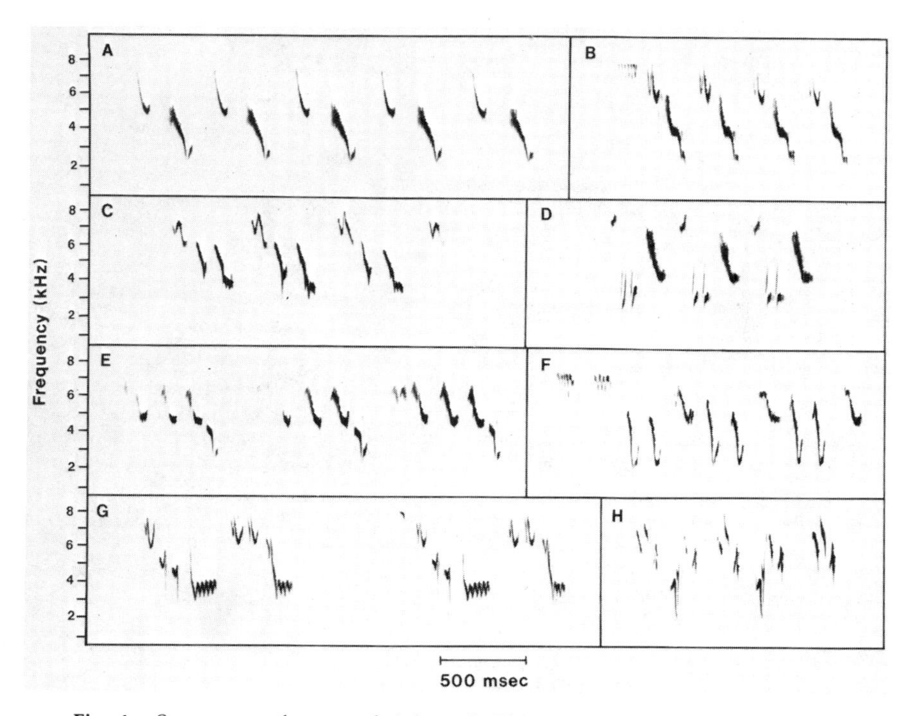

Fig. 4. Sonagrams of songs of eight male Blue Tits (*Parus caeruleus*) from Tenerife, Canary Islands. (Originals provided by H.-H. Bergmann; analyzing filter bandwidth, 300 Hz.)

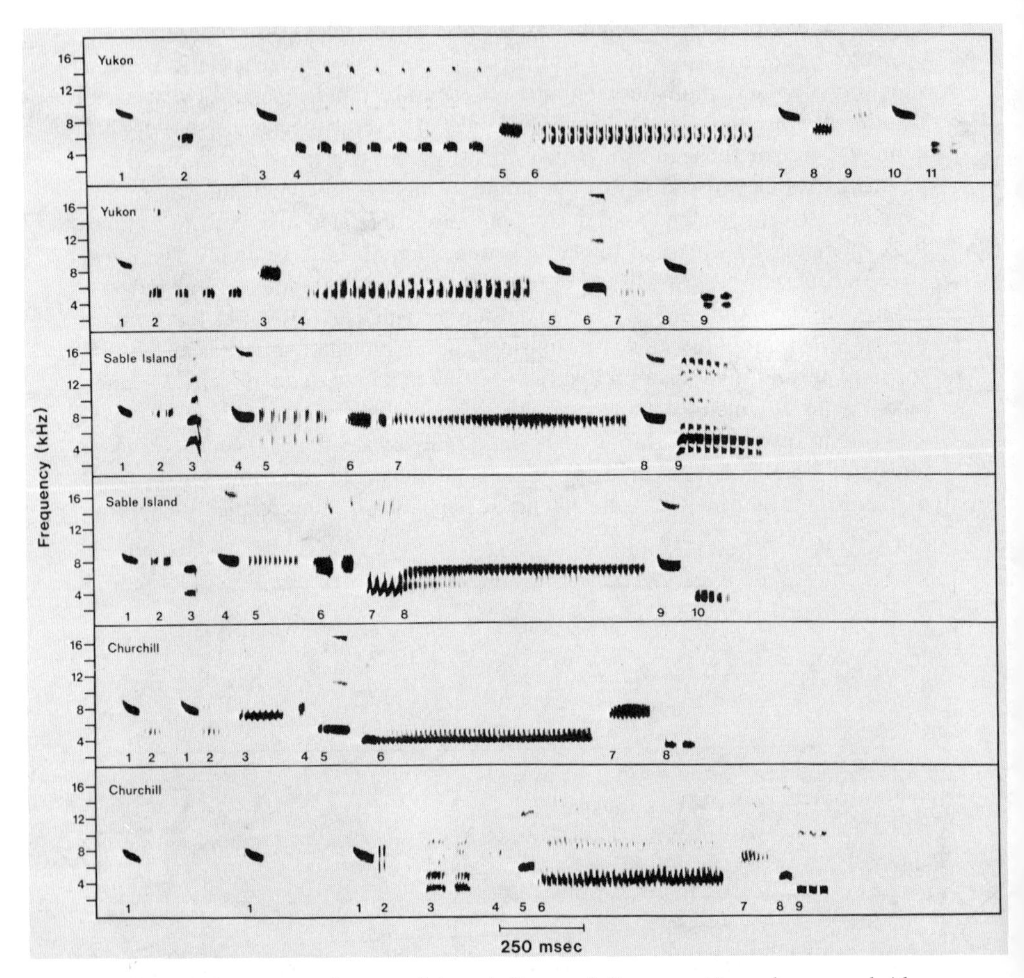

Fig. 5. Sonagrams of songs of six male Savannah Sparrows (*Ammodramus sandwichensis*) from Ogilvie Mountains (Yukon), Sable Island (Nova Scotia), and Churchill (Manitoba). Songs typically begin with several notes such as "1," as in the bottom panel. Only the last note in each introductory series is shown in the top five panels. "Note types" are single or compound notes which may be repeated immediately, and are distinguished by numbers. Subsequent repetitions are considered as separate note types (e.g., 1, 3, 7, and 10 in the top panel). Note types do not correspond across panels. (Analyzing filter bandwidth, 300 Hz.)

in *Parus caeruleus* on Tenerife and in the Dark-eyed Junco (*Junco hyemalis*) on Guadalupe Island, off Baja California (Becker *et al.*, 1980; Mirsky, 1976).[8] It may also explain some instances of interspecific song learning (e.g., Kroodsma, 1972).

It is clear from this and the preceding section that character and variance shift are not unitary concepts. Shift-like patterns can result from diverse processes, some related to hybridization and selection for enhancement of isolating mechanisms, and some not. Furthermore, there is no obvious reason to expect variance and character shift to covary.

The diversity of potential causes makes it difficult to explain which forces lead to shift and maintain it in any particular case. Nevertheless, two broad sets of forces can be distinguished, one concerned with general costs to interference, and the other with costs to species misidentification (including costs of hybridization).[9] Thus, selection promoting shift can stem from costs of lengthier, more frequent, or louder signaling in biologically noisy environments. Colonists or minority species in noisy environments may be under selective pressure to change simply because of the difficulty in being heard, irrespective of the similarity of their sounds to those of other residents, or the presence of related species. Change can also be promoted by the presence or absence of related species with similar sounds, because of danger of hybridization, or because time and energy are wasted through inappropriate responses by and to heterospecifics, or both.

Character shift can have diverse causes, but its mechanisms are probably quite uniform. Sound characteristics which encode species identity are usually important for recognition of conspecific mates or competitors. These characteristics are under normalizing selection *within* populations, for effective and efficient communication (see Carson, 1978; Hubbs, 1960; Manning, 1977; Paterson, 1978, 1980; Templeton, 1980). Thus, a male songbird with a deviant song may take a long time to attract a mate or may attract a low-quality mate, regardless of the ambient sound environment. In an altered sound environment, those males with only a slightly deviant but still recognizable song will be favored. Character shift should thus evolve gradually toward increased distinctiveness. Because most acoustic signals are complex and highly redundant, shift probably does not have to proceed very far before a new mode is established.

Variance shift should be uncommon, if intrapopulational normalizing selection on characteristics encoding species identity is widespread. This is because further reduction of population variance may be unnecessary even if a new sound

[8] "The aberrant insular song type . . . may be as *constant* . . . as the mainland song," however (Hansen, 1979, p. 44; italics added).

[9] A distinction between "reproductive" and "competitive" character displacement is often noted (e.g., Huey and Pianka, 1974; White, 1978), but their effects are usually inseparable, as these and earlier remarks indicate (Wilson, 1965).

environment favors it, and increased variance is selected against, for the reasons stated, even in a permissive sound environment.

3. *Social and Ecological Influences on Species Specificity*

The evolutionary importance of species specificity, and hence the variation and complexity of acoustic isolating mechanisms, is set by numerous interrelated factors. Some of these factors derive from the social functions of different sounds and the relative importance of sound in communication; others come from characteristics of the prevailing bioacoustic environment.

Sounds which are broadcast over large areas are more likely to come under pressure to diverge or contract from those of other species than are short-range sounds. Sound function also affects this likelihood. Thus, sounds emitted by unmated males to attract mates should diverge or contract more, or more quickly, than sounds with a similar broadcast area but which just function to space out males; and both of these kinds of sounds should respond more quickly than short-range sounds with equivalent functions. Such simple pair-wise examples could be multiplied but of course various factors operate simultaneously in any real situation, with inseparable effects. Consider a species in which unmated males broadcast over large areas to attract females. Following Marler (1957), we predict that such signals are more species-specific than: (i) short-range signals with different functions but equivalent importance to fitness; (ii) signals with the same broadcast area but with smaller effects on fitness; and (iii) signals with equivalent functions and effects on fitness in a different species, but with a smaller broadcast area. The latter prediction has several sources. First, acoustic interference from related and unrelated species is weak over small broadcast areas. Second, visual signals become increasingly important over short display distances, and these weaken the necessity for shifts in acoustic characteristics, or even make them unnecessary. Finally, intraspecific acoustic interference can be severe for species which display at high densities; this promotes evolutionary simplification of acoustic signals and increased reliance upon visual signals. Highly polygynous species often fall in this category, and sexual selection in such species is so strong that extreme conformity with this prediction is expected.

The Calidridinae (Scolopacidae) illustrate some of these points. The Ruff (*Philomachus pugnax*) is a traditional lek species. Males have complex and striking visual epigamic displays on lekking hills, but are utterly silent there (Hogan-Warburg, 1966; van Rhijn, 1973). The Buff-breasted Sandpiper (*Tryngites subruficollis*) is also a lek species, but leks are transient, males are separated on them by much greater distances than in *Philomachus*, and males display with flutter-jumps and wing extensions. These displays are visible over considerable distances because of the white undersurface of the wings. During courtship over short distances, males use elaborate visual displays and emit soft, simple "tick" sounds (Myers, 1979; Oring, 1964; Parmelee *et al.*, 1967; Sutton,

1967). Males of the polygynous Sharp-tailed and Pectoral sandpipers (*Calidris acuminata* and *C. melanotos*) are even more widely dispersed during courtship but otherwise have a social system similar to that of *Tryngites* (exploded lek). A major display of males is a low flight with characteristic flight path, wing action, throat expansion, and repeated trains of simple, loud calls (Flint and Kishchinskii, 1973; Pitelka, 1959). Males of most monogamous calidridine species are typically dispersed even more during courtship. Even their simplest long-distance calls are more complex than those of *C. melanotos* (analyses of sounds of *C. acuminata* are not available), and only some gross motor characteristics of display flights have potential signal value (height above ground and fluttering actions of the wings, mainly). Thus, calidridines exemplify some simple relationships among mating system, broadcast area, prominence of visual displays, and complexity of acoustic displays.

These observations of relationships are strengthened by observations on a related lek species, the Great Snipe (*Capella media*), which is nocturnal. The reduced emphasis on visual signaling in *C. media* is accompanied by complex and prominent vocal displays (Ferdinand, 1966; Ferdinand and Gensbøl, 1966; Lemnell, 1978).

Generalization is difficult, however. Complex species-specific song occurs in short-range courtship in various species of grassfinches (Estrildidae), in which visual signaling with plumage and posture is also important (Hall, 1962; Zann, 1976a,b).[10] Many species of birds that communicate frequently and importantly over short distances have complex graded systems of vocal communication, whether or not plumage is strongly modified for visual signaling (e.g., ducks quails, jays, phalaropes, jacanas, rails: Abraham, 1974; Anderson, 1978; Berger and Ligon, 1977; Hardy, 1979; Hope, 1980; Howe, 1972; Huxley and Wilkinson, 1977; Jenni *et al.*, 1974; Mace, 1981). Therefore it may be possible only to elucidate relationships among the factors discussed above, for closely related species.

Dispersion, density, and stability of species' spacing patterns all affect signal functions and are all strongly affected by ecological forces. Such forces include seasonality, spatial and temporal characteristics of resource distribution, predation, species richness, and trophic level. The effects of these factors on spacing patterns are intertwined with their effects on social structure, activity budgets, and so on, which influence signaling characteristics in turn. Numerous perspectives on such interrelationships are available, and further treatment here is unnecessary (e.g., Bradbury and Vehrencamp, 1977; Crook, 1965; Crook *et al.*, 1976; Emlen and Oring, 1977; Geist, 1978; Jarman, 1974; Lill, 1974; Waser and

[10]Hall (1962, p. 41) states that "The extent of the song differences . . . is . . . much less than might be predicted on the basis of what is known of the more familiar passerines," but sonagraphic evidence there and in Zann (1976a) is not supportive.

Wiley, 1979; Wittenberger, 1979; Wittenberger and Tilson, 1980, and references therein). Frequency and rate of occurrence, magnitude, direction, and predictability of ecological variations will determine the extent to which adaptive shifts occur. Variations of particular significance for shifts in acoustic signals lie in characteristics of spacing, of the overall bioacoustic environment (reflected in avifaunal composition), and of the presence of related species. All of these vary importantly with ecological factors like productivity. For example, spatial variations in productivity yield heterogeneity in a species' breeding density and spacing pattern. Highly productive areas may accommodate many species, including related species at high density; conversely, regions of low productivity may be inhabited by fewer species at lower densities (and probably distributed less uniformly). It is straightforward to consider comparable variations over time. Some of the most significant of these for adaptive shifts are year-to-year fluctuations in avifaunal composition and breeding density. Sustained shifts are unlikely to evolve where these and related ecological characteristics differ strongly from year to year. Additionally, the physical environment plays an important role in selecting for certain sound characteristics (see Bowman, 1979; Gish and Morton, 1981; Hunter and Krebs, 1979; James, 1981; Chapter 5, this volume).

In summary, character and variance shifts are most likely to evolve in long-distance signals of species which inhabit structurally simple environments that are temporally and spatially stable in physical characteristics, in those factors which set spacing patterns, and in high avifaunal complexity (including related species). Of course this is just a specific prediction related to general concepts about evolution in changing environments, and many other relevant considerations exist (Levins, 1968; Roughgarden, 1979).

In this section I have outlined some examples and principles which emphasize the complexity of selective forces favoring shift and the difficulty in documenting it. Some of the difficulty arises from the practical difficulties in estimating fitness differentials through field research, but most is due to the anticipated rarity of measurable shift because of variation in the kind and magnitude of selective pressures, and the importance of visual signaling.

B. Minimizing Interference and Achieving Species Specificity

1. *Behavioral and Ecological Adjustments*

Individuals can adaptively modify their acoustic signals, and this must impede evolutionary response to natural selection promoting contraction or divergent shift. For example, when isolated from mates, Common Quail (*Coturnix cotur-*

nix) emit more frequent separation crows in a noisy than in a quiet environment (Potash, 1975); male Ovenbirds (*Seiurus aurocapillus*) tend to sing incomplete songs when at moderate distances from their mates, but give full songs at greater distances (Lein, 1981); males of various songbird species avoid singing when neighbors are singing (Gochfeld, 1978; Kroodsma and Verner, 1978; Wasserman, 1977); White-rumped Shamas (*Copsychus malabaricus*) imitate one another during agonistic interactions, whereas mates avoid singing similar song structures (Kneutgen, 1969); subordinate male Long-billed Marsh Wrens (*Cistothorus palustris*) tend to follow and match song types of dominants during countersinging (Kroodsma, 1979); and so on. Given the widespread occurrence of these and other facultative adjustments in intraspecific communication, it is easy to envision responses which minimize acoustic interference with heterospecifics. Some data indicate simply that individuals tend to signal when heterospecifics are silent [Cody and Brown, 1969; Ficken *et al.*, 1974; for other taxa, see Littlejohn and Martin (1969), Samways (1977), and Samways and Broughton (1976)]. There are few comparable observations on facultative spatial adjustments. However, male gray treefrogs (*Hyla versicolor*) space out so that sound pressure levels at each calling position are no higher than about 93 dB (Fellers, 1979; see also Thiele and Bailey, 1980; Whitney and Krebs, 1975). A simple mechanism like this may operate widely, with minor ecological consequences.

Selection which favors unequivocal intraspecific signaling promotes improved discrimination by receivers, as well as increased species specificity in signals. In birds, improved discrimination is probably learned through unrewarding or aversive interaction with heterospecifics; it is unlikely to result directly from genetically determined tuning of the auditory system, as occurs in anurans and acoustic insects (Capranica, 1976; Elsner and Popov, 1978).

Altered responsiveness to sound signals has been documented in sympatric species of birds. Song discrimination by sympatric *Vermivora chrysoptera* and *V. pinus* in Michigan has been studied by Gill and Murray (1972a) (see Section II,A,1). *V. pinus* males responded to playbacks of heterospecific song type I on 4 of 18 trials, and *V. chrysoptera* responded on 2 of 14 (totalling $6/32$; see further). In allopatry, *V. pinus* (Maryland) responded on 7 of 9 trials, and *V. chrysoptera* (West Virginia) on 1 of 17 (Ficken and Ficken, 1969). *V. pinus* males thus responded to song type I of *V. chrysoptera* less often in sympatry than in allopatry. Other data that support the suggestion that responsiveness is less in sympatry come from playback experiments on song types I and II. Males of both species in sympatry responded more often to playbacks of heterospecific song type II than I, as predicted: *V. pinus* responded to II on 6 of 10 trials (versus $4/18$), and *V. chrysoptera* on 9 of 12 (versus $2/14$; Gill and Murray, 1972a, p. 287). Overall, then, there were $6/32$ (18.8%) responses to heterospecific I and $15/22$

TABLE II

Responsiveness of Dark-Phase Male Western Grebes (*Aechmophorus occidentalis*) to Playback of Advertising Calls of Dark- and Light-Phase Females, in Sympatry and Allopatry[a]

		Color phase of female whose stimulus call was used	
		Dark	Light
Percentage (and sample size) of males advertise–calling or approaching in response to stimulus calls	In sympatry (Oregon)	66(38)	3(33)
	In allopatry (Manitoba)	68(40)	48(40)

[a] Data are from Nuechterlein (1981).

(68.2%) to II in sympatry [see also Gill and Lanyon (1964) on allopatric *V. pinus*].

Western Grebes (*Aechmophorus occidentalis*) occur in dark and light phases, which act as good species (Ratti, 1979). Individuals emit simple advertising calls early in "courtship"; these are usually single calls in light-phase birds and double in dark-phase birds. Observational data and playback experiments reveal a strong tendency for birds to reply or otherwise respond to calls of their own color phase, particularly in sympatry (Nuechterlein, 1981) (Table II).

The trends discussed above should typify species which are not interspecifically territorial. Opposite trends are predicted for species which hold mutually exclusive territories. A few male Reed Warblers (*Acrocephalus scirpaceus*) which were interspecifically territorial with Marsh Warblers (*A. palustris*) also had mixed song. These males sang at the rapid tempo characteristic of *A. palustris* during interspecific singing encounters, and one uttered more *palustris* song while singing with them than when singing alone (Lemaire, 1977). In sympatry, males of *Passerina cyanea* and *P. amoena* respond equally strongly to playbacks of conspecific and heterospecific song (see Section II,A,1). Males in allopatric populations respond strongly to conspecific song, but very weakly to heterospecific song (Emlen, 1972; Emlen *et al.*, 1975; Thompson, 1969), a trend also approximated in some *Acrocephalus* species (Catchpole, 1978). Emlen *et al.* (1975, p. 172) "hypothesize that sympatric buntings 'learn' to misidentify congeners as a result of individual behavioral experiences," and that such "learned misidentification" is adaptive when it is more costly to share a territory with an ecologically, genetically, or morphologically similar heterospecific or hybrid than to exclude them. Similar considerations probably also apply to other pairs of related species (Becker, 1977; Falls and Szijj, 1959; Kroodsma, 1973; Thielcke, 1972).[11]

Individuals of most bird species can probably adjust to acoustic interference facultatively and (or) through learning. Local adjustments in aggressive and sexual behavior must be particularly advantageous for species that inhabit complex communities, are widespread or ecologically generalized, and have high vagility or weak philopatry. Mechanisms for such adjustments already exist at the level of intraspecific acoustic signaling. It is likely that these are just taken over for interspecific functions, and that responses in sympatry are simply weakened or strengthened through local contingencies of reinforcement (see Miller, 1967, 1968).

Acoustic interference among species can also be reduced by ecological differences in spatial or temporal characteristics of signaling. The permissive role played by ecological differences is illustrated by Australian lyrebirds (*Menura*), which mimic species that breed and vocalize at a different season (Robinson, 1973, 1974, 1975). It is improbable that acoustic interference ever effects substantial ecological differences (see also Orians and Collier, 1963). However, even ecologically similar species can reduce acoustic interference by using conventional encounter sites, or calling sites which differ in ecologically irrelevant ways, as in some anurans and insects (Littlejohn, 1977; MacNally, 1979; Parker, 1978; Salthe and Mecham, 1974). The two sibling species of Western Grebes may minimize interbreeding partly through local segregation of nesting colonies (Ratti, 1979; but see Nuechterlein, 1981), and different species of weaver finches (*Ploceus cucullatus* and *P. nigerrimus*) nesting in the same tree may likewise show clustering of nests (Crook, 1964). For species which advertise at low population densities, it may be unimportant to space out once a minimal distance from the nearest caller is exceeded. This probably applies mostly to large species, carnivorous species, and species in low-productivity environments (Harestad and Bunnell, 1979).

To summarize, high acoustic interference need not lead to character or variance shift in signal structure. Selection promoting shifts must often be blunted by nonevolutionary, facultative responses in signal form and reception, and by true evolutionary responses in spacing or timing which have minor ecological consequences.

[11]True evolutionary convergence in social signals can be achieved by convergent character shift or expanded character variance. Where individuals that react aggressively and display effectively to ecologically similar heterospecifics have increased fitness, convergent signals (or mimicry) and interspecific territoriality could evolve (Cody, 1969, 1973; Marler, 1960; Moynihan, 1968; Rice, 1978a,b). This seductive outline has few supporting data, and has been severely criticized (Becker, 1977; Brown, 1977; Murray, 1971, 1976). In any case, the significance of adequately documented examples of convergence for the concepts of character and variance shift is obvious.

2. *Shifts in Character Location (Character Shift)*

a. How Are Sound Characteristics Likely to Diverge? Species may be under pressure to diverge even in relatively dissimilar signals if the signals occupy the same frequency band or are emitted at the same time and place. Signals which occupy the same frequency band can become distinct through differentiation of temporal features. Such differentiation is probably only opposed when physiological limits to sound production and reception are approached (Ewing, 1979; Littlejohn, 1977), or when temporal chracteristics are closely adapted to environmental conditions. For example, the rate of trilling cannot increase freely in vegetated and other closed habitats because trills are so susceptible to reverberation (Richards and Wiley, 1980; Chapter 5, this volume). Despite such constraints, the evolution of sound distinctiveness based on temporal features is probably achieved fairly easily. The same is true of amplitude characteristics, though details of amplitude are unimportant in long-range communication (when acoustic interference among species is generally highest) because they degrade so quickly over distance (Wiley and Richards, 1978; Chapter 5, this volume).[12] In contrast, divergence in the frequency domain is opposed by several factors.

First, attenuation of a signal is closely tied to its frequency spectrum, and this relationship is strongly affected by many environmental characteristics (Wiley and Richards, 1978; Chapter 5, this volume). Thus, the frequency spectrum of calls (especially long-distance calls) is generally closely adapted to optimal (often maximal) transmission in the prevailing environment, and there is probably strong selection against changes in it (Bowman, 1979; Hunter and Krebs, 1979; James, 1981). Indeed, adaptive differences in the frequency spectrum of song exist even among local populations of White-throated Sparrows (*Zonotrichia albicollis*) in habitats with different sound-transmission properties (Wasserman, 1979). Further evidence for the selective importance of effective transmission lies in the prevalence of dawn chorusing in numerous bird species, when conditions are best for long-distance acoustic signaling but when acoustic interference is greatest (Henwood and Fabrick, 1979). In addition, there is no evidence of partitioning of the audible frequency spectrum in complex anuran communities, despite their relatively simple calls (Littlejohn, 1977).

Second, calls should occupy a frequency band in which conspecifics have high auditory sensitivity. This conservative force may be strong in lower vertebrates (Hopkins and Bass, 1981; but see Myrberg and Spires, 1980), but may not be particularly important for homeotherms, where most sound processing occurs in

[12]Adaptation of amplitude characteristics to local environments is suggested by Gish and Morton (1981), however. If individuals are under selection to conceal their distance from conspecifics while singing, it can be done most effectively if callers can retain *degradable* source characteristics in long-distance calls (see also Chapter 6, this volume).

the central nervous system (Nottebohm, 1976): auditory sensitivity functions are remarkably similar across bird species and typically correspond only roughly to frequency spectra of the same species' sounds (Cohen *et al.*, 1978; Dooling, 1980; Dooling *et al.*, 1978; Sachs *et al.*, 1978; Chapter 4, this volume).

Finally, effecting a shift in frequency of a sound signal must usually entail a change in the size of sound-producing structures, which varies directly with body size across species in most taxa [see Bowman (1979), Eisenberg (1976), Hutterer and Vogel (1977), McGrath *et al.* (1972), Martin (1972), and Würdinger (1970); but not across *Vireo solitarius* subspecies: see James (1981)]. Frequency shifts may also necessitate changes in tracheal length, which can influence resonance frequency (Hinsch, 1972; Chapter 2, this volume). The ability of a population to respond to selective pressures favoring a shift in the frequency spectrum of its sounds is impeded further by the relationships among age, size, sound frequency, and auditory sensitivity. Within species, sound frequency declines with age (Schleidt and Shalter, 1973; Schubert, 1976a,b; Würdinger, 1970; Ehret, 1980) and bears an inverse relationship to body size among adults (Lanyon, 1978; Schubert, 1976a). In addition, cochlear microphonics for precocial and altricial species suggest increasing sensitivity to higher frequencies with age (Gates *et al.*, 1975; Golubeva, 1978; see also Brown and Grinnell, 1980; Brzoska *et al.*, 1977, Fig. 7).

When a species exhibits shift in acoustic signals, temporal features and time-varying characteristics of frequency are likely to change first. Major changes in frequency spectra of sounds are probably harder to effect, especially for long-distance signals, and changes in amplitude characteristics are easily achieved but are relevant mainly for short-range signals. The precise way in which shift comes about will be hard to predict. This depends on which characters are undergoing shift, and the effect of their change on receivers. But a general prediction can be made. Consider two species which are under equal pressure to diverge in a song character Y, and which differ slightly in Y to begin with. In general, the evolutionary cost or difficulty of change is proportional to ($\Delta Y/\bar{Y}$), so the species with the largest measure on Y should diverge most.

b. Analysis of Character Shift. Most putative examples of divergent character shift in signals rest on a few variables for two related species. Some clear examples of such shift are in Section I, and more will be documented, especially for simple, special-purpose, low-dimensional signals. There are some general problems with this approach. First, low-dimensional sound signals are uncommon in general, and are certainly uncommon among birds. Second, sounds are probably adapted to the overall acoustic environment, and not just to sounds of related species; in any case, the joint effect of low community complexity with respect to related species and (acoustic) niche complementarity renders the likelihood of simple compensation among related species very low. Finally, compensation can be achieved through reduced variance as well as by divergent character shift (Fig. 6C). Indeed, the potential for evolutionary change of variance may profoundly affect the possibility of character shift (Slatkin, 1980). All these considerations make the simultaneous examination of location and variance essential, except for extremely simple sounds (Section II,B,3,b).

3. *Shifts in Character Complexity and Dispersion (Variance Shift)*

a. Measurement of Variation and Complexity. The Fox Sparrow (*Zonotrichia iliaca*) is a monogamous species whose breeding range extends across North America. It consequently breeds in bird communities with very different levels of complexity, and its song characteristics also vary: in the simple terrestrial avifauna of Newfoundland, males average only 1 song type per male (Blacquiere, 1979); in Utah and Idaho, they average 3 (Martin, 1977). Syllable diversity per song (no. of syllable types/no. of syllables) varies accordingly, averaging 0.78 in Newfoundland and 1.00 in Utah and Idaho [based on data in Blacquiere (1979) and Martin (1977); see also Martin (1979)]. These observations suggest that the species has simpler emissions in simple sound environments. However, the trends are opposite for song complexity as estimated by song length (number of syllables) and number of syllable types, which are, respectively, 11.7 and 9.1 in Newfoundland, and 8.2 and 8.2 in Utah and Idaho (data sources are as above). Clearly, opposite interpretations could be placed on these trends.

The preceding example illustrates the importance of the way in which complexity is defined in assessing variance shift. There are numerous ways to construe and define complexity, not all of which are applicable when studying variance shift. Two different questions must be addressed: in what characters should complexity be estimated?, and, how should complexity be estimated? (See also Chapter 5, Volume 2.)

Characteristics chosen for measurement depend on the level of integration in which a worker is interested. In bird songs, the smallest unit recognized is usually a temporally continuous utterance (syllable). Even at this level, simple and compound measures on frequency (F), amplitude (A), and time (T) must be evaluated. For example, the simple song of the Grasshopper Warbler (*Locustella naevia*) is a sequence of rapid, rhythmical repetitions of two alternative types of pulses, and may last up to 2 min. The second pulse in each couplet is about 6 msec long and has a characteristic amplitude envelope and signs of frequency modulation (Brackenbury, 1978). Given this typical sort of situation for even very simple syllables, it is clear that a single, uniquely derived measure of syllable complexity cannot exist: any such measure will reflect complexity in several domains (F, A, and T) among which interaction can occur. Thus, similar estimates of complexity for two syllable types may be due to different contributions from the three domains, different patterns of covariation, etc. Complexity at higher levels of integration can obviously be even more varied, and a single comprehensive measure of complexity (e.g., song complexity) will be even more elusive. In practice, measures of sound complexity must be based upon chosen characteristics. It is imperative that the level of integration is explicit (e.g., syllable, phrase, ... , song, song repertoire), and desirable that complexity be assessed in various ways and for each domain separately, as far as possible.

There are numerous ways to view, measure, and represent variation and complexity (see also Chapter 5, Volume 2). For continuous variables, conventional measures are standard deviation (or standard error, or variance) and coefficient of variation (CV). The latter is used when variables being compared have different means, on the assumption that the magnitude of the CV is independent of the mean (this assumption is rarely tested in ethology).[13] There have been numerous univariate statistical treatments of continuous variation in bird sounds (e.g., Bergmann, 1976; Johnson, 1980; Konishi, 1964; Marler and Isaac, 1960a,b, 1961; Payne and Budde, 1979; Shiovitz, 1975; Smith *et al.*, 1980).[14] Multivariate statistics of variation are also available (van Valen, 1974, 1978) but none has been used in ethological studies, to my knowledge.

Variation and complexity can also be expressed through various diversity and information–theoretic measures, for both continuous and discrete data. These are treated in many texts and review papers (e.g., Colgan, 1978; Hazlett, 1977; Pielou, 1975, 1977). Some (notably the Shannon–Wiener information measure, H') have been widely used in the analysis of signaling, and may prove useful in studying variance shift.[15] For example, predictability and redundancy are important attributes of animal sounds which may be modified in sympatry, and both have formal expressions derivable from H'. Other approaches can be applied to sequential complexity: recurrence intervals, time-series analyses, formal grammars, etc.

There is a plethora of quantitative methods available for documenting and analyzing structural variation, and none is natural and "best." The variation and complexity in sound properties being assessed may be expressed differently according to the measure chosen. For example, "diversity" indices are not all

[13]Barlow (1977, p. 101) suggests a converse measure of stereotypy ST = $[\bar{Y}/(SD + 0.01\bar{Y})]$, which has the undesirable property of being a curvilinearly decreasing function of CV.

[14]Reliability and repeatability of measurements on sonagrams are low. Measurement error and bias are particularly likely to lead to spurious differences in variation between frequency and temporal measures, and between brief and long calls in temporal measures. This matter merits close scrutiny, since sonagrams are the main display used for measurement. Alternative forms of sound analysis are probably necessary for adequate study of continuous variation in sound characteristics, especially for brief sounds with rapid frequency shifts.

[15]Routledge (1980) suggests that the Simpson index is preferable to the Shannon–Wiener index for estimating ecological α diversity and for enabling comparison of diversity estimates.

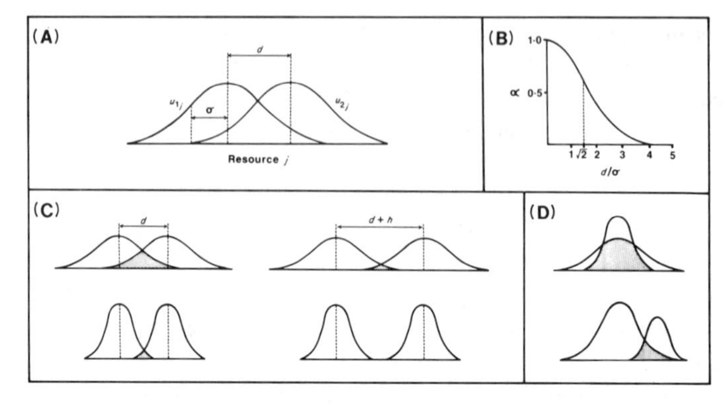

Fig. 6. (A) Symbolism for utilization curves (u) on resource axis j, each with standard deviation σ, for two species (1 and 2) separated by distance d (after Fig. 2–4 of MacArthur, 1972). (B) Relationship of overlap (α) to the ratio of (d/σ) (after Fig. 2–5 of MacArthur, 1972). (C) Relationship of overlap between resource utilization curves to their variance and to the distance between them. Equal overlap occurs between curves with low variance (lower left), and more widely separated curves with greater variance (upper right). (D) Relationship of overlap between resource utilization curves, to their unequal variance (upper) and unequal areas (lower).

concordant and emphasize or measure different properties of a collection (Hurlbert, 1971; Routledge, 1979). Similarly, "redundancy" may be evaluated in a correlational or a sequential sense. The most important procedure in evaluating variation and complexity is the specification of their level and nature appropriate to the hypothesis being tested.

b. Ecological Analogs. It is easy to analogize sound characteristics with ecological roles, and hence acoustic with ecological space. The analogy provides conceptual and analytic approaches from ecology for studying overlap and variation in sound signals.

Hypothetical utilization functions u for resource j, by species 1 and 2 are shown in Fig. 6A. The means of u are separated by distance d, and the overlap between the species is determined by this value and the standard deviation σ (estimated by w). The unidimensional overlap value, α_{12}, is estimated as C exp $[-d^2{}_{12}/2\,(w_1^2 + w_2^2)]$, where C is a normalization constant $[= f(w_1, w_2)]$ (May, 1974).[16] It is straightforward to extend this to multidimensional overlap, assuming a common covariance matrix (Harner and Whitmore, 1977). This assumption, however, is probably rarely met with for acoustic variables, and more

[16]This expression is not generally used in ecology, where multistate categorical resources are recognized and overlap across them is measured (Hurlbert, 1978; May, 1975).

complex computations involving the substitution of α into the formula for the p-variate normal density are necessary (G. Sugihara, personal communication). In any case, overlap as measured in these ways obviously depends simultaneously on location and dispersion, and in the simplest case varies inversely with the ratio d/w (Fig. 6B and C). Overlap (α) declines as d/w increases, and declines most rapidly at $d/w = \sqrt{2}$ (Fig. 6B). Even where $d = 0$, α is less than 1.0 if $w_1 \neq w_2$ (Fig. 6D, upper).

This approach is applicable to sound characteristics, but two further points must be made. First, overlap must generally have unequal effects on species. Thus, the rarer of two species will be affected most by their overlap (Fig. 6D, lower). The same effect is present if two species are equally abundant, but their overlapping sound signals have different effects on fitness (e.g., the signals have different functions, or they have equivalent functions with different quantitative effects on fitness, or some combination of these). Second, overlap *per se* may not be necessary for acoustic signals to be under pressure to diverge or contract. Sounds which are similar to one another (small d) but are nonoverlapping may still cause some confusion in receivers, and lead to reduced fitness of similar-sounding emitters.

Overlap between most bird sounds cannot be realistically assessed by univariate analysis, because of their complexity. Consider another simple ecological analogy, where species 1 and 2 overlap on resource axes i and j. The species' two-dimensional overlap is determined by the direction and strength of the correlations between the separate utilization functions (Fig. 7).

Overlap and possible interference between heterospecifics cannot be predicted on the basis of overlap between the species to which they belong. This is because a species may have small or large σ (corresponding to a "specialist" or "generalist," in ecological terms), which has no fixed relationship to how specialized or generalized individuals are. Overlap in sound signals and resource use affects individual fitness, so it is clearly necessary to consider within- and among-individual components of variation [cf. within- and between-phenotype components of variance recognized by Roughgarden (1972)]. In general, we predict that competing species in sympatry will show reduced population and individual variances (the latter assumes that some level of individuality is favored, and that a species' variance is reduced; see Chapter 8, Volume 2).

To interpret the relationships of individual and population variation to sympatry, it is necessary to establish baseline measures. I do this here by analyzing variation in simple calls of two species. The Black-legged Kittiwake (*Rissa tridactyla*) emits a loud, stereotyped, and individualistic call when at the nest site, or when approaching and landing there. The species is densely colonial, and individuality in the landing call is important in communication within mated pairs. Kittiwake colonies include few other species, so landing calls need not encode species identity. These conditions provide for high individuality which

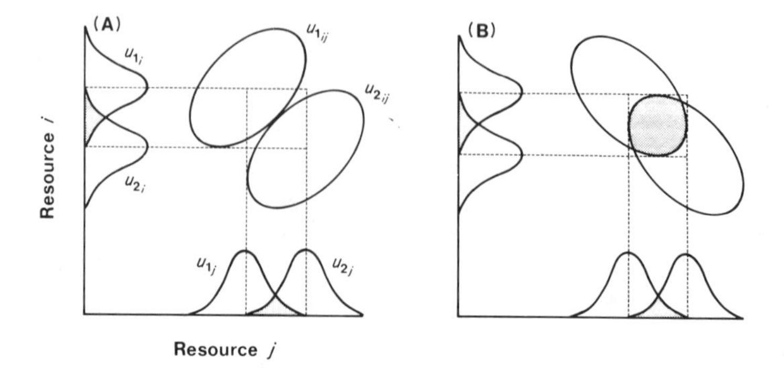

Fig. 7. Relationship of one-dimensional overlap between utilization curves of two species (1, 2), on resource axes *i* and *j*, to two-dimensional overlap. The direction (and strength, not varied in this figure) of correlations between separate resource utilization functions determines whether higher-dimensional overlap is low (A) or high (B). (After Fig. 7.8 of Pianka, 1978.)

can be estimated by the percentage variation among groups, $100s_A^2/(s^2 + s_A^2)$, where s_A is the added variance component among groups (here, among birds) in one-way analysis of variance (model II) [this approach has been used extensively by Jenssen (1979); see also Chapter 8, Volume 2].

Kittiwake calls include five distinct temporal components ($a - e$), and show strong differences among individuals in tonality, noise, and other spectral features (Fig. 8) (Wooller, 1978). Analysis of variance on the durations of temporal components reveals a range in percentage variation among groups of 30–68 (30–58 for males, 37–68 for females; Table III). These figures point to components b and d as having the highest variation among birds (for each sex), hence as being likely variables to encode individuality. Furthermore, the analysis establishes estimates against which others can be compared. Landing calls are fairly complex and occur at short range, so they are supplemented by visual sources of information. A logical comparison is with a species which differs on both points.

The American Woodcock (*Scolopax minor*) presents a simpler situation. Like kittiwakes, the American Woodcock faces no problems of species misidentification. Males emit nasal *peent* calls during mate-attraction displays at dusk (Mendall and Aldous, 1943; Sheldon, 1967). The calls are structurally simple, and because of their long transmission distance and crepuscular emission, are not accompanied by visual display components (Fig. 8). In brief, the *peent* call is a very low dimensional display, so would be expected to be more individualistic than the landing call of the kittiwake. Here though, advantages to individuality lie in attraction of females and repulsion of competing males. Individuality in *peent* calls may enable woodcock to be accurately censused, so considerable research on this matter has been done (Beightol and Samuel, 1973; Bourgeois,

1977; Bourgeois and Couture, 1977; Samuel and Beightol, 1972, 1973; Thomas and Dilworth, 1980; Weir, 1979; Weir and Graves, 1981). Data of Bourgeois (1977) indicate substantially higher percentage variation among male woodcock than among kittiwakes, as predicted (range 65–90%; Table III). Again, some variables are much more individualistic than others.

Approaches which address questions about both character location and dispersion may prove to be valuable in exploring shift. A simple situation which can be

TABLE III

Percentage Variation among Groups for Landing Calls of Black-legged Kittiwake (*Rissa tridactyla*) and Peent Calls of Male American Woodcock (*Scolopax minor*)

Variable (duration of components)[a]	*R. tridactyla*[b]	*S. minor*[c]	Variable[d]
e, male	29.8		
e, female	36.8		
a, male	43.4		
c, male	48.9		
c, female	56.0		
d, male	56.8		
b, male	57.6		
a, female	58.6		
d, female	60.8		
		65.3	*F*3
		66.8	*F*G
b, female	68.0		
		68.1	*F*C
		71.6	*F*D
		86.6	total no. of pulses
		88.7	total duration
		89.6	*T*
		90.4	rate of FM (= 1/*DP*)

[a] Letters a to e represent temporal components recognized by Wooller (1978) (see Fig. 8).

[b] Analyses were based on 50 calls by each of 25 males and 25 females, raw data on which were provided by R. D. Wooller.

[c] Analyses were based on seven to ten calls by each of eight males, from data in Tables 1 to 4, and Appendix Tables 1 to 5 of Bourgeois (1977).

[d] Measurements were taken on sonagrams for which the bandpass filter was 300 Hz, and intensity contours differed by 6 dB (area of maximal intensity, 36–42 dB). F_3, minimal frequency at the start of the 24- to 30-dB contour; F_G, frequency 5 mm after the start of the 36- to 42-dB contour, and centered vertically; F_C, frequency in the center of the 36- to 42-dB band; *FD*, frequency 5 mm before the end of the 36- to 42-dB contour, and centered vertically; *T*, duration of the 36- to 42-dB band; and *DP*, mean duration of ten pulses in the center of a call [see Fig. 5 of Bourgeois (1977) and Fig. 1 of Bourgeois and Couture (1977)].

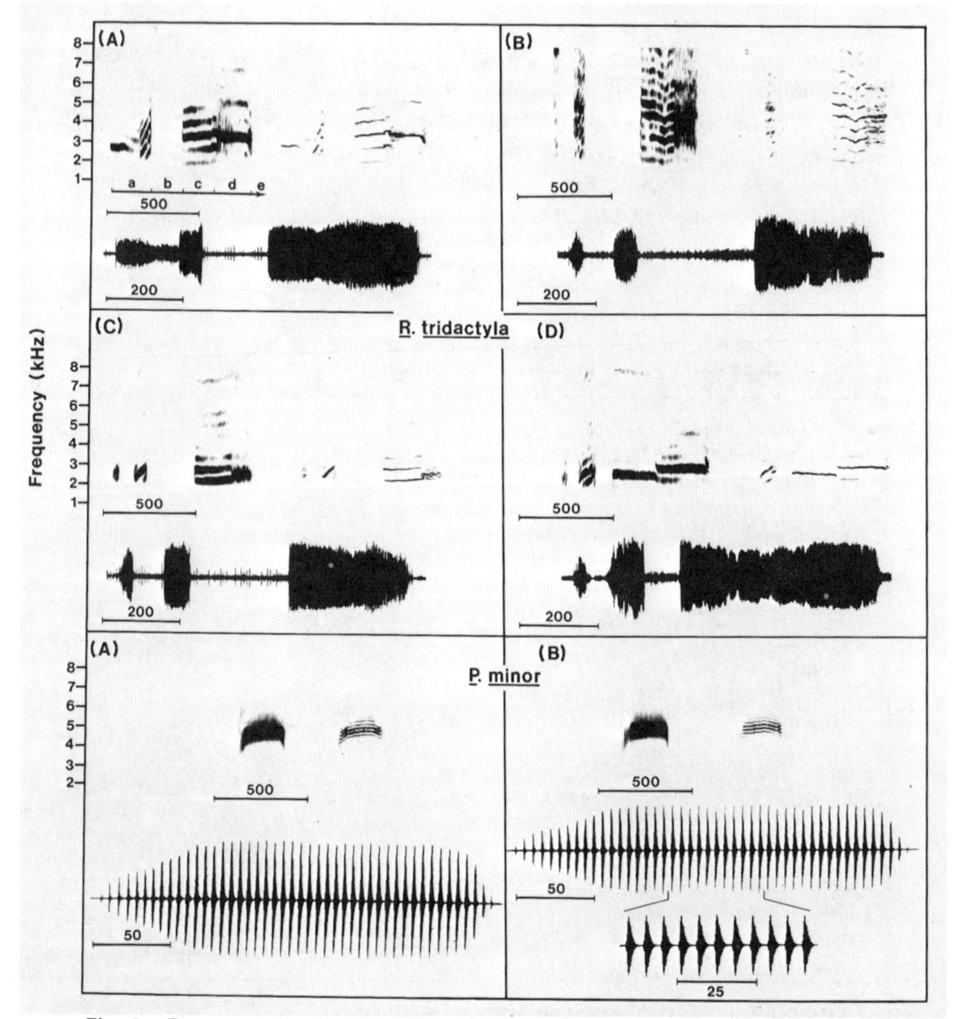

Fig. 8. Representative landing calls of four Black-legged Kittiwakes (*Rissa tridactyla;* upper, A–D) and peent calls of two male American Woodcock (*Scolopax minor;* lower, A and B). Each call is shown as wide band (300 Hz, on left) and narrow band (45 Hz, on right) sonagrams, and as an oscillogram (below). Time markers below the representations are expressed in milliseconds. Lettering (a–e) for *R. tridactyla* indicates temporal measurements discussed in the text and Table III (e = interval to next call; after Fig. 4 of Wooller, 1978). This figure was made from tape recordings of *R. tridactyla* in England, by R. D. Wooller, and of *S. minor* in Quebec, by J.-C. Bourgeois.

analyzed more comprehensively is needed. I do not know of an adequate data set for birds, but one exists for two sibling species of fiddler crabs, *Uca panacea* and *U. pugilator*. These species are syntopic in parts of Florida, and *U. pugilator* is in the minority. *Uca panacea* occurs from Florida to Texas, and *U. pugilator* occurs from Florida to Massachusetts. Courting males attract females to their burrows by waving the major cheliped. When a female is near, males vibrate the claw rapidly against the substrate (*U. pugilator*) or first walking leg (*U. panacea*) (Pawlik *et al.*, 1980; Salmon *et al.*, 1978).

The pulse repetition rate in *U. panacea* is about 24 Hz in allopatry and sympatry, but in *U. pugilator* is 9.4 in sympatry and 11.6 in allopatry, suggestive of divergent shift [analyses discussed here and below are based on Tables 1 and 2 of Salmon *et al.* (1978) and raw data of M. Salmon)]. *Uca pugilator* shows slightly less variation among males in sympatry than in allopatry, as predicted (54.5 versus 58.7%), but the trend is strongly reversed for the majority species, *U. panacea* (51.1 versus 40.9%; based on one-way analyses of variance). Other analyses yielded the following results:

(1) *Uca pugilator* exhibits a reduced population variance in sympatry, as predicted, but *U. panacea* does not.

(2) Individual males are not less variable in sympatry than in allopatry, for either species, contrary to prediction. The data plotted in Fig. 9 are in the predicted direction, but are not statistically significant.

In summary, there is a suggestion of divergent character shift and reduced population variance in sympatry for the minority species, *U. pugilator*, but no evidence of reduced individual variation in either species. This example was discussed at some length because it involves sibling species with limited learning

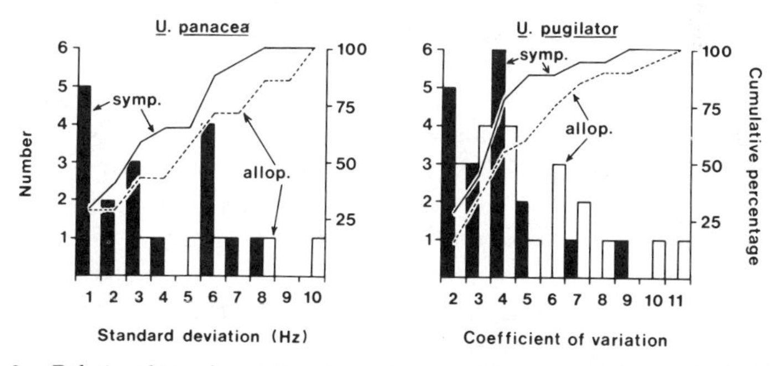

Fig. 9. Relationship of variation in pulse repetition rate within male fiddler crabs (*Uca*), to sympatry (symp.) and allopatry (allop.). The right vertical scales in each diagram refer to the cumulative frequency plots. The coefficient of variation (CV) was used for *U. pugilator* because pulse repetition rate differed between allopatry and sympatry; the CV's are statistically independent of means for this data set. (Based on Tables 1 and 2 of Salmon *et al.*, 1978, and raw data of M. Salmon.)

abilities and simple acoustic signals, all of which increase the likelihood of adaptive adjustments of variances and means. The analytic results of this and the preceding example are promising; research on variance components in chosen sounds across communities could yield insight into the bioacoustic community structure of birds.

III. CONCLUDING COMMENTS

In this chapter I have suggested that shift is not a unitary phenomenon, that two components of shift be recognized and dealt with analytically, and that shift be appreciated as a special case of coevolution. Shift in acoustic signals of birds is probably less common and important than generally thought: high dimensionality of most avian sound signals and multimodality of much avian signaling reduce the necessity for divergent shift or contraction; the physical environment is important in selecting for physical characteristics of sounds, and the social environment is important in selecting for various forms of sound complexity; spatial and temporal fluctuations in ecological factors yield variations in avifaunal characteristics to which close adaptation of sound signals is unlikely; facultative responses must often effectively reduce acoustic competition and interference; and sounds of syntopic species are usually sufficiently different anyway, adaptive adjustment being unnecessary.

Based on these considerations, we predict that measurable shift is most likely in low-dimensional, long-range, unlearned sound signals of species in fairly simple and constant environments, where richness and evenness of the avifauna are high. Shift should be especially pronounced in long-range sounds used in mate attraction and (secondarily) territorial advertisement, in avifaunas which include many related species. Temporal features and characteristics of frequency modulation are most likely to undergo shift.

The mere documentation of species differences which conform with predictions in the last paragraph is inadequate for the inference of shift, however (Green and Marler, 1979). Consider the Red-headed and Downy woodpeckers (*Melanerpes erythrocephalus* and *Picoides pubescens*), which are broadly sympatric and simply drum with the bill on a substrate in long-distance signaling. Some of the likely important dimensions of this simple sound are duration, number of pulses, pulse repetition rate, and pulse interval, several of which differ strongly between the species (Table IV) (Crusoe, 1980). The simplest explanation of these trends is that the species evolved differently; there is no basis for invoking shift or other evolutionary accommodation. In any case, considering variation in avian community characteristics, examination of only two species might generally be pointless. The study of signal structure at the community level has begun for anurans (Hödl, 1977; Rand and Drewry, 1972) and mormyrid fish (Hopkins, 1980, 1981). Steiner (1979) notes that those spe-

TABLE IV

Characteristics of a Low-Dimensional Sound Signal (Drumming) in the Broadly Sympatric Red-Headed and Downy Woodpeckers (*Melanerpes erythrocephalus* and *Picoides pubescens*)[a]

Variable	*M. erythrocephalus*	*P. pubescens*
Pulses per drum	14.7 ± 0.18	14.8 ± 0.21
Pulse train duration (dsec)	6.4 ± 0.08	8.7 ± 0.14
Pulse repetition rate (Hz)	20.8 ± 0.04	16.0 ± 0.09
Pulse interval (msec)[b]	47.9 ± 0.04	63.0 ± 0.11
Sample size	559	306

[a] Data are listed as $\bar{Y} \pm$ SE, and are based on data in Crusoe (1980).
[b] Data are based on 7655 pulse intervals from 559 drums of *M. erythrocephalus* and 4217 intervals from 306 drums of *P. pubescens*.

cies of odontocete whales which have the widest distribution in the western North Atlantic have the most different calls, and those with narrow distributions (and which presumably encounter fewer other species) have less different calls (Table V). Similar approaches at the community level should also be applied to birds (Leroy, 1978, 1979).

A major hindrance to understanding the adaptedness of sounds to the bioacoustic environment is our ignorance of the behavioral significance of sounds and sound variants in the lives of different species. Without detailed studies of form, function and context, we cannot assume that sound qualities like constancy, predictability, and complexity have similar evolutionary significance across species or even populations [e.g., see Morton's discussion of song of Carolina Wrens (*Thryothorus ludovicianus*) at different latitudes, in Chapter 6, this volume]. Nor can we assume that even major classes of sounds like song have comparable significance among related species. The study of character and vari-

TABLE V

Summary of Mahalanobis' Distance between Whistles of Five Species of Delphinidae (Odontoceti) from the Western North Atlantic[a]

Lagenorhynchus acutus	—				
Stenella plagiodon	3.1	—			
Stenella longirostris	2.8	1.8	—		
Tursiops truncatus	7.6	6.2	7.1	—	
Globicephala melaena	10.3	12.4	16.2	20.0	—
	L.a.	*S.p.*	*S.l.*	*T.t.*	*G.m.*

[a] Mahalanobis' distance D^2 is the square of the Euclidean distance in D- space. Here, $D = 6$ for the variables beginning frequency, terminal frequency, maximal frequency, minimal frequency, duration, and number of changes between increasing and decreasing frequency. After Steiner (1979).

ance shift in bird sounds should begin with information on sound morphology, on the communicative contexts and significance of different sound types, and on the relationships of sound types and variants to fitness budget (e.g., Beer, 1975, 1976, 1980; Green, 1975; Lein, 1978, 1980, 1981; Smith, 1977). These matters should be examined across classes (age, sex, reproductive status, etc.), and considered against variations in species' ecological characteristics. The dynamics of acoustic interaction among species are diverse in kind, strength, predictability, and importance, and shifts are of correspondingly varied form and significance.

I feel that much evolutionary discussion surrounding character and variance shift, reproductive isolation, and species-specific attributes pays insufficient attention to the diversity of nonselective and natural selective processes underlying such patterns. For example, we have detailed knowledge of species-specific characteristics of short-range calls in some species of Anatidae and Phasianidae, through the fine research by Gottlieb and his colleagues (Gottlieb, 1979, and references therein; Miller, 1978; Miller and Gottlieb, 1976; see also Shapiro, 1980). Yet, we have no comprehensive understanding of whether or how species specificity in such calls (and responsiveness to them) has been selected for as such, why it is encoded in the characteristics it is, and so on. Character and variance shift, and related concepts, have been viewed as good evolutionary explanations for many observed natural patterns, but "the trouble about good answers is that people tend to look at problems in term of answers they already know and which they expect or hope to find" (Birch, 1979, p. 197).

ACKNOWLEDGMENTS

I am indebted to many people for generously providing advice, critical comments, references, and data, not all of which could be incorporated in the final version of this chapter: G. W. Barlow, J. C. Barlow, M. D. Beecher, H.-H. Bergmann, R. Blacquiere, J.-C. Bourgeois, R. I. Bowman, K. B. Bryan, G. L. Chew, D. Crusoe, D. W. Davidson, T. Fenchel, M. J. Fouquette, Jr., F. B. Gill, J. B. Gollop, P. R. Grant, R. D. James, D. E. Kroodsma, R. M. May, B. A. Menge, A. S. Rand, K. Rohde, M. Salmon, L. J. Shapiro, C. T. Snowdon, W. Steiner, G. Sugihara, J. K. Waage, E. O. Willis, and R. D. Wooller.

REFERENCES

Abraham, R. L. (1974). Vocalizations of the Mallard (*Anas platyrhynchos*). *Condor* **76,** 401–420.

Alexander, R. D. (1968). Acoustical communication in arthropods. *Annu. Rev. Entomol.* **12,** 495–526.

Anderson, W. L. (1978). Vocalizations of Scaled Quail. *Condor* **80,** 49–63.

Baird, J. (1967). Some courtship displays of the Golden-winged Warbler. *Wilson Bull.* **79,** 301–306.

Banks, R. C., and Johnson, N. K. (1961). A review of North American hybrid hummingbirds. *Condor* **63,** 3–28.

Barlow, G. W. (1977). Modal action patterns. *In* "How Animals Communicate" (T. A. Sebeok, ed.), pp. 98–134. Indiana Univ. Press, Bloomington.

Becker, P. H. (1977). Verhalten auf Lautäusserungen der Zwillingstart, interspezifische Territorialität und Habitatansprüche von Winter- und Sommergoldhähnchen (*Regulus regulus, R. ignicapillus*). *J. Ornithol.* **118,** 233–260.

Becker, P. H., Thielcke, G., and Wustenberg, K. (1980). Versuche zum angenommenen Kontrastverlust im Gesang der Blaumeise (*Parus caeruleus*) auf Teneriffa. *J. Ornithol.* **121,** 81–95.

Beer, C. G. (1975). Multiple functions and gull displays. *In* "Function and Evolution in Behaviour—Essays in Honour of Professor Niko Tinbergen, F.R.S.," (G. P. Baerends, C. G. Beer, and A. Manning, eds.), pp. 16–54. Oxford Univ. Press (Clarendon), London and New York.

Beer, C. G. (1976). Some complexities in the communication behavior of gulls. *Ann. N.Y. Acad. Sci.* **280,** 413–432.

Beer, C. G. (1980). The communication behavior of gulls and other seabirds. *Behav. Mar. Anim.* **4,** 169–205.

Beightol, D. R., and Samuel, D. E. (1973). Sonagraphic analysis of the American Woodcock's *peent* call. *J. Wildl. Manage.* **37,** 470–475.

Berger, A. J. (1958). The Golden-winged–Blue-winged warbler complex in Michigan and the Great Lakes area. *Jack-Pine Warbler* **36,** 37–71.

Berger, L. R., and Ligon, J. D. (1977). Vocal communication and individual recognition in the Piñon Jay, *Gymnorhinus cyanocephalus*. *Anim. Behav.* **25,** 567–584.

Bergmann, H.-H. (1976). Inseldialekte in den Alarmrufen von Weibert- und Samtkopfgräsmucke (*Sylvia cantillans* und *S. melanocephala*). *Vogelwarte* **28,** 245–257.

Bigelow, R. S. (1965). Hybrid zones and reproductive isolation. *Evolution* **19,** 449–458.

Birch, L. C. (1979). The effect of species of animals which share common resources on one another's distribution and abundance. *Fortschr. Zool.* **25,** 197–221.

Blacquiere, J. R. (1979). Some aspects of the breeding biology and vocalizations of the Fox Sparrow (*Passerella iliaca* Merrem) in Newfoundland. M.S. Thesis, Memorial Univ. of Newfoundland, St. John's.

Bondesen, P. (1977). "North American Bird Songs—A World of Music." Scand. Sci. Press, Klampenborg, Denmark.

Bourgeois, J.-C. (1977). Contribution à l'étude des problèmes relatifs à l'interprétation des recensements de populations de Bécasses d'Amérique (*Philohela minor*). M.S. Thesis, Univ. Québec à Trois-Rivières, Trois-Rivières.

Bourgeois, J.-C., and Couture, R. (1977). A method for identifying American Woodcock males based on peent call sonagraphic analysis. *Proc. Woodcock Symp.* **6,** 171–184.

Bowman, R. I. (1979). Adaptive morphology of song dialects in Darwin's finches. *J. Ornithol.* **120,** 353–389.

Brackenbury, J. H. (1978). A comparison of the origin and temporal arrangement of pulsed sounds in the songs of the Grasshopper and Sedge warblers, *Locustella naevia* and *Acrocephalus schoenobaenus*. *J. Zool.* **184,** 187–206.

Bradbury, J. W., and Vehrencamp, S. L. (1977). Social organization and foraging in emballonurid bats III. Mating systems. *Behav. Ecol. Sociobiol.* **2,** 1–17.

Bradley, R. A. (1977). Geographic variation in the song of Belding's Savannah Sparrow (*Passerculus sandwichensis beldingi*). *Bull. Fla. State Mus. Biol. Sci.* No. 22, 57–100.

Brown, P. E., and Grinnell, A. D. (1980). Echolocation ontogeny in bats. *In* "Animal Sonar Systems" (R. G. Busnel and J. F. Fish, eds.), pp. 355–377. Plenum, New York.

Brown, R. N. (1977). Character convergence in bird song. *Can. J. Zool.* **55,** 1523–1529.

Brown, W. L., and Wilson, E. O. (1956). Character displacement. *Syst. Zool.* **5,** 49–64.

Brzoska, J., Walkowiak, W., and Schneider, H. (1977). Acoustic communication in the grass frog (*Rana t. temporaria*): Calls, auditory thresholds and behavioral responses. *J. Comp. Physiol.* **118,** 173–186.

Bush, G. L. (1975). Modes of animal speciation. *Annu. Rev. Ecol. Syst.* **6,** 339–364.

Capranica, R. R. (1976). The auditory system. *In* "Physiology of the Amphibia" (B. Lofts, ed.), Vol. 3, pp. 443–466. Academic Press, New York.

Carson, H. L. (1978). Speciation and sexual selection in Hawaiian *Drosophila. In* "Ecological Genetics: The Interface" (P. F. Brussard, ed.), pp. 93–107. Springer-Verlag, Berlin and New York.

Catchpole, C. K. (1978). Interspecific territorialism and competition in *Acrocephalus* warblers as revealed by playback experiments in areas of sympatry and allopatry. *Anim. Behav.* **26,** 1072–1080.

Chew, G. L. (1979). Species geographic and individual trends of variation in the song of the Savannah Sparrow (*Passerculus sandwichensis*). M.A. Thesis, Dalhousie Univ., Halifax, Nova Scotia.

Chew, L. (1981). Geographic and individual variation in the morphology and sequential organization of the song of the Savannah Sparrow (*Passerculus sandwichensis*). *Can. J. Zool.* **59,** 702–713.

Cody, M. L. (1969). Convergent characteristics in sympatric species: A possible relation to interspecific competition and aggression. *Condor* **71,** 222–239.

Cody, M. L. (1973). Character convergence. *Annu. Rev. Ecol. Syst.* **4,** 189–211.

Cody, M. L., and Brown, J. H. (1969). Song asynchrony in neighboring bird species. *Nature (London)* **222,** 778–780.

Cohen, S. M., Stebbins, W. C., and Moody, D. B. (1978). Audibility thresholds of the Blue Jay. *Auk* **95,** 563–568.

Colgan, P. W., ed. (1978). "Quantitative Ethology." Wiley (Interscience), New York.

Confer, J. L., and Knapp, K. (1981). Golden-winged Warblers and Blue-winged Warblers: The relative success of a habitat specialist and a generalist. *Auk* **98,** 108–114.

Crook, J. H. (1964). The evolution of social organization and visual communication in the weaver birds (Ploceinae). *Behaviour, Suppl.* No. 10.

Crook, J. H. (1965). The adaptive significance of avian social organizations. *Symp. Zool. Soc. London* **14,** 181–218.

Crook, J. H., Ellis, J. E., and Goss-Custard, J. D. (1976). Mammalian social systems: Structure and function. *Anim. Behav.* **24,** 261–274.

Crusoe, D. A. (1980). Acoustic behavior and its role in social relations of the Red-headed Woodpecker (Picidae: *Melanerpes erythrocephalus*). M.S. Thesis, Univ. of Illinois at Chicago Circle.

Davidson, D. W. (1978). Size variability in the worker caste of a social insect (*Veromessor pergandei* Mayr) as a function of the competitive environment. *Am. Nat.* **112,** 523–532.

den Boer, P. J. (1980). Exclusion or coexistence and the taxonomic or ecological relationship between species. *Neth. J. Zool.* **30,** 278–306.

Dobzhansky, T. (1937). "Genetics and the Origin of Species." Columbia Univ. Press, New York.

Dobzhansky, T. (1951). "Genetics and the Origin of Species," 3rd ed., rev. Columbia Univ. Press, New York.

Dobzhansky, T. (1970). "Genetics of the Evolutionary Process." Columbia Univ. Press, New York.

Dobzhansky, T., Ayala, F. J., Stebbins, G. L., and Valentine, J. W. (1977). "Evolution." Freeman, San Francisco, California.

Dooling, R. J. (1980). Behavior and psychophysics of hearing in birds. *In* "Comparative Studies of Hearing in Vertebrates" (A. N. Popper and R. R. Fay, eds.), pp. 261–288. Springer-Verlag, Berlin and New York.

Dooling, R. J., Zoloth, S. R., and Baylis, J. R. (1978). Auditory sensitivity, equal loudness, temporal resolving power, and vocalizations in the House Finch (*Carpodacus mexicanus*). *J. Comp. Physiol. Psychol.* **92,** 867–876.

Ehret, G. (1980). Development of sound communication in mammals. *Adv. Study Behav.* **11,** 179–225.

Eisenberg, J. F. (1976). Communication mechanisms and social integration in the black spider monkey, *Ateles fusciceps robustus,* and related species. *Smithson. Contrib. Zool.* No. 213.

Elsner, N., and Popov, A. V. (1978). Neuroethology of acoustic communication. *Adv. Insect Physiol.* **13,** 229–355.

Emlen, S. T. (1972). An experimental analysis of the parameters of bird song eliciting species recognition. *Behaviour* **41,** 130–171.

Emlen, S. T., and Oring, L. W. (1977). Ecology, sexual selection, and the evolution of mating systems. *Science* **197,** 215–223.

Emlen, S. T., Rising, J. D., and Thompson, W. L. (1975). A behavioral and morphological study of sympatry in the Indigo and Lazuli buntings of the Great Plains. *Wilson Bull.* **87,** 145–177.

Ewing, A. W. (1979). Complex courtship songs in the *Drosophila funebris* species group: Escape from an evolutionary bottleneck. *Anim. Behav.* **27,** 343–349.

Falls, J. B., and Szijj, L. J. (1959). Reactions of Eastern and Western meadowlarks in Ontario to each others' vocalizations. *Anat. Rec.* **134,** 560.

Fellers, G. M. (1979). Aggression, territoriality, and mating behaviour in North American treefrogs. *Anim. Behav.* **27,** 107–119.

Fenchel, T. (1975). Character displacement and coexistence in mud snails (Hydrobiidae). *Oecologia* **20,** 19–32.

Ferdinand, L. (1966). Display of the Great Snipe (*Gallinago media* Latham). *Dan. Ornithol. For. Tidsskr.* **60,** 14–34.

Ferdinand, L., and Gensbøl, B. (1966). Maintenance of territory in the Great Snipe (*Gallinago media* Latham) on the display grcund. *Dan. Ornithol. For. Tidsskr.* **60,** 35–43.

Ficken, M. S., and Ficken, R. W. (1962). The comparative ethology of the wood warblers: a review. *Living Bird* **1,** 103–122.

Ficken, M. S., and Ficken, R. W. (1967). Singing behaviour of Blue-winged and Golden-winged warblers, and their hybrids. *Behaviour* **28,** 149–181.

Ficken, M. S., and Ficken, R. W. (1968a). Reproductive isolating mechanisms in the Blue-winged Warbler–Golden-winged Warbler complex. *Evolution* **22,** 166–179.

Ficken, M. S., and Ficken, R. W. (1968b). Ecology of Blue-winged Warblers, Golden-winged Warblers and some other *Vermivora. Am. Midl. Nat.* **79,** 311–319.

Ficken, M. S., and Ficken, R. W. (1968c). Territorial relationships of Blue-winged Warblers, Golden-winged Warblers, and their hybrids. *Wilson Bull.* **80,** 442–451.

Ficken, M. S., and Ficken, R. W. (1969). Responses of Blue-winged Warblers and Golden-winged Warblers to their own and the other species' song. *Wilson Bull.* **81,** 69–74.

Ficken, R. W., Ficken, M. S., and Hailman, J. P. (1974). Temporal pattern shifts to avoid acoustic interference in singing birds. *Science* **183,** 762–763.

Fisher, R. A. (1958). "The Genetical Theory of Natural Selection," 2nd ed. Dover, New York.

Flint, V. E., and Kishchinskii, A. A. (1973). Data on the biology of the Siberian Pectoral Sandpiper. *In* "Fauna and Ecology of Waders," pp. 100–105. Moscow Soc. Naturalists, Moscow.

Fouquette, M. J., Jr. (1975). Speciation in chorus frogs. I. Reproductive character displacement in the *Pseudacris nigrita* complex. *Syst. Zool.* **24,** 16–23.

Gates, G. R., Perry, D. R., and Coles, R. B. (1975). Cochlear microphonics in the adult Domestic Fowl (*Gallus domesticus*). *Comp. Biochem. Physiol. A* **51,** 251–252.

Geist, V. (1978). On weapons, combat, and ecology. *In* "Aggression, Dominance, and Individual Spacing" (L. Krames, P. Pliner, and T. Alloway, eds.), pp. 1–30. Plenum, New York.

Gill, F. B. (1980). Historical aspects of hybridization between Blue-winged and Golden-winged warblers. *Auk* **97,** 1–18.

Gill, F. B., and Lanyon, W. E. (1964). Experiments on species discrimination in Blue-winged Warblers. *Auk* **81,** 53–64.

Gill, F. B., and Murray, B. G., Jr. (1972a). Discrimination behavior and hybridization of the Blue-winged and Golden-winged warblers. *Evolution* **26,** 282–293.

Gill, F. B., and Murray, B. G., Jr. (1972b). Song variation in sympatric Blue-winged and Golden-winged warblers. *Auk* **89,** 625–643.

Gish, S. L., and Morton, E. S. (1981). Structural adaptations to local habitat acoustics in Carolina Wren songs. *Z. Tierpsychol.* **56,** 74–84.

Gochfeld, M. (1978). Intraspecific social stimulation and temporal displacement of songs of the Lesser Skylark, *Alauda gulgula. Z. Tierpsychol.* **48,** 337–344.

Golubeva, T. B. (1978). Heterochronous development of hearing in birds during ontogenesis. *J. Evol. Biochem. Physiol.* **14,** 589–596.

Gottlieb, G. (1979). Development of species identification in ducklings: V. Perceptual differentiation in the embryo. *J. Comp. Physiol. Psychol.* **93,** 831–854.

Grant, P. R. (1972). Convergent and divergent character displacement. *Biol. J. Linn. Soc.* **4,** 39–68.

Grant, P. R. (1975). The classical case of character displacement. *Evol. Biol.* **8,** 237–337.

Green, S. (1975). Variation of vocal pattern with social situation in the Japanese Monkey (*Macaca fuscata*): A field study. *Primate Behav.* **4,** 1–102.

Green, S., and Marler, P. (1979). The analysis of animal communication. *Hand. Behav. Neurobiol.* **3,** 73–158.

Güttinger, H. R., and Nicolai, J. (1973). Struktur und Funktion der Rufe bei Prachtfinken (Estrildidae). *Z. Tierpsychol.* **33,** 319–334.

Hall, M. F. (1962). Evolutionary aspects of estrildid song. *Symp. Zool. Soc. London* **8,** 37–55.

Hansen, P. (1979). Ecological adaptations in bird sounds. *Nat. Jutland.* **20,** 33–53.

Hardy, J. W. (1979). Vocal repertoire and its possible evolution in the Black and Blue jays (*Cissilopha*). *Wilson Bull.* **91,** 187–201.

Harestad, A. S., and Bunnell, F. L. (1979). Home range and body weight—a re-evaluation. *Ecology* **60,** 389–402.

Harner, J. E., and Whitmore, R. C. (1977). Multivariate measures of niche overlap using discriminant analysis. *Theor. Pop. Biol.* **12,** 21–36.

Hazlett, B. A., ed. (1977). ''Quantitative Methods in the Study of Animal Behavior.'' Academic Press, New York.

Henwood, K., and Fabrick, A. (1979). A quantitative analysis of the dawn chorus: Temporal selection for communicatory optimization. *Am. Nat.* **114,** 260–274.

Heth, G., and Nevo, E. (1981). Origin and evolution of ethological isolation in subterranean mole rats. *Evolution* **35,** 259–274.

Hinsch, K. (1972). Akustische Gesangs analyse beim Fitis (*Phylloscopus trochilus*) zur Untersuchung der Rolle der Luftröhre bei der Stimmerzeugung der Singvögel. *J. Ornithol.* **113,** 315–322.

Hödl, W. (1977). Call differences and calling site segregation in anuran species from central Amazonian floating meadows. *Oecologia* **28,** 351–363.

Hogan-Warburg, A. J. (1966). Social behavior of the Ruff, *Philomachus pugnax* (L.). *Ardea* **54,** 109–225.

Hope, S. (1980). Call form in relation to function in the Steller's Jay. *Am. Nat.* **116,** 788–820.

Hopkins, C. D. (1980). Evolution of electric communication channels of mormyrids. *Behav. Ecol. Sociobiol.* **7,** 1–13.

Hopkins, C. D. (1981). On the diversity of electric signals in a community of mormyrid electric fish in West Africa. *Am. Zool.* **21,** 211–222.

Hopkins, C. D., and Bass, A. H. (1981). Temporal coding of species recognition signals in an electric fish. *Science* **212,** 85–87.

Howe, M. A. (1972). Pair bond formation and maintenance in Wilson's Phalarope, *Phalaropus tricolor*. Ph.D. Thesis, Univ. of Minnesota, Minneapolis.

Hubbs, C. L. (1960). Isolating mechanisms in the speciation of fishes. *In* ''Vertebrate Speciation'' (W. F. Blair, ed.), pp. 5–23. Univ of Texas Press, Austin.

Huey, R. B., and Pianka, E. R. (1974). Ecological character displacement in a lizard. *Am. Zool.* **14,** 1127–1136.

Hunter, M. L., Jr., and Krebs, J. R. (1979). Geographical variation in the song of the Great Tit (*Parus major*) in relation to ecological factors. *J. Anim. Ecol.* **48,** 759–785.

Hurlbert, S. H. (1971). The non-concept of species diversity: A critique and alternative parameters. *Ecology* **52,** 577–586.

Hurlbert, S. H. (1978). The measurement of niche overlap and some relatives. *Ecology* **59,** 67–77.

Hutterer, R., and Vogel, P. (1977). Abwehrlaute afrikanischer Spitzmäuse der Gattung *Crocidura* Wagler, 1832 und ihre systematische Bedentung. *Bonn. Zool. Beitr.* **28,** 218–227.

Huxley, C. R., and Wilkinson, R. (1977). Vocalizations of the Aldabra White-throated Rail *Dryolimnas cuvieri aldabranus*. *Proc. R. Soc. London, Ser. B.* **197,** 315–331.

James, R. D. (1981). Factors affecting variation in the primary song of North American Solitary Vireos (Aves: Vireonidae). *Can. J. Zool.* **59,** 2001–2009.

Jarman, P. J. (1974). The social organisation of antelope in relation to their ecology. *Behaviour* **48,** 215–267.

Jenni, D. A., Gambs, R. D., and Betts, B. J. (1974). Acoustic behavior of the Northern Jacana. *Living Bird* **13,** 193–210.

Jenssen, T. A. (1979). Display behaviour of male *Anolis opalinus* (Sauria, Iguanidae): A case of weak display stereotypy. *Anim. Behav.* **27,** 173–184.

Johnson, N. K. (1963). Biosystematics of sibling species of flycatchers in the *Empidonax hammondii-oberholseri- wrightii* complex. *Univ. Calif.* (*Berkeley*), *Publ. Zool.* **66,** 79–238.

Johnson, N. K. (1980). Character variation and evolution of sibling species in the *Empidonax difficilis–flavescens* complex (Aves: Tyrannidae). *Univ. Calif.* (*Berkeley*), *Publ. Zool.* **112,** 1–151.

Kneutgen, J. (1969). ''Musikalische'' Formen im Gesang der Schamadrossel (*Kittacincla macroura* Gm.) und ihre Funktionen. *J. Ornithol.* **110,** 250–285.

Konishi, M. (1964). Song variation in a population of Oregon Juncos. *Condor* **66,** 423–436.

Kroodsma, D. E. (1972). Variations in songs of Vesper Sparrows in Oregon. *Wilson Bull.* **84,** 173–178.

Kroodsma, D. E. (1973). Coexistence of Bewick's Wrens and House Wrens in Oregon. *Auk* **90,** 341–352.

Kroodsma, D. E. (1979). Vocal dueling among male Marsh Wrens: Evidence for ritualized expressions of dominance/subordinance. *Auk* **96,** 505–515.

Kroodsma, D. E. (1981). Geographical variation and functions of song types in warblers (Parulidae). *Auk* **98,** 743–751.

Kroodsma, D. E., and Verner, J. (1978). Complex singing behaviors among *Cistothorus* wrens. *Auk* **95,** 703–716.

Lack, D., and Southern, H. N. (1949). Birds on Tenerife. *Ibis* **91,** 607–626.

Lanyon, W. E. (1978). A revision of the *Myiarchus* flycatchers of South America. *Bull. Am. Mus. Nat. Hist.* **161,** 427–628.

Lanyon, W. E., and Gill, F. B. (1964). Spectrographic analysis of variation in the songs of a population of Blue-winged Warblers (*Vermivora pinus*). *Am. Mus. Novit.* No. 2176, 1–18.

Lein, M. R. (1978). Song variation in a population of Chestnut-sided Warblers (*Dendroica pensylvanica*): its nature and suggested significance. *Can. J. Zool.* **56,** 1266–1283.

Lein, M. R. (1981). Display behavior of Ovenbirds (*Seiurus aurocapillus*) II. Song variation and singing behavior. *Wilson Bull.* **93,** 21–41.

Lein, M. R. (1981). Display behavior of Ovenbirds. (*Seirus aurocapillus*) II. Song variation and singing behavior. *Wilson Bull.* **93,** 21–41.

Lemaire, F. (1977). Mixed song, interspecific competition and hybridisation in the Reed and Marsh Warblers (*Acrocephalus scirpaceus* and *palustris*). *Behaviour* **63,** 215–240.

Lemnell, P. A. (1978). Social behaviour of the Great Snipe *Capella media* at the arena display. *Ornis Scand.* **9,** 146–163.
Leroy, Y. (1978). Analysis of the acoustic environment. *Biophon* **6**(2), 6–8.
Leroy, Y. (1979). "L'univers Sonore Animal. Rôles et Evolution de la Communication Acoustique." Gauthier-Villars, Paris.
Levins, R. (1968). "Evolution in Changing Environments. Some Theoretical Explorations." Princeton Univ. Press, Princeton, New Jersey.
Liley, N. R. (1966). Ethological isolating mechanisms in four sympatric species of poeciliid fishes. *Behaviour, Suppl.* No. 13.
Lill, A. (1974). Lek behavior in the Golden-headed Manakin, *Pipra erythrocephala* in Trinidad (West Indies). *Z. Tierpsychol., Suppl.* No. 18.
Littlejohn, M. J. (1977). Long-range acoustic communication in anurans: An integrated and evolutionary approach. *In* "The Reproductive Biology of Amphibians" (D. H. Taylor and S. I. Guttman, eds.), 263–294. Plenum, New York.
Littlejohn, M. J., and Martin, A. A. (1969). Acoustic interaction between two species of leptodactylid frogs. *Anim. Behav.* **17,** 785–791.
MacArthur, R. H. (1972). "Geographical Ecology. Patterns in the Distribution of Species." Harper, New York.
MacArthur, R. H., and Wilson, E. O. (1967). "The Theory of Island Biogegraphy." Princeton Univ. Press, Princeton, New Jersey.
Mace, T. R. (1981). Causation, function, and variation of the vocalizations of the Northern Jacana, *Jacana spinosa*. Ph.D. Thesis, Univ. of Montana, Missoula.
McGrath, T. A., Shalter, M. D., Schleidt, W. M., and Sarvella, P. (1972). Analysis of distress calls of Chicken × Pheasant hybrids. *Nature (London)* **237,** 47–48.
McLaren, I. A. (1972). Polygyny as the adaptive function of breeding territory in birds. *Trans. Conn. Acad. Arts Sci.* **44,** 191–210.
MacNally, R. C. (1979). Social organisation and interspecific interactions in two sympatric species of *Ranidella* (Anura). *Oecologia* **42,** 293–306.
Manning, A. (1977). Aspects génétiques et évolutifs des mécanismes éthologiques d'isolement. *In* "Les Mécanismes Ethologiques de l'Evolution" (J. Médioni and E. Boesiger, eds.), pp. 70–78. Masson, Paris.
Marler, P. (1955). Characteristics of some animal calls. *Nature (London)* **176,** 6–7.
Marler, P. (1957). Specific distinctiveness in the communication signals of birds. *Behaviour* **11,** 13–39.
Marler, P. (1960). Bird songs and mate selection. *In* "Animal Sounds and Communication" (W. E. Lanyon and W. N. Tavolga, eds.), Publ. No. 7, pp. 348–367. Am. Inst. Biol. Sci., Washington D.C.
Marler, P., and Boatman, D. J. (1951). Observations on the birds of Pico, Azores. *Ibis* **93,** 90–99.
Marler, P., and Hamilton, W. J., III (1966). "Mechanisms of Animal Behavior". Wiley, New York.
Marler, P., and Isaac, D. (1960a). Physical analysis of a simple bird song as exemplified by the Chipping Sparrow. *Condor* **62,** 124–135.
Marler, P., and Isaac, D. (1960b). Song variation in a population of Brown Towhees. *Condor* **62,** 272–283.
Marler, P., and Isaac, D. (1961). Song variation in a population of Mexican Juncos. *Wilson Bull.* **73,** 193–206.
Martens, J. (1975). Akustische Differenzierung verwandtschaftlicher Beziehungen in der *Parus* (*Periparus*)—Gruppe nach Untersuchungen im Nepal–Himalaya. *J. Ornithol.* **116,** 369–433.
Martin, D. J. (1977). Songs of the Fox Sparrow. I. Structure of song and its comparison with song in other Emberizidae. *Condor* **79,** 209–221.

Martin, D. J. (1979). Songs of the Fox Sparrow. II. Intra- and interpopulation variation. *Condor* **81,** 173–184.
Martin, W. F. (1972). Evolution of vocalization in the genus *Bufo. In* "Evolution in the Genus *Bufo*" (W. F. Blair, ed.), pp. 279–309. Univ. of Texas Press, Austin.
May, R. M. (1974). On the theory of niche overlap. *Theor. Popul. Biol.* **5,** 297–332.
May, R. M. (1975). Patterns of species abundance and diversity. *In* "Ecology and Evolution of Communities" (M. L. Cody and J. M. Diamond, eds.), pp. 81–120. Belknap Press, Cambridge, Massachusetts.
Mayr, E. (1942). "Systematics and the Origin of Species from the Viewpoint of a Zoologist." Columbia Univ. Press, New York.
Mayr, E. (1963). "Animal Species and Evolution." Belknap Press, Cambridge, Massachusetts.
Mayr, E. (1976). Darwin, Wallace, and the origin of isolating mechanisms. *In* "Evolution and the Diversity of Life. Selected Essays," pp. 129–134. Belknap Press, Cambridge, Massachusetts.
Mecham, J. S. (1960). Isolating mechanisms in anuran amphibians. *In* "Vertebrate Speciation" (W. F. Blair, ed.), pp. 24–61. Univ. of Texas Press, Austin.
Mendall, H. L., and Aldous, C. M. (1943). "The Ecology and Management of the American Woodcock." Maine Coop. Wildl. Res. Unit, Orono.
Meyerriecks, A. J., and Baird, J. (1968). Agonistic interactions between Blue-winged and "Brewster's" warblers. *Wilson Bull.* **80,** 150–160.
Miller, D. B. (1978). Species-typical and individually distinctive acoustic features of crow calls of Red Jungle Fowl. *Z. Tierpsychol.* **47,** 182–193.
Miller, D. B., and Gottlieb, G. (1976). Acoustic features of Wood Duck (*Aix sponsa*) maternal calls. *Behaviour* **57,** 260–280.
Miller, R. S. (1967). Pattern and process in competition. *Adv. Ecol. Res.* **4,** 1–74.
Miller, R. S. (1968). Conditions of competition between Redwings and Yellowheaded Blackbirds. *J. Anim. Ecol.* **37,** 43–62.
Mirsky, E. N. (1976). Song divergence in hummingbird and junco populations on Guadalupe Island. *Condor* **78,** 230–235.
Moynihan, M. (1968). Social mimicry; character convergence versus character displacement. *Evolution* **22,** 315–331.
Müller, H. J. (1942). Isolating mechanisms, evolution and temperature. *Biol. Symp.* **6,** 71–125.
Murray, B. G., Jr. (1971). The ecological consequences of interspecific territorial behavior in birds. *Ecology* **52,** 414–423.
Murray, B. G., Jr. (1976). A critique of interspecific territoriality and character convergence. *Condor* **78,** 518–525.
Murray, B. G., Jr., and Gill, F. B. (1976). Behavioral interactions of Blue-winged and Golden-winged warblers. *Wilson Bull.* **88,** 231–254.
Myers, J. P. (1979). Leks, sex, and Buff-breasted Sandpipers. *Am. Birds* **33,** 823–825.
Myrberg, A. A., Jr., and Spires, J. Y. (1980). Hearing in damselfishes: An analysis of signal detection among closely related species. *J. Comp Physiol.* **140,** 135–144.
Nottebohm, F. (1976). Vocal tract and brain: A search for evolutionary bottlenecks. *Ann. N.Y. Acad. Sci.* **280,** 643–649.
Nuechterlein, G. L. (1981). Courtship behavior and reproductive isolation between Western Grebe color morphs. *Auk* **98,** 335–349.
Orians, G. H., and Collier, G. (1963). Competition and blackbird social systems. *Evolution* **17,** 449–459.
Oring, L. W. (1964). Displays of the Buff-breasted Sandpiper at Norman, Oklahoma. *Auk* **81,** 83–86.
Parker, G. A. (1978). Evolution of competitive mate searching. *Annu. Rev. Entomol.* **23,** 173–196.
Parkes, K. C. (1951). The genetics of the Golden-winged × Blue-winged warbler complex. *Wilson Bull.* **63,** 5–15.

Parkes, K. C. (1961). Intergeneric hybrids in the family Pipridae. *Condor* **63,** 345–350.

Parkes, K. C. (1978). Still another parulid intergeneric hybrid (*Mniotilta* × *Dendroica*) and its taxonomic and evolutionary implications. *Auk* **95,** 682–690.

Parmelee, D. F., Stephens, H. A., and Schmidt, R. H. (1967). The birds of southeastern Victoria Island and adjacent small islands. *Bull. Nat. Mus. Can.* No. 222.

Paterson, H. E. H. (1978). More evidence against speciation by reinforcement. *S. Afr. J. Sci.* **74,** 369–371.

Paterson, H. E. (1980). A comment on "mate recognition systems." *Evolution* **34,** 330–331.

Pawlik, J., Lewis, R., and Barnwell, F. H. (1980). Different mechanisms of sound production in two closely related species of fiddler crabs. *Am. Zool.* **20,** 957.

Payne, R. B., and Budde, P. (1979). Song differences and map distances in a population of Acadian Flycatchers. *Wilson Bull.* **91,** 29–41.

Pianka, E. R. (1978). "Evolutionary Ecology," 2nd ed. Harper, New York.

Pielou, E. C. (1975). "Ecological Diversity." Wiley (Interscience), New York.

Pielou, E. C. (1977). "Mathematical Ecology." Wiley (Interscience), New York.

Pitelka, F. A. (1959). Numbers, breeding schedule, and territoriality in Pectoral Sandpipers of northern Alaska. *Condor* **61,** 233–264.

Potash, L. M. (1975). An experimental analysis of the use of location calls by Japanese Quail, *Coturnix coturnix japonica*. *Behaviour* **54,** 153–180.

Rand, A. S., and Drewry, G. E. (1972). Communication strategies in Puerto Rican frogs of the genus *Eleutherodactylus*. Unpublished manuscript. Cited in Smith (1977, pp. 369–370).

Ratti, J. T. (1979). Reproductive separation and isolating mechanisms between sympatric dark- and light-phase Western Grebes. *Auk* **96,** 573–586.

Rice, J. (1978a). Ecological relationships of two interspecifically territorial vireos. *Ecology* **59,** 526–538.

Rice, J. (1978b). Behavioural interactions of interspecifically territorial vireos. I. Song discrimination and natural interactions. *Anim. Behav.* **26,** 527–549.

Rice, J. O., and Thompson, W. L. (1968). Song development in the Indigo Bunting. *Anim. Behav.* **16,** 462–469.

Richards, D. G., and Wiley, R. H. (1980). Reverberations and amplitude fluctuations in the propagation of sound in a forest: Implications for animal communication. *Am. Nat.* **115,** 381–399.

Robinson, F. N. (1973). Vocal mimicry and bird song evolution. *New Sci.* **58,** 742–743.

Robinson, F. N. (1974). The function of vocal mimicry in some avian displays. *Emu* **74,** 9–10.

Robinson, F. N. (1975). Vocal mimicry and the evolution of bird song. *Emu* **75,** 23–27.

Roughgarden, J. (1972). Evolution of niche width. *Am. Nat.* **106,** 683–718.

Roughgarden, J. (1979). "Theory of Population Genetics and Evolutionary Ecology: An Introduction." Macmillan, New York.

Routledge, R. D. (1979). Diversity indices: Which ones are admissible? *J. Theor. Biol.* **76,** 503–515.

Routledge, R. D. (1980). Bias in estimating the diversity of large, uncensused communities. *Ecology* **61,** 276–281.

Sachs, M. B., Sinnott, J. M., and Heinz, R. D. (1978). Behavioral and physiological studies of hearing in birds. *Fed. Proc., Fed. Am. Soc. Exp. Biol.* **37,** 2329–2335.

Salmon, M., Hyatt, G., McCarthy, K., and Costlow, J. D., Jr. (1978). Display specificity and reproductive isolation in the fiddler crabs, *Uca panacea* and *U. pugilator*. *Z. Tierpsychol.* **48,** 251–276.

Salthe, S. N., and Mecham, J. S. (1974). Reproductive and courtship patterns. *In* "Physiology of the Amphibia" (B. Lofts, ed.), Vol. 2, pp. 309–521. Academic Press, New York.

Samuel, D. E., and Beightol, D. R. (1972). Monitoring Woodcock singing sites through sonagrams. *Proc. SE Assoc. Game Fish Commis.* **26,** 301–305.

Samuel, D. E., and Beightol, D. R. (1973). The vocal repertoire of male American Woodcock. *Auk* **90,** 906–909.

Samways, M. J. (1977). Bush cricket interspecific acoustic interactions in the field (Orthoptera, Tettigoniidae). *J. Nat. Hist.* **11,** 155–168.

Samways, M. J., and Broughton, W. B. (1976). Song modification in the Orthoptera II. Types of acoustic interaction between *Platycleis intermedia* and other species of the genus (Tettigoniidae). *Physiol. Entomol.* **1,** 287–297.

Schleidt, W. M., and Shalter, M. D. (1973). Stereotypy of a fixed action pattern during ontogeny in *Coturnix coturnix coturnix. Z. Tierpsychol.* **33,** 35–37.

Schubert, M. (1976a). Uber die Variabilität von Lockrufen des Gimpels *Pyrrhula pyrrhula. Ardea* **64,** 62–71.

Schubert, M. (1976b). Das akustische Repertoire des Fitislaubsängers (*Phylloscopus t. trochilus*) und seine erblichen und durch Lernen erworbenen Bestandteile. *Beitr. Vogelkd.* **22,** 167–200.

Selander, R. K., and Giller, D. R. (1961). Analysis of sympatry of Great-tailed and Boat-tailed grackles. *Condor* **63,** 29–86.

Shalter, M. D. (1978). Localization of passerine seeet and mobbing calls by Goshawks and Pygmy Owls. *Z. Tierpsychol.* **46,** 260–267.

Shapiro, L. J. (1980). Species identification in birds: a review and synthesis. *In* "Species Identity and Attachment" (A. Roy, ed.), pp. 69–111. Garland STPM Press, New York.

Sheldon, W. G. (1967). "The Book of the American Woodcock." Univ. of Massachusetts Press, Amherst.

Shiovitz, K. A. (1975). The process of species-specific song recognition by the Indigo Bunting, *Passerina cyanea,* and its relationship to the organization of avian acoustical behavior. *Behaviour* **55,** 128–179.

Short, L. L., Jr. (1962). The Blue-winged Warbler and Golden-winged Warbler in central New York. *Kingbird* **12,** 59–67.

Short, L. L., Jr. (1963). Hybridization in the wood warblers *Vermivora pinus* and *V. chrysoptera. Proc. Int. Ornithol. Congr.* **13,** 147–160.

Short, L. L., Jr. (1969). "Isolating mechanisms" in the Blue-winged Warbler—Golden-winged Warbler complex. *Evolution* **23,** 355–356.

Sibley, C. G., and Short, L. L., Jr. (1959). Hybridization in the buntings (*Passerina*) of the Great Plains. *Auk* **76,** 443–463.

Slatkin, M. (1980). Ecological character displacement. *Ecology* **61,** 163–177.

Smith, D. G., Reid, F. A., and Breen, C. B (1980). Stereotypy of some parameters of Red-winged Blackbird song. *Condor* **82,** 259–266.

Smith, W. J. (1977). "The Behavior of Communicating. An Ethological Approach." Harvard Univ. Press, Cambridge, Massachusetts.

Stein, R. C. (1958). The behavioral, ecological and morphological characteristics of two populations of the Alder Flycatcher, *Empidonax traillii* (Audubon). *N.Y. State Mus. Sci. Ser., Bull.* No. 371.

Stein, R. C. (1963). Isolating mechanisms between populations of Traill's Flycatchers. *Proc. Am. Philos. Soc.* **107,** 21–50.

Steiner, W. W., II (1979). A comparative study of the pure tonal whistle vocalizations from five western north Atlantic dolphin species. Ph.D. Thesis, Univ. of Rhode Island, Kingston.

Stobo, W. T., and McLaren, I. A. (1975). The Ipswich Sparrow. *Proc. N. S. Inst. Sci.* **27,** Suppl. No. 2.

Sutton, G. M. (1967). Behaviour of the Buff-breasted Sandpiper at the nest. *Arctic* **20,** 3–7.

Templeton, A. R. (1980). Once again, why 300 species of Hawaiian *Drosophila? Evolution* **33,** 513–517.

Thielcke, G. (1970). Die sozialen funktionen der Vogelstimmen. *Vogelwarte* **25,** 204–229.

Thielcke, G. (1972). Waldbaumläufer (*Certhia familiaris*) ahmen artfremdes Signal nach und reagieren darauf. *J. Ornithol.* **113,** 287–296.

Thielcke, G. (1973). On the origin of divergence of learned signals (songs) in isolated populations. *Ibis* **115,** 511–516.

Thielcke, G., Wustenberg, K., and Becker, P. H. (1978). Reaktionen von Zilpzalp und Fitis (*Phylloscopus collybita, Ph. trochilus*) auf verschiedene Gesangsformen des Zilpzalps. *J. Ornithol.* **119,** 213–226.

Thiele, D., and Bailey, W. J. (1980). The function of sound in male spacing behaviour in bush-crickets (Tettigoniidae, Orthoptera). *Aust. J. Ecol.* **5,** 275–286.

Thomas, D. W., and Dilworth, T. G. (1980). Variation in peent calls of American Woodcock. *Condor* **82,** 345–347.

Thompson, W. L. (1968). The songs of five species of *Passerina*. *Behaviour* **31,** 261–287.

Thompson, W. L. (1969). Song recognition by territorial male buntings (*Passerina*). *Anim. Behav.* **17,** 658–663.

Thompson, W. L. (1970). Song variation in a population of Indigo Buntings. *Auk* **87,** 58–71.

Thompson, W. L. (1972). Singing behavior of the Indigo Bunting, *Passerina cyanea*. *Z. Tierpsychol.* **31,** 39–59.

Thompson, W. L. (1976). Vocalizations of the Lazuli Bunting. *Condor* **78,** 195–207.

Thomson, J. D. (1980). Implications of different sorts of evidence for competition. *Am. Nat.* **116,** 719–726.

van Rhijn, J. G. (1973). Behavioural dimorphism in male Ruffs, *Philomachus pugnax* (L.). *Behaviour* **47,** 153–229.

van Valen, L. (1965). Morphological variation and width of ecological niche. *Am. Nat.* **99,** 377–390.

van Valen, L. (1974). Multivariate structural statistics in natural history. *J. Theor. Biol.* **45,** 235–247.

van Valen, L. (1978). The statistics of variation. *Evol. Theory* **4,** 33–43.

Waage, J. K. (1975). Reproductive isolation and the potential for character displacement in the damselflies, *Calopteryx maculata* and *C. aequabilis* (Odonata: Calopterygidae). *Syst. Zool.* **24,** 24–36.

Waage, J. K. (1979). Reproductive character displacement in *Calopteryx* (Odonata: Calopterygidae). *Evolution* **33,** 104–116.

Waser, P. M., and Wiley, R. H. (1979). Mechanisms and evolution of spacing in animals. *Hand. Behav. Neurobiol.* **3,** 159–223.

Wasserman, F. E. (1977). Intraspecific acoustical interference in the White-throated Sparrow (*Zonotrichia albicollis*). *Anim. Behav.* **25,** 949–952.

Wasserman, F. E. (1979). The relationship between habitat and song in the White-throated Sparrow. *Condor* **81,** 424–426.

Watson, G. F., and Littlejohn, M. J. (1978). The *Litoria ewingi* complex (Anura: Hylidae) in south-eastern Australia V. Interactions between northern *L. ewingi* and adjacent taxa. *Aust. J. Zool.* **26,** 175–195.

Weatherhead, P. J. (1977). Ecology and behavior of reproduction in a tundra population of Savannah Sparrows. Ph.D. Thesis, Queen's Univ., Kingston, Ontario.

Weatherhead, P. J. (1979). Ecological correlates of monogamy in tundra-breeding Savannah Sparrows. *Auk* **96,** 391–401.

Weir, N. L. (1979). The use of sonagraphic analysis of the American Woodcock (*Philohela minor*) peent call as a method for identifying individual males. M.S. Thesis, Pennsylvania State Univ., University Park.

Weir, N. L., and Graves, H. B. (1981). Discriminant analysis of the peent call for identification of individual male American Woodcock. *In* "Woodcock Ecology and Management," pp. 34–39. *U.S. Fish and Wildl. Serv., Wildl. Res. Rep.* No. 14.

Welsh, D. A. (1975). Savannah Sparrow breeding and territoriality on a Nova Scotia dune beach. *Auk* **92,** 235–251.

White, M. J. D. (1978). "Modes of Speciation." Freeman, San Francisco, California.

Whitney, C. L., and Krebs, J. R. (1975). Spacing and calling in Pacific tree frogs, *Hyla regilla*. *Can. J. Zool.* **53,** 1519–1527.

Wiley, R. H., and Richards, D. G. (1978). Physical constraints on acoustic communication in the atmosphere: Implications for the evolution of animal vocalizations. *Behav. Ecol. Sociobiol.* **3,** 69–94.

Wilson, E. O. (1965). The challenge from related species. *In* "The Genetics of Colonizing Species" (H. G. Baker and G. L. Stebbins, eds.), pp. 7–24. Academic Press, New York.

Wittenberger, J. F. (1979). The evolution of mating systems in birds and mammals. *Hand. Behav. Neurobiol.* **3,** 271–349.

Wittenberger, J. F., and Tilson, R. L. (1980). The evolution of monogamy: hypotheses and evidence. *Annu. Rev. Ecol. Syst.* **11,** 197–232.

Wooller, R. D. (1978). Individual vocal recognition in the Kittiwake Gull, *Rissa tridactyla* (L.). *Z. Tierpsychol.* **48,** 68–86.

Würdinger, I. (1970). Erzeugung, Ontogenie und Funktion der Lautäusserungen bei vier Gänsearten (*Anser indicus, A. caerulescens, A. albifrons* und *Branta canadensis*). *Z. Tierpsychol.* **27,** 257–302.

Zann, R. A. (1976a). Variation in the songs of three species of estrildine grassfinches. *Emu* **76,** 97–108.

Zann, R. (1976b). Inter- and intraspecific variation in the courtship of three species of grassfinches of the subgenus *Poephila* (Gould) (Estrildidae). *Z. Tierpsychol.* **41,** 409–433.

9

The Evolution of Bird Sounds in Relation to Mating and Spacing Behavior

CLIVE K. CATCHPOLE

I. INTRODUCTION

Birds have evolved an extremely sophisticated system of acoustic communication which they use for both species (Becker, Chapter 7, this volume) and individual recognition (Falls, Chapter 8, Volume 2). When a more detailed functional interpretation of a particular sound is attempted, serious problems can arise. For example, it is relatively simple to record and analyze the potential information transmitted in a particular sound but much more difficult to determine exactly what it means to the receiving individual. It was Smith (1965) who

ISBN 0-12-426801-3

first clarified this important distinction between message and meaning in animal communication.

Another problem is the immense variety of bird sounds, and in the past it has been traditional to divide them into calls and songs. Calls are generally short and simple and are uttered by both sexes at all times of year. Songs are longer, more complex, and generally restricted to the male bird during the breeding season. This division, although largely arbitrary and artificial, is still retained by most modern workers today. Although there is no clear gradation in sound complexity from primitive to advanced birds, songbirds are thought to be the most recently evolved group, and have the most complex vocalizations. The highly vocal passerines occupy a dominant position in most avifaunas, and it is hardly surprising that their complicated songs have attracted so much attention. Most passerines also have a large vocabulary of relatively simple calls which serve a wide variety of functions. Because calls occur in particular contexts and have an immediate effect upon the behavior of conspecifics, it has been relatively easy to arrive at functional interpretations. Marler (1956), in a pioneering study on the Chaffinch (*Fringilla coelebs*), was the first to combine sonagraphic analysis with such a functional interpretation for some 14 different calls. For Chaffinch song he concluded "Unmated reproductive females are attracted by it, and all reproductive males are repelled by it."

Thorpe (1961) has traced some of the first functional interpretations of bird song back as far as Aristotle and Pliny, and in 1871 Charles Darwin wrote "Naturalists are much divided with respect to the object of the singing of birds." Darwin (1871) suggested that songs function to "charm" the females and also used bird song as an example when developing his theory of sexual selection. He was nevertheless aware that earlier English naturalists, such as Gilbert White (1789) of Selborne, had emphasized male rivalry for available space as the most important function of song. With the publication of Eliot Howard's (1920) book the concept of territory became widely known, and ever since then the view that bird song has developed primarily as a territorial proclamation to rival males has held sway among most ornithologists. In his major review Armstrong (1973) unhesitatingly uses the term "territorial song" although he also pointed out that such song might have several different functions, including female attraction. Tinbergen (1939), working on the Snow Bunting (*Plectrophenax nivalis*), was one of the first to emphasize the probable dual function of "territorial" song as a sound "that serves to attract a sex partner, ward off a bird of the same sex or both." Tinbergen proposed the term "advertising song," which makes no *a priori* assumptions about its precise function.

Any attempt to arrive at a broad perspective concerning the functions of bird sounds is beset with a number of major obstacles. Comparison is hindered by the lack of any standard system for the description and classification of bird sounds. Whereas older studies may lack analytical data, modern studies which include sonagraphic analysis are often too specialized to include details of territorial or mating behavior. Taxonomic and geographical coverage is strongly biased toward north temperate passerines, and as Lack (1968) pointed out in his review, there are certain dangers inherent in comparing adaptations between major and widely disparate taxonomic groups. Given these problems, and the bewildering variety of bird sounds, it may well be that functional comparisons are best made between closely related species, populations, or even individuals.

Attracting females and repelling rival males have undoubtedly been two of the most important factors in shaping the various sounds which male birds have evolved. While these two functions are certainly not mutually exclusive, it may well be that the emphasis and pressures vary among different bird groups and may also change with different forms of mating and spacing behavior. Three different techniques will be used in order to examine what evidence there is, if any, to support this view. The approaches are certainly not mutually exclusive any more than the evolutionary forces they attempt to reveal at work, and wherever possible they should be used together as an integrated approach. They involve contextual correlations within individual species, correlations with mating and spacing systems between closely related species, and finally direct effects upon males and females. To some extent they are also historical, starting with observations, moving on to comparative studies, and finishing off by testing any hypotheses with laboratory or field experiments.

II. CONTEXTUAL CORRELATIONS

In a collection of edited contributions published over a decade ago, Hinde (1969) wrote "The first indication of function often comes from contextual information." Early descriptive studies (well reviewed in Armstrong, 1973) relied heavily on this effective if simplistic approach, relating different sounds to the immediate or long-term context in which they were given. There are several criticisms which might be leveled at this kind of study. They were often anecdotal, lacked analytical and quantitative data, and were often unaware of Smith's distinction between message and meaning.

A. Context, Message, and Meaning

Smith (1963) first outlined his message-meaning approach in a paper dealing with vocal communication in birds, and also developed the approach extensively in his recent book (Smith, 1977). In brief, a signal which can be analyzed by sonagraph contains one or more messages, the meaning of which may vary depending upon the context in which they are transmitted or received. Smith often uses the example of the short "kitter" vocalization made by the Eastern

Kingbird (*Tyrannus tyrannus*). The sound is made in a variety of contexts associated with tendencies to approach a perch, a mate, or other individuals. Considered in each particular context it could have a number of different meanings. For example, given by a male, it could signify his unmated status to any females, but to a rival male nearby it could signify that he is prepared to defend his territory if challenged. The sound is also given within the mated pair as an appeasement signal, and even given to predators which kingbirds often attempt to drive away. It is clearly a vocalization which conveys quite different messages in each of these different contexts.

As Smith points out, such a system of communication is certainly economical, for it would enable a repertoire of relatively few different sounds to communicate a great deal of information. However, there clearly must be some lower limitation since selection would also discourage undue overlap which could lead to ambiguities and confusion in some areas of communication. This may explain why the call repertoires of most passerines contain signals which are used only in certain restricted contexts, such as alarm calls. More elaborate and ambiguous sounds, such as passerine songs, may well prove to be much more appropriate for this type of analysis.

B. The Seasonal Context

In most bird species there is a simple correlation between seasonal breeding activity and either certain call (Thielcke, 1976) or song production (Armstrong, 1973). Even though this relationship is well known, there is also a surprising lack of data from quantitative studies which explore it in any detail. In European *Acrocephalus* warblers peak song production by males in a population is closely followed by egg-laying in females (Fig. 1). The Sedge Warbler (*A. schoenobaenus*) and Reed Warbler (*A. scirpaceus*) are migrants and the initial increase in singing reflects male arrival. Females arrive later and the decline in singing reflects pairing, for males which have attracted a female then cease to sing. Sedge Warbler arrival and pairing appears more synchronized, and breeding activity, as indicated by egg-laying, peaks earlier in the season. Slagsvold (1977) has now applied a similar technique to over 20 woodland species of passerines, and investigated several variables using multiple regression analysis. One of the strongest relationships to emerge was song production, reaching a peak several days before the onset of egg-laying.

It would obviously be premature to conclude that seasonal song production in passerines has solely a sexual function. In most cases a male also occupies a territory throughout the breeding season and so song is also produced in a territorial context. In such cases the dual function suggested by Tinbergen seems extremely likely, unless there is some conflict of interests. To tease apart the relative importance of the sexual and territorial function of seasonal song needs a

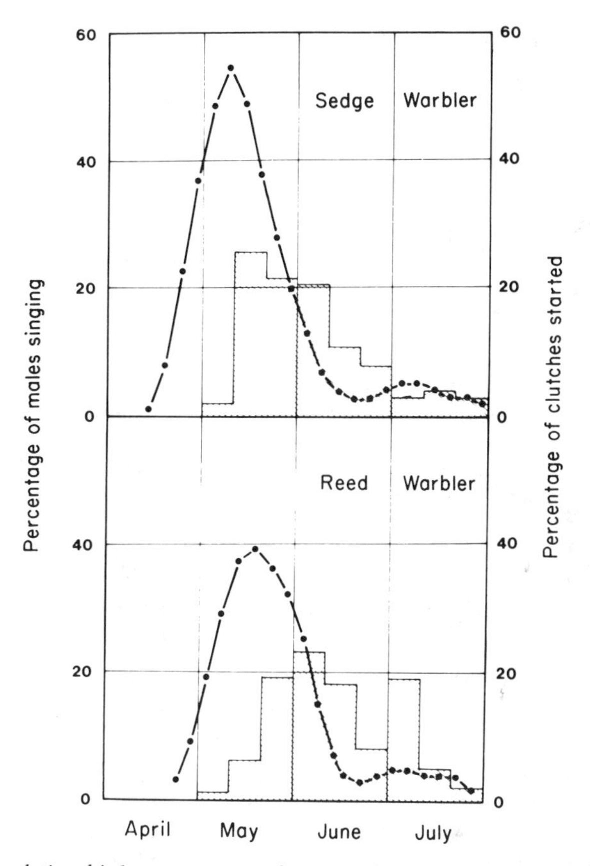

Fig. 1. The relationship between seasonal song production and breeding activity in populations of two *Acrocephalus* species. The percentage of males in song has been graphed at 5-day intervals and the percentage of clutches started in 10-day periods are shown as histograms. (From Catchpole, 1973.)

more detailed contextual approach, such as monitoring song output from individual birds over full 24-hr periods.

This technique has been applied to the same two *Acrocephalus* species to check whether individual males really do decrease or even cease song production after pairing. The latter case would be powerful evidence in favor of a sexual attraction function, since after a female has been obtained the territory must still be defended throughout the breeding season. Claims for song cessation after pairing have often been made (reviewed in Catchpole, 1973) but are not usually backed up by quantitative evidence obtained from detailed monitoring before and after pairing. The Sedge Warbler exhibits true song cessation, where the characteristic diurnal rhythm ceases completely after pairing (Fig. 2) and is not even

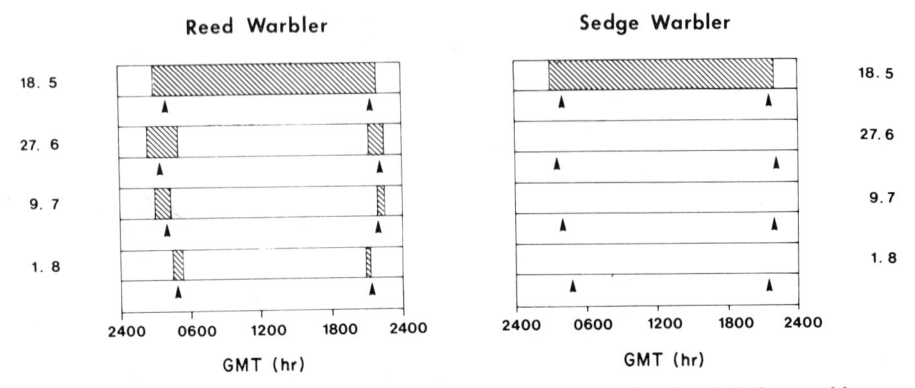

Fig. 2. Changes in the diurnal rhythm of song in male Reed and Sedge warblers monitored over 24-hr periods throughout the breeding season. Both males were unpaired at May 18.5 but paired soon after. Shading indicates singing and arrows the approximate times of sunrise and sunset. (From Catchpole, 1973.)

elicited by playback experiments (Catchpole, 1973, 1977). The Reed Warbler also loses its persistent diurnal rhythm of song after pairing, but retains a vestigial dawn and dusk chorus throughout the remainder of the breeding season. It also responds with song during playback experiments after pairing. These data suggest that the Sedge Warbler uses its advertising song primarily if not solely for sexual attraction. Sedge Warbler territories, in fairly open habitat, are still defended throughout the remainder of the breeding season, not by song but by visual threat displays and overt aggression. Reed Warblers nest much closer together but in dense reed beds. This may have placed an additional premium on vocal rather than visual defense of territory, and explains why they still retain song as an important part of their territorial behavior.

C. Different Songs for Different Functions?

It seems likely that in many species the same song may function in both mate attraction and territorial behavior. When different song-types are found within a species, correlations with different functions have been looked for but with limited success. As Krebs (1977a) has pointed out, in most cases different song types within a repertoire appear to occur within the same context and carry the same message. It has been extremely difficult to demonstrate any major differences in function, although one exception appears to be the Grasshopper Sparrow (*Ammodramus savannarum*) studied by Smith (1959). In a detailed contextual study he described different song-types used in territorial defence and mate attraction, but presented no sonagraphic evidence. Morse (1970) did reveal two structurally distinct songs in the Black-throated Green Warbler (*Dendroica virens*), type A used chiefly against rival males and type B in the presence of females. In addition he attempted to reduce the territorial context by comparing males in groups with those isolated on small islands. Song A, the territorial song, occurred much less frequently in the isolated males when compared to the grouped controls. However, Morse's functional interpretation has been questioned by Lein (1972). In a study on the Chestnut-sided Warbler (*D. pensylvanica*), Lein (1978) found five different song types which formed a graded series of signals associated with territorial behavior, but none with sexual behavior. Smith *et al.* (1978) reported similar results for the eight different song types of the Yellow-throated Vireo (*Vireo flavifrons*) which again appear to be associated with different aspects of territorial behavior. Jarvi *et al.* (1980) have also found out that the normal, spontaneous song of the Willow Warbler (*Phylloscopus trochilus*) changes in structure during territorial disputes. The songs become shorter, change in amplitude and frequency, and also contain a special syllable type at the start of each song. One other example of a species with separate song types for sexual and territorial functions has recently been reported by Baptista (1978) working on the Melodious Grassquit (*Tiaris canora*). Baptista used captive males to demonstrate that the frequency of the sexual song increased in the presence of females, whereas territorial song increased in the presence of males.

III. CORRELATIONS WITH DIFFERENT MATING AND SPACING SYSTEMS

With contextual correlations, sounds made in some immediate and temporary situations can be used to construct hypotheses regarding function. Another approach is to consider the possible influence of more constant factors that may be acting over longer periods of time. With respect to mating and spacing behavior, the obvious candidates are the different territorial and mating systems which birds have evolved.

A. Different Territorial Systems

Aggression related to space is a common characteristic of bird behavior, and all bird groups have developed special sounds that are aggressively motivated and function in territorial defense. For this reason a strictly phylogenetic approach to the study of territorial sounds may well be futile. Bird territorial systems are classified largely on the basis of size as well as what resources they may contain (e.g., Hinde, 1956). As distances between signaling individuals may well have had a considerable effect upon the evolution of sounds, the different territorial systems will be briefly considered in increasing order of size.

It would be difficult to produce convincing evidence that there are birds that have no territory at all, but some species defend only an extremely small area around the nest. Colonial seabirds such as Gannets (*Morus bassanus*) produce short, relatively simple calls which function in mate recognition (White, 1971). One of the most thorough comparative studies relating small territory size to sounds comes from Hall's (1962) work on colonial estrildine weaver finches. Hall analyzed the songs of 24 species and found that they were short, simple and given only in a sexual context, a view since confirmed by Immelmann (1969). Miller (1979) claims that the Zebra Finch (*Poephila guttata*) has no territorial behavior at all, and that the songs of individual males are recognized and responded to by their mates. It may be that being in close proximity in a breeding colony has placed more of an emphasis on the evolution of visual signaling, and that where sounds are used they are relatively short, simple, and more related to sexual behavior and mate recognition.

Typical passerine territories are much larger and it is here that we find considerable emphasis on both sound signaling and territorial function. When individuals are separated by greater distances, sound signaling clearly becomes advantageous. Passerines are relatively small, may also be cryptic, and often nest in dense vegetation where visibility is reduced. Such conditions also place a premium upon vocal advertising for a mate, so there are certain theoretical grounds for expecting a duality of function in passerine songs. Also, in terms of complexity, the evolution of birdsong certainly appears to have reached its peak in passerines breeding under these conditions.

But the correlation with distance holds good only to a certain extent, and at the other extreme we find species such as most raptors which have enormous territories. Most species seem to produce relatively simple sounds primarily in a sexual context (e.g., the Harpy Eagle, *Harpia harpyja*, Gochfeld and Kleinbaum, 1978). Extremely long distances between individuals produce obvious problems in sound transmission as well as reducing the need for territorial behavior. Flightless ratites also have very large territories and their rather simple sounds are restricted to parent–young communication in species such as the Greater Rhea (*Rhea americana*) (Beaver, 1978).

B. Different Mating Systems

Darwin (1871) was the first to discuss the evolution of different mating systems in birds. In developing his theory of sexual selection he also pointed out that sexual selection pressure would be more intense on polygynous males leading to such extreme elaborations as the peacock's tail. Bird sounds can be considered as the acoustic equivalent of visual sexual dimorphism, and similar elaboration in passerine songs might also be the result of sexual selection pressure. Lack (1968) estimated that over 90% of all birds are monogamous, with polygyny the most common alternative mating system. Polyandry is the least common system and in many ways is merely the ecological equivalent of polygyny with sex role reversal. Lek systems are also rare, and close proximity has resulted in strong sexual selection pressure favoring visual rather than vocal elaboration among males. Workers studying polyandry or lek species have therefore devoted very little attention to vocal behavior and there are no analytical data available for detailed comparison. For reasons outlined earlier, gross comparisons across widely disparate groups will be avoided in favor of studies which compare closely related species. Ideally, these should agree quite closely in their main breeding adaptations and be sexually monomorphic and even cryptic in order to minimize as many confounding variables as possible. There are relatively few passerine groups which contain a mixture of monogamous and polygynous species, and only two have so far been studied in this way. They are the North American wrens studied by Kroodsma (1977) and European warblers of the genus *Acrocephalus** (Catchpole, 1980).

Kroodsma studied nine species of North American wrens, of which four are regularly polygynous to varying degrees, and five monogamous. One very fundamental measure of complexity is the repertoire of basic building-blocks (syllables or elements) from which songs are constructed. In Table I it can be seen that the four polygynous wren species all have high estimated repertoire ranges when compared with the monogamous species. Kroodsma also used other measures of song complexity such as length, and these combine to make the general trend even more obvious. Male Winter Wrens (*Troglodytes troglodytes*) are highly polygynous (up to 50%) and have the longest and most complicated songs of all. The Long-billed Marsh Wren (*Cistothorus palustris*) is also highly polygynous, has a similarly high repertoire range and also the highest rate of presentation of new song-types. Among the monogamous species, the Rock Wren (*Salpinctes obsoletus*) does have a high repertoire range, but when it sings it merely repeats one syllable as a short, simple song. Overall, there is a strong positive correlation between polygyny and the evolution of more complicated songs among North American wrens.

However, a completely opposite trend emerges from the study on European *Acrocephalus* warblers (Table I). Here the two polygynous species have smaller repertoire ranges, and it is the monogamous species which have evolved the most elaborate songs. The polygynous species also have shorter songs lasting only about 3 sec. Among the monogamous species, the Sedge Warbler has the smallest estimated repertoire range, but from it constructs a seemingly endless variety of long, complicated songs sometimes lasting over 1 min in duration. The other three monogamous species have even larger repertoires and produce continuous songs with no clear intersong intervals. Mimicry is also a common feature in

*The Moustached Warbler is currently placed in the genus *Lusciniola* by some authorities.

TABLE I

The Relationship between Song Complexity (Estimated Repertoire Ranges) and Mating Systems within Two Different Groups of Passerine Birds

North American wrens, Troglodytidae (Kroodsma, 1977)			European *Acrocephalus* warblers, Muscicapidae, Sylviinae (Catchpole, 1980)
	Polygynous		
Long-billed Marsh *Cistothorus palustris*	44–118	10–20	Aquatic *A. paludicola*
Short-billed Marsh *Cistothorus platensis*	112	10–20	Great Reed *A. arundinaceus*
Winter *Troglodytes troglodytes*	89–95		
House *Troglodytes aedon*	37–90		
	Monogamous		
Rock *Salpinctes obsoletus*	69–119	35–55	Sedge *A. schoenobaenus*
Bewick's *Thryomanes bewickii*	25–65	60–80	Moustached *A. melanopogon*
Carolina *Thryothorus ludovicianus*	22	70–90	Reed *A. scirpaceus*
Canyon *Salpinctes mexicanus*	9	80–100	Marsh *A. palustris*

three of the monogamous species, and in the Marsh Warbler Dowsett-Lemaire (1979) has shown that most of the syllables are acquired through mimicry of a large number of other species from both the European and African avifaunas. Mimicry seems to be another way of increasing even further the song complexity of monogamous warbler species.

These results appear to run contrary to the orthodox view of more intense sexual selection pressure acting on polygynous species, but the explanation can be found in the ecology of these warblers, particularly in relation to the feeding ecology of individual species. *Acrocephalus* warblers are primarily inhabitants of marshland and breed in and around reed beds. The territories of monogamous species are quite small and most insect food is collected outside the territory and involves flights of considerable distances. Both parents are obviously needed to bring food to the young and female choice must be clearly influenced by this. Females should ensure they select a male who is capable of helping to bring in food, and pay less attention to territory quality. The very converse applies to female choice in polygynous species. Here the female must ensure that the territory can supply enough food for her to feed the young alone, as the male may well desert her to attract other females. Territory quality rather than male quality should be her prime criterion in mate choice. Such a situation may have resulted in the apparent paradox of more intense sexual selection upon monogamous males, where females select males directly on a number of cues including the quality of their songs.

There are other differences between the two studies which may also be important. North American wrens are a larger group found in a wider variety of different habitats from marsh to desert. As Kroodsma has pointed out, they may have evolved their songs in widely differing avian communities with varying species diversities. *Acrocephalus* warblers are migrants, often arriving in Europe relatively late from Africa. The Marsh Warbler, with the most elaborate song of all, is the latest to arrive back and has a short, compressed breeding season. The ability to attract females quickly in order to breed successfully may have been another important factor leading to intense sexual selection pressure on the songs of monogamous warbler males.

IV. DIRECT EFFECTS ON MALES AND FEMALES

Contextual studies and correlations with different mating and spacing systems can give important clues as to the functions of bird sounds. It may emerge that the song of a particular species appears to be particularly important in the establishment of territory or in the attraction of females. If so, the next step should be to subject these hypotheses to experimental testing, a technique which poses obvious problems, particularly in the field. However, in the last few years a number of useful techniques have been developed which are now providing much needed direct experimental evidence.

A. Direct Effects on Males

Although there are countless observations in the literature that passerine songs are associated with the acquisition and defense of territory, until recently there has been no experimental evidence to confirm that song itself does have a direct effect upon males. The main problem in an experimental design is to separate song itself from the visual stimulus of the singing bird. One solution is to remove any possible effects of song by muting territorial males, and another is to remove the males completely and substitute playback recordings of their songs.

The latter approach is the most direct method and was first attempted by Göransson *et al.* (1974) working on the Thrush Nightingale (*Erithacus luscinia*). They simulated the presence of territorial males in a wood by playback of songs through loudspeakers, and appeared to demonstrate some aversive effect upon real males. This technique was further developed and refined by Krebs (1977b) working on the Great Tit (*Parus major*). He removed all eight territorial males from a small wood and replaced them with a sophisticated system of loudspeakers (Fig. 3). Three of the original territories were occupied by several speakers and a tape recorder containing a continuous loop recording of normal Great Tit songs. Multiway switches ensured that the normal pattern of a Great Tit singing in different parts of his territory could be faithfully reproduced. The wood also had two control areas, one being left completely empty and silent, while the other area contained similar equipment and a control noise from a tin whistle. The first experiment took place in February and it can be seen from Fig. 3 that after a few hours both control areas were occupied by prospecting Great Tit males, whereas the experimental area was avoided. As a final precaution, the experimental and control areas were changed and the experiment repeated in March, with much the same result. In both experiments the experimental area was eventually reoccupied in about 2 days.

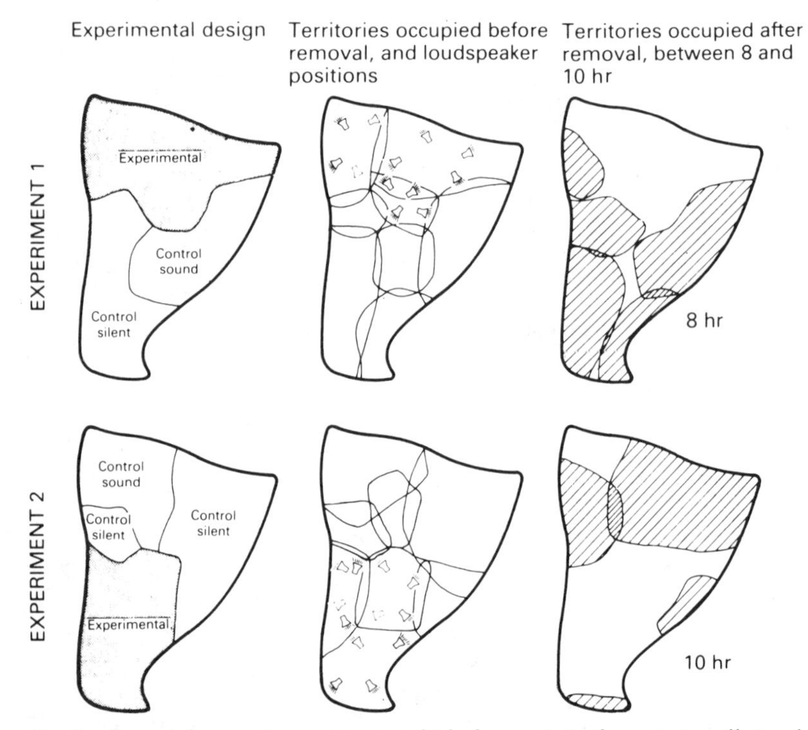

Fig. 3. Two replacement experiments which demonstrate the aversive effects of recorded songs upon male Great Tits prospecting for territories in spring. Occupied territories at 8 and 10 hr in Experiments 1 and 2, respectively, are indicated by cross-hatching. (From Krebs, 1977b.)

The other approach of muting territorial males and then studying their ability to maintain territory against rival males was adopted by Peek (1972) working on Red-winged Blackbirds (*Agelaius phoeniceus*). He trapped territorial males, anesthetized them, and removed a small portion of the hypoglossal nerves to stop normal song production. The operated birds quickly recovered and when released back into territory appeared to be normal in all other respects. Control males were operated on in a similar fashion, but the hypoglossal nerves were left intact. He found that the muted males had their territories invaded much more frequently than the controls. Smith (1975) repeated Peek's experiments in another location and obtained less convincing results, but then later (Smith, 1979) changed his muting technique. Instead of severing the hypoglossal nerves, which could also lead to respiratory problems, he cut the membrane of the interclavicular air sac. This operation seems not only to produce more effective muting, but also has fewer additional side effects which could affect a male's ability to defend territory. The mutes eventually recover their vocal ability after 2 or 3 weeks, and so in a sense serve as their own controls as well. Smith confirmed that invasions by rival males into territories of muted males increased in both frequency and amount. The muted males were also forced to resort to more visual signaling during territorial defense. When they recovered their ability to sing, they regained any ground which had been lost after initial muting.

In both the Red-winged Blackbird and Great Tit experiments, song appears to have a strong aversive effect upon rival males. But it is only effective as a first line of defense, and must normally be backed up by the visual stimulus of a real male or actual fighting. Of the two approaches song replacement is the most direct test and also has certain experimental advantages since it removes most other variables. Any aversive effects can be due only to the songs themselves, and these can be controlled, varied, and manipulated by the experimenter.

Indeed, Krebs has now gone on to use this technique to investigate whether recordings of Great Tits with song-type repertoires repel prospecting males more effectively than recordings of a single song type (Krebs *et al.*, 1978). As before, the wood was "occupied" by systems of loudspeakers divided into three areas. The control area was left empty, another area occupied by loudspeakers broadcasting recordings of a single song type, and the remaining area by those broadcasting a repertoire of song types. It can be seen (Fig. 4) that whereas the control area was occupied almost immediately, and the single song-type area soon afterward, the repertoirc area was occupied last of all. This experiment again demonstrates the aversive effects of song upon prospecting males, but also suggests that one of the functions of song-type repertoires may be to enhance the effect even further. This was one of the predictions which stemmed from Kreb's Beau Geste hypothesis (Krebs, 1977a) in which he suggested that resident males might deceive prospecting males as song-type repertoires create the impression that the area is already crowded. Yasukawa (1981) has now performed speaker replace-

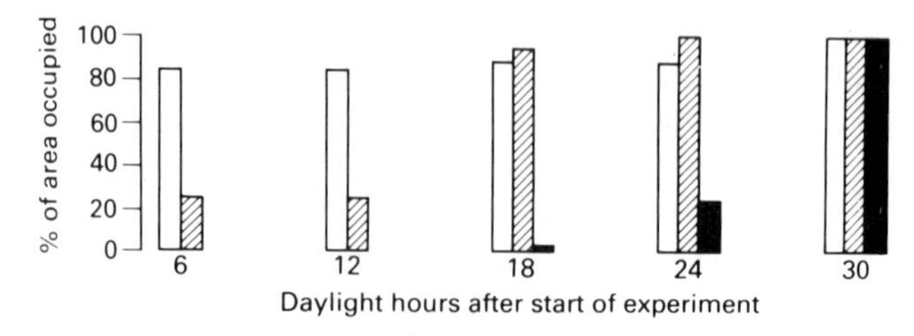

Fig. 4. Occupation by male Great Tits of unoccupied experimental areas defended by no song recordings (white), one recorded song type (shaded), or a repertoire of recorded song types (black). (From Krebs *et al.*, 1978.)

ment experiments on the Red-winged Blackbird, and his results also support the hypothesis.

B. Direct Effects on Females

Territorial conflicts between rival males are much more conspicuous than interactions between males and females. One reason is that after pairing a male may stop whatever behavior he has used to initially attract a female. Although cessation of song may be highly suggestive that the sound signal has attracted a female, the hypothesis needs to be tested experimentally. It could be that cessation of song is merely the result of the female interfering with production of territorial song by placing new time and energy demands on the male. One way to test this is by conventional playback experiments on males before and after pairing, and this has been done for the Sedge Warbler (Catchpole, 1977). Before pairing, all ten males responded by approaching and spending time near the playback speaker; eight produced some song, and six produced threat calls (Fig. 5). After pairing, six still approached the speaker, although on average much less, and eight produced even more threat calls. Although they all had the time, energy, and motivation to respond in this way, not one produced song. Such an experiment provides strong, but still indirect evidence, that the primary function of song in such species is female attraction.

For more direct evidence we must look to studies which have attempted to play back songs to females. Males are usually the first occupants of territories and a conventional playback experiment may do little except attract any resident males. Waiting for females to arrive could take days or even weeks, and any response may be slight and difficult to observe or quantify. For these reasons, female replacement experiments have yet to be seriously attempted, but some progress has been made with laboratory studies on captive females. However, none of the experiments so far has dealt with female attraction, but rather with the longer-term effects of sounds upon female reproductive physiology or behavior. Brockway (1965) showed that playback of male vocalizations alone stimulated ovarian development and egg-laying in captive female Budgerigars (*Melop-*

sittacus undulatus). Hinde and Steel (1976), working on captive Canaries (*Serinus canaria*), used nest-building behavior as an index of female reproductive behavior. Nest-building is known to be estrogen-dependent and ovariectomized females were given a standard dose of exogenous estrogen before being exposed to playback of either Canary or Budgerigar songs. Females exposed to song from males of their own species built far more actively than controls exposed to Budgerigar song or no song at all.

Kroodsma (1976) used a similar technique, but manipulated the structure and complexity of the Canary songs used in playback. He divided his females into two groups. To one group he played relatively simple songs containing only 5 syllable-types, and to the other he played more normal, complicated songs containing about 35 syllable-types. After several days the females exposed to the more complicated songs were building much more actively and using more strings for nesting (Fig. 6) than those exposed to the simpler songs. By day 36 of the experiment the females in the complex song group had laid 15 eggs compared to only 5 in the simple song groups. Whether more complicated songs actually attract females for pairing more than simpler songs is really another question, but one which can be tested under natural conditions by working with a wild population.

Male Mockingbirds (*Mimus polyglottos*) have extremely elaborate songs and Howard (1974) set out to investigate whether those with more elaborate songs acquired better territories, attracted females earlier, or both. He found a strong inverse correlation between repertoire size and pairing date. He also measured territory size and quality, ranked the territories, and found that territory rank also showed a strong correlation with pairing date. In an attempt to separate the

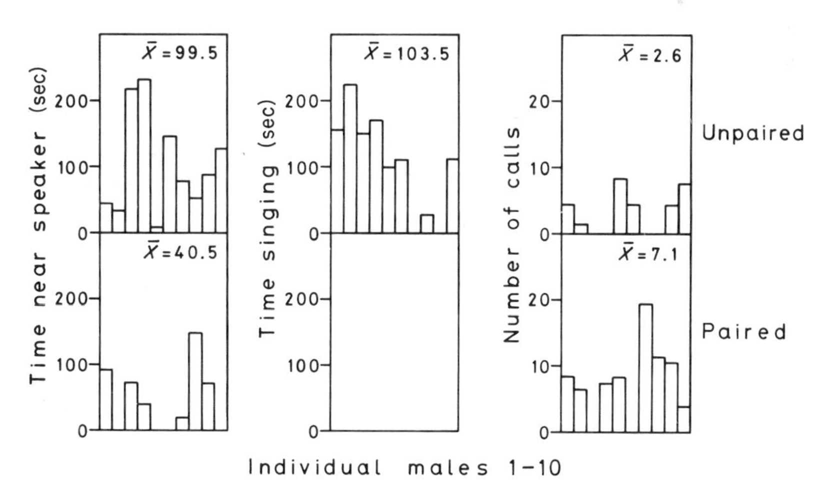

Fig. 5. Responses of ten male Sedge Warblers to playback of species song before (April 29–May 5) and after (June 4–11) pairing. (From Catchpole, 1977.)

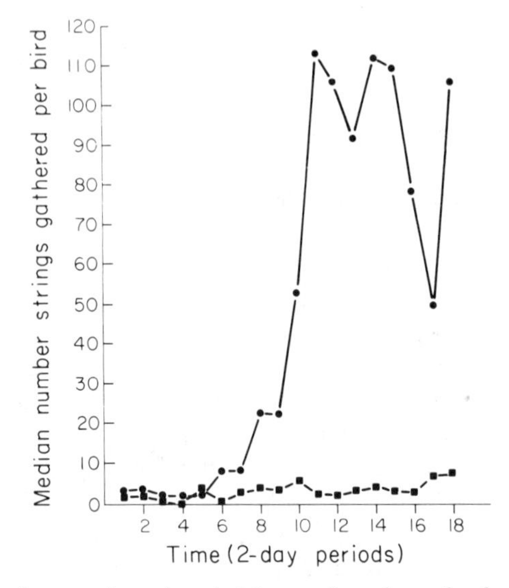

Fig. 6. The median number of nest strings gathered per day by 24 female Canaries exposed to either large (dots) or small (squares) male song repertoires. (From Kroodsma, 1976. Copyright 1976 by the American Association for the Advancement of Science.)

effects of male song and territory quality he used partial correlation analysis, holding each of the two variables constant in turn. His results suggested that both repertoire size and territory rank influence female choice, but that territory is the more important of the two.

I applied a similar technique to a population of Sedge Warblers (Catchpole, 1980), and also obtained a strong inverse correlation between repertoire size and pairing date (Fig. 7). Again it seems clear that males with more complicated songs are attracting females earlier than their rivals, but in this case territory quality may be less important. As mentioned earlier, Sedge Warblers feed primarily outside their territories and a female should be more concerned about a male's ability to help with foraging rather than his territory quality. In practice it was difficult to rank these small, quite uniform, nonfeeding territories on any other criteria of quality, although male birds may be more discriminating. If there are important differences in territory quality, then the first males back should select the best territories, and if females are selecting males primarily on territory quality, then they should follow a similar settling pattern. However, when male arrival date (male choice) is plotted against pairing date (female choice), no significant relationship emerges. Additional checks using partial correlation analysis confirmed that repertoire size appears to be the most significant variable affecting female choice in this species.

Krebs (1977a) has pointed out that such studies have so far ignored the possible effects of age as another confounding variable. It could be that age is also

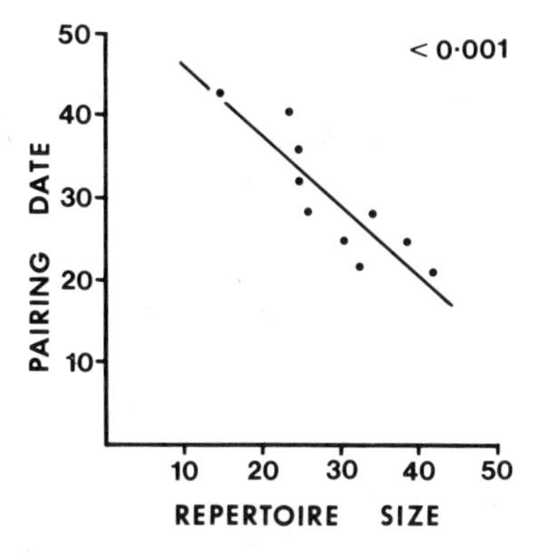

Fig. 7. The relationship between pairing date and syllable repertoire size in ten male Sedge Warblers during 1979. Earliest (22) and latest (43) pairing dates were May 10 and June 1 respectively. (From Catchpole, 1980.)

correlated with repertoire size and that females prefer to mate with older more experienced males who just happen to have larger repertoires. So far the evidence for this view is rather slight, in that most species appear to acquire their song structures during a relatively short period within the first year of life (Chapter 1, Volume II). One of the few documented cases of males adding to their repertoire after the first year is in captive Canaries (Nottebohm and Nottebohm, 1978). In the wild there are even fewer data, although Yasukawa *et al.* (1980) have some evidence that male Red-winged Blackbirds continue to develop new song-types. They found that older males had larger repertoires, mated earlier, and acquired more females. They attempted to control for age by restricting a second analysis to first year males only. Repertoire size was no longer correlated with earlier breeding or attracting more females, but there are a number of other reasons for variation among inexperienced breeders. What is really needed is a study on experienced males of known ages in a species where increasing repertoire size is significantly correlated with age. Such species are far from common and in most cases any significant variation seems more related to normal ontogeny during the first year only. At present it seems unlikely that variation in repertoire size forms the basis for a reliable index of male age in most species, but much more likely that an impoverished repertoire merely indicates the presence of an inexperienced first year bird which females would do well to avoid.

V. SEXUAL SELECTION AND BIRD SOUNDS

Charles Darwin (1871) not only originated the theory of sexual selection, but also suggested that it may have played an important part in the evolution of the elaborate sounds produced by many male birds during courtship. He recognized that male competition for females could occur in two rather different ways. In the first case, males might fight with one another and the winner acquire either the available female or a breeding territory which is an essential prerequisite in acquiring a female later. Because males first compete among themselves, there will be strong intrasexual selection favoring traits which increase success in male–male contests. In the second case, males attempt to influence female choice more directly by producing increasingly elaborate or conspicuous signals. This assumes that females do find elaborate signals more attractive, and if so, there will be strong intersexual selection on signals to increase their complexity.

Intersexual selection has always been the more controversial part of the sexual selection hypothesis, as Darwin failed to suggest how it might originate or be maintained in natural populations. Although the advantage to the male seems fairly obvious in that he will attract more females and leave behind more offspring, what possible advantage does the female obtain? It was Fisher (1958) who pointed out that if females do select the most attractive males, their male offspring will presumable inherit this ability, also attract more females, and so leave behind more second generation offspring of the original female. This would lead to the characteristic runaway process which tends to produce such bizarre elaborations as the peacock's tail. Fisher also pointed out that although the attractive feature might initially be linked to some aspect of male fitness, once established it might later continue to be selected simply for its attractiveness.

Although the complex sounds produced by birds have been the subject of much debate, the theory of sexual selection has rarely been emphasized and is often relegated to a minor role. There are several main reasons for this, and one is that many writers appear to equate "sexual selection" with intersexual selection for which there is rather less evidence. The often dominant aspects of male–male vocal interactions are then regarded as something quite unrelated to sexual selection, and interpreted solely in terms of proximate territorial function. Such an approach is also one of the reasons for the apparent dichotomy between the territorial and sexual functions of song so prevalent in the early literature (reviewed in Armstrong, 1973). In many ways there is a good argument for regarding this dichotomy as a false and misleading one. First, the two proximate functions are not mutually exclusive, and there is good evidence that most species maintain some duality of function. Second, both functions are ultimately related to sexual selection, since acquiring a territory through song is merely an initial and essential stage in attracting a female for breeding.

Nottebohm (1972) was one of the first to rekindle the idea that sexual selection is an important force in shaping the more elaborate bird sounds. The Sedge Warbler, which uses its song almost solely for sexual attraction (Catchpole, 1973) and also has unusually long, complicated songs (Catchpole, 1976), led me to suggest that sexual selection had played a significant role in the evolution of song structure. But it was Howard (1974) who was the first to emphasize and clarify the important distinction between intra- and intersexual selection, and their effects upon the evolution of Mockingbird songs. He made the point that sexual selection must be the major selective force, and that in proximate terms song has a duality of function in both territorial defense and mate attraction.

Whether song evolves primarily for mate attraction, territorial defense, or both may depend to some extent upon the relationship between parental investment in polygynous or monogamous species. In his review Trivers (1972) concluded that when each sex invests nearly equally, then sexual selection affects both equally; but when, for example, females invest more heavily than males, then there will be increased competition among males for females. This is just what happens in polygynous males, who invest very little and after copulation may even desert one female for another. The key point in the argument is that it is male–male competition (i.e., intrasexual selection) that is the main pressure affecting their songs. This should produce songs designed primarily for territorial repulsion of rival males, rather than female attraction. It is intersexual selection pressure which produces elaborate songs designed to attract females (Fig. 8). Furthermore, females selecting potentially polygynous males who may soon desert them must ensure that they select a territory of sufficient quality to enable them to feed their young alone. Females of monogamous species should select males more directly on the basis of their physical attributes, to help in feeding young, and it may be that song is one important cue. It is difficult to separate completely male quality and territory quality, and species may also have subtle differences in feeding ecology which render territory quality more or less important, as in the *Acrocephalus* warblers discussed earlier.

It seems clear that we can and should distinguish between the two types of sexual selection, for in the evolution of bird songs the distinction is both impor-

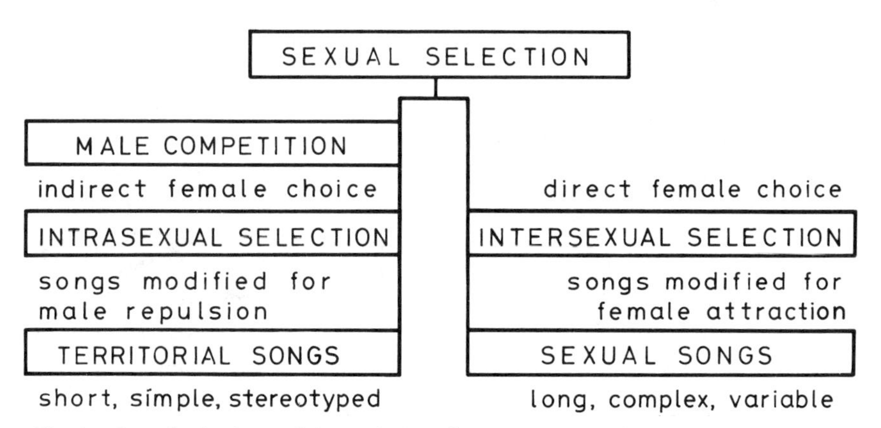

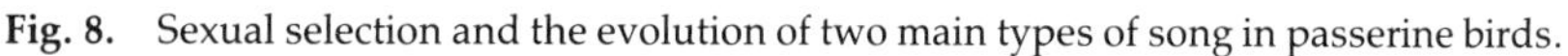

Fig. 8. Sexual selection and the evolution of two main types of song in passerine birds.

TABLE II

A Tentative Diagnostic Checklist for the Probable Evolution of Two Main Types of Song in Male Passerine Birds

Main proximate function		
	Female attraction (e.g., Sedge Warbler)	Male repulsion (e.g., Great Tit)
Song structure	Large syllable repertoire	Small syllable repertoire
	Variable sequencing	Stereotyped sequencing
	Songs not repeated	Song types repeated
	Continuous singers	Discontinuous singers
Contextual correlations	Sings in territory before pairing and stops after	Sings in territory after pairing
	Does not sing in response to playback after pairing	Sings in response to playback after pairing
	More likely to be migratory	More likely to be resident
Direct effects on males	No matched countersinging occurs	Matched countersinging occurs
	Speaker replacement experiments have no significant effect	Speaker replacement experiments repel rival males and are more effective with larger repertoires
	Males with larger syllable repertoires do not obtain better territories	Males with larger song type repertoires obtain better territories
Direct effects on females	Males with larger syllable repertoires attract females first	Males with larger song type repertoires do not attract females first
Main selection pressure involved	Intersexual selection	Intrasexual selection

tant and useful. Intrasexual selection appears to be more common and must produce song structures adapted mainly for territorial defense. Intersexual selection may be less common (or more difficult to detect) and must produce song structures adapted primarily for the direct attraction of females. Therefore we might expect the evolution of two main types of song whose very structure should reflect any differences imposed by the particular problems of female attraction or male repulsion. As noted earlier, the vast majority of bird species are monogamous, and I have left the complicating factor of mating system out of the simple model (Fig. 8). Where cases of polygyny do occur, we have seen that an increase in song complexity may or may not occur, depending upon the ecology and behavior of the species concerned. In order to clarify the argument and indicate the many factors involved, I have drawn up a tentative list summarizing the main differences (Table II). The list follows much the same approach as the whole chapter, in that it emphasizes the importance of initial contextual work followed up by later experiments and tests in the field. I have also selected two specific examples with which to illustrate the comparison, the Sedge Warbler and the Great Tit. These two species are in many ways opposite extremes, and have also been subjected to many of the techniques advocated.

The Great Tit is a fairly typical example of a species where males have a song-type repertoire. As Krebs (1977b) has shown, Great Tit songs function primarily in territorial defense, although some duality of function cannot be ruled out (Krebs *et al.*, 1981). The factors which control the size of song-type repertoires are not known, but McGregor *et al.* (1981) have shown that male Great Tits with larger repertoires breed more successfully because they occupy better territories. Features which seem designed to enhance a territorial function include repeating relatively short, stereotyped song-types which could facilitate the positional learning of neighbors as well as permit individual recognition. Being resident rather than migratory also allows more time for such learning to occur. The existence of song-type repertoires may not enhance individual recognition per se, and indeed the possible functions of song repertoires remain controversial. However, it does now seem that they might confer a number of advantages in territorial behavior, such as in the Beau Geste effect, in matched countersinging against neighbors, and acting as graded threat signals. In contrast, the variability and complexity of Sedge Warbler song seem to preclude any of these more sophisticated territorial functions, and indeed all the available evidence points to female attraction. The species is migrant and has a comparatively short time in which to breed, and this may well have placed an additional premium upon quick and effective mate attraction with little time to indulge in long-term spatial relationships with other males. Increasing elaboration and more effective attraction may have lead to runaway intersexual selection resulting in the acoustic equivalent of such visual extravagances as the peacock's tail. These two species may be extreme forms, but the same basic trends can probably be detected in many other passerines which have either song-type repertoires or very elaborate songs. As pointed out earlier, the two trends are not mutually exclusive, and other species may have retained considerable duality of function and evolved particular song structures subtly shaped to their particular ecological and social requirements.

REFERENCES

Armstrong, E. A. (1973). "A Study of Bird Song." Dover, New York.

Baptista, L. (1978). Territorial, courtship and duet songs of the Cuban Grassquit (*Tiaris canora*). *J. Ornithol.* **119,** 91–101.

Beaver, P. W. (1978). Ontogeny of vocalization in the Greater Rhea. *Auk* **95,** 382–388.

Brockway, B. F. (1965). Stimulation of ovarian development and egg laying by male courtship vocalization in Budgerigars (*Melopsittacus undulatus*). *Anim. Behav.* **13,** 575–578.

Catchpole, C. K. (1973). The functions of advertising song in the Sedge Warbler (*Acrocephalus schoenobaenus*) and the Reed Warbler (*A. scirpaceus*). *Behaviour* **46,** 300–320.

Catchpole, C. K. (1976). Temporal and sequential organisation of song in the Sedge Warbler (*Acrocephalus schoenobaenus*). *Behaviour* **59,** 226–246.

Catchpole, C. K. (1977). Aggressive responses of male Sedge Warblers (*Acrocephalus schoenobaenus*) to playback of species song and sympatric species song, before and after pairing. *Anim. Behav.* **25,** 489–496.

Catchpole, C. L. (1980). Sexual selection and the evolution of complex songs among European warblers of the genus *Acrocephalus*. *Behaviour* **74,** 149–166.

Darwin, C. (1871). "The Descent of Man and Selection in Relation to Sex." Murray, London.

Dowsett-Lemaire, F. (1979). The imitative range of the song of the Marsh Warbler *Acrocephalus palustris,* with special reference to imitations of African birds. *Ibis* **121,** 453–468.

Fisher, R. A. (1958). "The Genetical Theory of Natural Selection," 2nd ed. Dover, New York.

Gochfeld, M., and Kleinbaum, M. (1978). Observations on behavior and vocalizations of a pair of wild Harpy Eagles. *Auk* **95,** 192–194.

Göransson, G., Högstedt, G., Karlsson, J., Källander, H., and Ulfstrand, S. (1974). Sångensroll för revirhållandet hos näktergal *Luscinia luscinia* några experiment med playbackteknik. *Var Fagelvarld* **33,** 201–209.

Hall, M. F. (1962). Evolutionary aspects of estrildid song. *Symp. Zool. Soc. London* **8,** 37–55.

Hinde, R. A. (1956). The biological significance of the territories of birds. *Ibis* **98,** 240–369.

Hinde, R. A., ed. (1969). "Bird Vocalizations." Cambridge Univ. Press, London and New York.

Hinde, R. A., and Steel, E. (1976). The effect of male song on an estrogen-dependent behavior pattern in the female Canary (*Serinus canarius*). *Horm. Behav.* **7,** 293–304.

Howard, E. (1920). "Territory in Bird Life." Murray, London.

Howard, R. D. (1974). The influence of sexual selection and interspecific competition on Mockingbird song (*Mimus polyglottos*). *Evolution* **28,** 428–438,

Immelmann, K. (1969). Song development in the Zebra Finch and other estrildid finches. *In* "Bird Vocalizations" (R. A. Hinde, ed.), pp. 61–74. Cambridge Univ. Press, London and New York.

Jarvi, T., Radesater, T., and Jakobsson, S. (1980). The song of the Willow Warbler *Phylloscopus trochilus* with special reference to singing behaviour in agonistic situations. *Ornis Scand.* **11,** 236–242.

Krebs, J. R. (1977a). The significance of song repertoires: The Beau Geste hypothesis. *Anim. Behav.* **25,** 475–478.

Krebs, J. R. (1977b). Song and territory in the Great Tit. *In* "Evolutionary Ecology" (B. Stonehouse and C. Perrins, eds.), pp. 47–62. Macmillan, New York.

Krebs, J., Ashcroft, R., and Webber, M. (1978). Song repertoires and territory defence in the Great Tit. *Nature (London)* **271,** 539–542.

Krebs, J., Avery, M., and Cowie, R. (1981). Effect of removal of mate on the singing behaviour of Great Tits. *Anim. Behav.* **29,** 635–637.

Kroodsma, D. E. (1976). Reproductive development in a female songbird: Differential stimulation by quality of male song. *Science* **192,** 574–575.

Kroodsma, D. E. (1977). Correlates of song organisation among North American wrens. *Am. Nat.* **111,** 995–1008.

Lack, D. (1968). "Ecological Adaptations for Breeding in Birds." Methuen, London.

Lein, M. R. (1972). Territorial and courtship songs of birds. *Nature (London)* **237,** 48–49.

Lein, M. R. (1978). Song variation in a population of Chestnut-sided Warblers: Its nature and suggested significance. *Can. J. Zool.* **56,** 1266–1283.

McGregor, P., Krebs, J., and Perrins, C. (1981). Song repertoires and lifetime reproductive success in the Great Tit *Parus major*. *Am. Nat.* **118,** 149–159.

Marler, P. (1956). The voice of the Chaffinch and its function as a language. *Ibis* **98,** 231–261.

Miller, D. (1979). The acoustic basis of mate recognition by female Zebra Finches (*Taeniopygia guttata*). *Anim. Behav.* **27,** 376–380.

Morse, D. H. (1970). Territorial and courtship songs of birds. *Nature* (*London*) **226,** 659–661.

Nottebohm, F. (1972). The origins of vocal learning. *Am. Nat.* **106,** 116–140.

Nottebohm, F., and Nottebohm, M. (1978). Relationship between song repertoire and age in the Canary *Serinus canarius*. *Z. Tierpsychol.* **46,** 298–305.

Peek, F. W. (1972). An experimental study of the territorial function of vocal and visual display in the male Red-winged Blackbird (*Agelaius phoeniceus*). *Anim. Behav.* **20,** 112–118.

Slagsvold, T. (1977). Bird song activity in relation to breeding cycle, spring weather, and environmental phenology. *Ornis Scand.* **8,** 197–222.

Smith, D. G. (1975). An experimental analysis of the function of Red-winged Blackbird song. *Behaviour* **46,** 136–156.

Smith, D. G. (1979). Male singing ability and territorial integrity in Red-winged Blackbirds (*Ageliaus phoeniceus*). *Behaviour* **68,** 191–206.

Smith, R. L. (1959). The songs of the Grasshopper Sparrow. *Wilson Bull.* **71,** 141–152.

Smith, W. J. (1963). Vocal communication of information in birds. *Am. Nat.* **97,** 117–125.

Smith, W. J. (1965). Message, meaning and context in ethology. *Am. Nat.* **99,** 405–409.

Smith, W. J. (1977). "The Behavior of Communicating." Harvard Univ. Press, Cambridge, Massachusetts.

Smith, W., Pawlukiewicz, J., and Smith, S. (1978). Kinds of activities correlated with singing patterns of the Yellow-throated Vireo. *Anim. Behav.* **26,** 862–884.

Thorpe, W. H. (1961). "Bird Song. The Biology of Vocal Communication and Expression in Birds," Monographs in Experimental Biology, No. 12. Cambridge Univ. Press, London and New York.

Thielcke, G. (1976). "Bird Sounds." Ann Arbor, Michigan.

Tinbergen, N. (1939). The behaviour of the Snow Bunting in spring. *Trans. Linn. Soc. N.Y.* **5,** 1–94.

Trivers, R. L. (1972). Parental investment and sexual selection. *In* "Sexual Selection and the Descent of Man" (B. Campbell, ed.), pp. 136–179. Aldine, Chicago, Illinois.

White, G. (1789). "The Natural History and Antiquities of Selborne." London.

White, S. (1971). Selective responsiveness by the Gannet (*Sula bassana*) to played-back calls. *Anim. Behav.* **19,** 125–131.

Yasukawa, K. (1981). Song repertoires in the Red-winged Blackbird (*Agelaius phoeniceus*): a test of the Beau Geste hypothesis. *Anim. Behav.* **29,** 114–125.

Yasukawa, K., Blank, J., and Patterson, C. (1980). Song repertoires and sexual selection in the Red-winged Blackbird. *Behav. Ecol. Sociobiol.* **7,** 233–238.

Taxonomic Index

Roman numerals in boldface refer to volume numbers.

H

N

O

X

Subject Index

Roman numerals in boldface refer to volume numbers.

J

Q

T

U

V

W

Contents of Volume 2

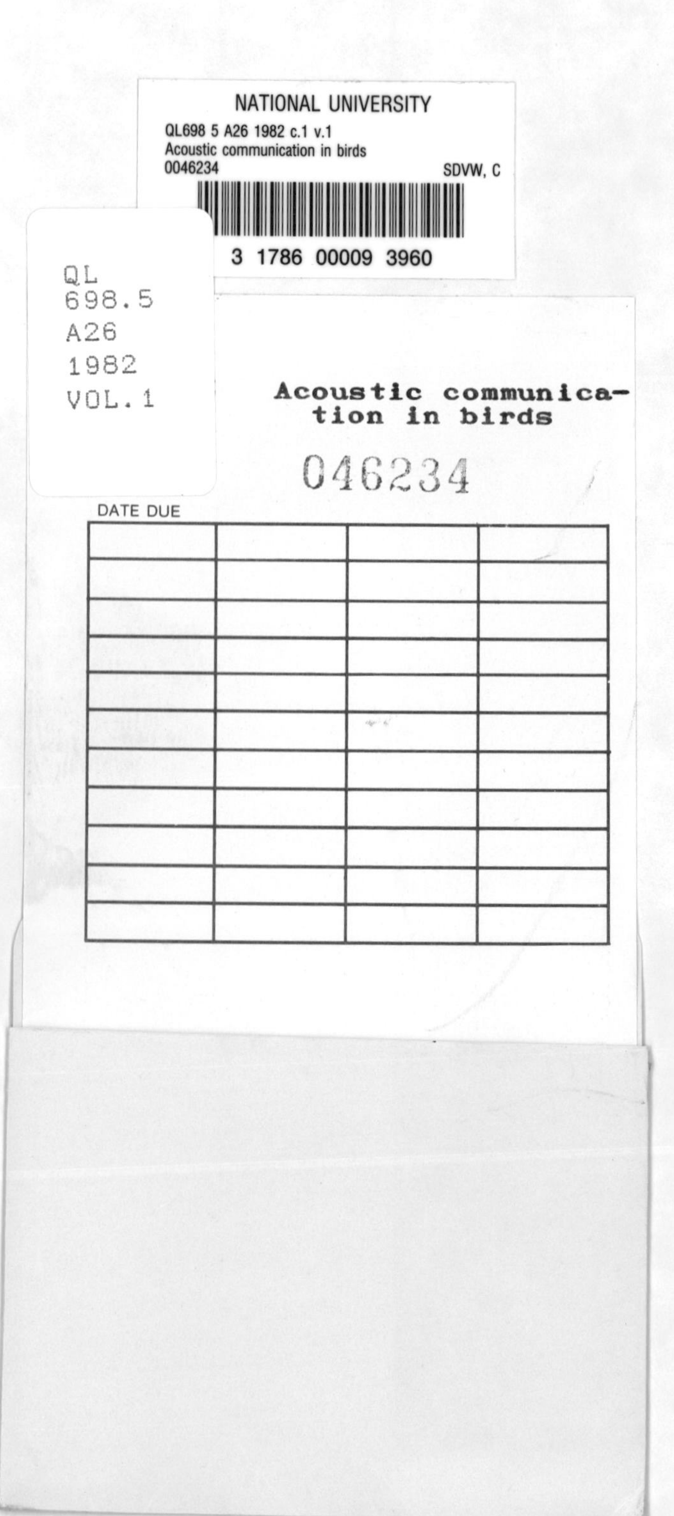